KEY SCIENCE

Chemistry

Eileen Ramsden BSc PhD DPhil
Formerly of Wolfreton School, Hull

Stanley Thornes (Publishers) Limited

First published in 1994 by:
Stanley Thornes (Publishers) Ltd
Ellenborough House
Wellington Street
CHELTENHAM GL50 1YD
England

A catalogue record of this book is available from the British
Library.

ISBN 0-7487-1675-0

Related titles
Key Science: Biology (0-7487-1676-9)
Key Science: Physics (0-7487-1674-2)
Key Science: Chemistry Teacher's Guide (0-7487-1721-8)
Key Science Resource Bank (0-7487-1719-6)

Design and artwork by Cauldron Design Studio, Auchencrow,
Berwickshire. Additional artwork by Barking Dog Art,
Nailsworth, Gloucestershire.
Typeset by Word Power, Auchencrow, Berwickshire and
Tech-Set, Gateshead, Tyne & Wear.
Printed and bound in Spain by Mateu Cromo.

Acknowledgements

I would like to express my thanks to Mr Gavin Cameron, Mr
Steve Hewitt and Mrs Sheila Rogers for their constructive
comments on the Earth Science topic and to Mr David Haslam
for his advice on some of the chemical topics.

Dr Jerry Wellington and Mr Jon Scaife have provided some
excellent material on information technology, and I thank them
for their contributions to the text.

I thank the following organisations and people who have
supplied photographs.
Action Plus Photographic: 3.1A (Tony Henshaw), p. 72
Alfred McAlpine plc: 3.6A
Allsport: 26.1B (Gray Mortimore), 30.3D (Bob Martin)
Ardea: 18.6C (Francois Gahier)
Ariel Plastics: 3.6D
British Canoe Union: 3.6C
British Rail: 19.6A
British Steel: 5.4A
Bruce Coleman Ltd: 31.4F (Jane Burton)
Camera Press: 8.5B, 8.5C (L Smillie), p. 92
Casio Electronics Co. Ltd: 20.2A
Collections: 21.7B (Keith Pritchard), 28.3A (Ben Boswell)
Diamond Information Council: 5.5A

Fisons plc: 13.4A
Gavin Cameron: 2.2A, 2.2C
Geological Museum: 2.5B
ICI Chemicals and Polymers Ltd: 3.2A, 3.5D, 9.6C, 11.2A, 11.8B,
29.11B
Image Bank: p. ix (top) (Bill Varie), p. ix (bottom) (Jaime
Villaseca)
International Centre for Conservation Education: 17.4A, 18.4A
(Andy Purcell), 18.6D (Mark Tasker), 18.6E, 25.1B (Mark Tasker),
26.1A
Mark Boulton: 28.1A, 28.1E, 30A, 30.3C
Mary Evans Picture Library: p. 70, p. 71
Metropolitan Police: 31.1D
Michael Holford: 5.3A
Mountain Camera: 3.5G
NASA: p. 1, 14.3A, 28.3D
Northern Irish Tourist Board: 2.1E
Novosti Press Agency (APN): p. 94, 8.9B
Oxford Scientific Films Ltd: p. 27 (Peter Gould), 14.5A
(G I Bernard), 14.10A (L D Zell), 14.10B (Jack Dermid)
Panos Pictures: 18.3A(a) (J Hartley), 18.3A(b) (Neil Cooper),
31.1G (David Hall)
Panasonic (UK) Ltd: 20.2B, 20.2E
Peter Newark's Western Americana: 5.2A
Pilkington Glass plc: 28.2B
Popperfoto: 18.6A, 18.6B (Erik Hill), 18.6C (David Aka)
Quadrant Picture Library: 9.6A
Science Photo Library: cover (Charlotte Raymond), 2.3D (Sinclair
Stammers), 2.5E (Prof Stewart Lowther), 4.4B (Ben Johnson),
4.4C (Dr Mitsuo Ohtsuki), 4.4D (Dr M B Hursthouse), 5.1C (Dr
G M Rackham), 8.1B (Hank Morgan), 8.1G (Lawrence Berkeley
Laboratory), 8.3C (Philippe Plailly), 8.4A(a) (Alexander Tsiaras),
11.3B, 14.1A (Susan Leavines), 14.4B (Alex Bartel), 17.11B, p. 213
(Martin Bond), 19.1A (John Ross), p. 267 (Geoff Tompkinson),
28.1D (Martin Bond), 31.6B, 31.6C
Shell Photographic Library: p. 123
Sotherbys: 5.2B
Sue Boulton: 30B, 31.1A
Swiss National Tourist Board: 2.3C
The Environmental Picture Library: 17.10A (David Townend),
18.1A (Alan Greig)
United Kingdom Atomic Energy Authority: 8.1D, 8.3D,
8.4A(b), (c)
Woodmansterne Ltd: 2.1D, 2.2F
Wookey Hole Caves: 16.8A
Portrait of Michael Faraday (p. 103) reproduced by courtesy of
the National Portrait Gallery

I thank the following examining groups for permission to
reproduce questions from examination papers:
University of London Examinations and Assessment Council,
Midland Examining Group, Northern Examinations and
Assessment Board, Northern Ireland Schools Examinations and
Assessment Council, Southern Examining Group, Welsh Joint
Education Committee.

The examining groups bear no responsibility for the answers
to questions taken from their question papers contained in this
publication.

The production of a science text book from manuscripts
involves considerable effort, energy and expertise from many
people, and we wish to acknowledge the work of all the
members of the publishing team. We are particularly grateful to
Penelope Barber (Senior Publisher), Adrian Wheaton (Science
Publisher), Lorna Godson, Helen Roberts and John Hepburn
(Editors), and Malcolm Tomlin (Senior Editor) for their advice
and support.

Finally we thank our families for the tolerance which they
have shown and the encouragement which they have given
during the preparation of this book.

Eileen Ramsden

Contents

Preface

Key Science: Chemistry has been written for GCSE Science: Chemistry and, with two other books, for all science syllabuses under the National Curriculum at GCSE. The related books are:

Key Science: Biology, David Applin
Key Science: Physics, Jim Breithaupt.

Topics are differentiated into core material for Double/Single Science and extension material for Science: Chemistry (red margin). The examining groups have chosen different extension topics in addition to the Programme of Study for Key Stage 4; therefore, teachers and their students will need to familiarise themselves with the specific requirements of the syllabus which they are following.

Key Science: Chemistry has features, often located in the margin, such as **It's a fact**, **Science at work** and **Who's behind the science**?, which enhance the text and improve the accessibility of the text to all students.

• **Look at Links** identify related topics in this book and in *Key Science: Biology* and *Key Science: Physics*. Their role is to help students to integrate the different areas of chemistry, to appreciate the structure of the subject and its relationship to other branches of science. For the teacher, they have another benefit in that they make it easy to depart from the order of topics and take a different route through the text. The links with the other texts are optional and allow each book to stand alone.

• **Summaries** and **Checkpoints** at regular intervals in the text help students to review their work and test their knowledge and understanding. More taxing questions which appear at the end of each theme will exercise their ability to apply their knowledge to novel situations, to analyse and interpret data and to solve problems. These longer questions co-ordinate the topics in the preceding theme and link up with other themes.

• **Information technology** 'IT' appears in the National Curriculum both as a cross-curricular skill (analogous to literacy and numeracy) and as an element of the separate subjects. The value of information technology, however, extends through all of the attainment targets in science. The 'IT' entries in the book reflect the wide applications of information technology. These 'IT' features show how computer-assisted learning (perhaps of a direct tutorial nature), simulations using 'IT', databases, videodisc, datalogging and other educational uses of information technology can be used throughout science at Key Stage 4.

Consultants in information technology, Dr Jerry Wellington and Mr Jon Scaife, both of the Educational Research Centre, University of Sheffield, have provided expert guidance for students (alongside the text) and for teachers (in the teacher's guides).

Key Science: Chemistry is supported by a detailed Teacher's Guide and Resource Bank of photocopiable activities and assignments – some of which can be used for teacher-based assessment of Attainment target 1 for Science, Sc1. Details are available from the publisher.

Student's Note

I hope that this book will provide all the information you need and enough practice to enable you to do well in the tests and examinations in your GCSE Chemistry course. However, examinations are not the most important part of your course. I hope also that you will develop a real interest in science, which will continue after your course is over. Finally, I hope that you enjoy using the book. If you do, all the effort will have been worth while!

Eileen Ramsden

Adventures in science

The Mirror of Galadriel

In the book *Lord of the Rings*, amazing things happen to Gandalf and the hobbits. They risk their lives on a perilous journey to save their world. They journey through secret tunnels and unknown lands and they witness astonishing events. In the early part of their travels, they are invited to look into a magic mirror – a silver bowl filled with water. It shows 'things that were, and things that are, the things that yet may be.' But the viewer can't tell which it is showing!

Imagine you have travelled in time from two centuries ago to the present day. What would you make of television – a silver bowl showing events from the past and the present? Television is a product of the scientific age in which we live.

Science has many more amazing things to reveal. Imagine food 'grown' in a factory or round-the-world flight in a few hours or 'intelligent' computers that need no programming. These and many more projects are the subject of intense scientific research now. Fact is stranger than fiction.

Figure A ● What can you see?

Can you believe what you see?

Look at the picture in Figure A. *What can you see?* Some people see an old lady; others see a young woman. Different people looking at the same picture see different images.

Witnesses of an event often report totally different versions of the event. Invent a harmless incident and try it out in front of unsuspecting witnesses without warning them. Then ask them to say what happened. Each person will probably have a different story to tell.

In science, there are countless events to observe. The story or 'account' of an event in science should not differ from one observer to another. Otherwise, who could you believe?

How science works

Science involves finding out how and why things happen. Natural curiosity sometimes provides the starting point for a scientific investigation. Another investigation might set out to solve a certain problem and therefore have a definite aim. Even then, unexpected results may emerge to stimulate our natural curiosity.

The history of science has many examples where an important discovery or invention came about unexpectedly. The electric motor was invented when an electric generator was wired up wrongly at the Vienna Exhibition of 1873. X-rays, radioactivity and penicillin are further examples of unexpected scientific discoveries.

Discoveries in science are unpredictable. No-one can predict which research projects will lead to exciting new developments because no-one knows what new areas of science lie waiting to be discovered beyond the frontiers of knowledge.

Chemistry as a key branch of science

Thank chemistry for a comfortable life. You can see the contribution which chemistry makes to our well-being in the materials that we use to make the clothes we wear, the houses we live in, the cars, boats and planes which transport us and the fuels which they burn. Fertilisers manufactured by the chemical industry help farmers to grow enough crops. Pesticides ensure that those crops survive to be harvested, and chemical preservatives ensure that the foods we eat are safe and appetising. Even the water we drink would not be safe without chemical treatment.

Thank chemistry for health. Chemotherapy (the use of chemicals in the treatment of diseases) allows us to live longer, healthier lives, more free from pain than in any previous century. Most of us have occasion to use pain-killers such as aspirin and antiseptics such as TCP. For more serious complaints we have antibiotics such as penicillin, and surgery is made safe by anaesthetics such as fluothane. Thanks to the development of new materials, surgeons can now do 'spare part surgery'. Someone who needs a replacement part can be fitted with a teflon earbone, a dacron polyester artery, a hip joint of polymer and alloy, a duralumin artificial leg, a polyethene heart valve or a complete artificial heart of polyurethane plastic.

Thank chemistry for leisure activities. Chemists have developed new, cheaper materials to replace natural materials such as metals, wood and ox-gut and allow many people to acquire equipment that otherwise would have been beyond their means. When you surf on your polystyrene and fibre glass surfboard, when you use your tennis racquet of polymer fibre strings on a frame of epoxy resin with graphite fibres, when you coast downhill on your skis of acrylic foam and polyester laminates on a polyurethane base, and when you simply go for a jog in your polyurethane trainers, think of the chemists who make these things happen.

A scientific race

On 18 March 1987, more than two thousand scientists crammed into a conference room in New York to listen to progress reports on research into superconductivity. The conference had to be relayed to hundreds of scientists in adjacent rooms.

Superconducting materials have no electrical resistance. Electric cables and electric motors made from superconductors would be far more efficient than those made from ordinary conductors. Superconducting computers and magnets would be much more powerful. However, before 1987, most scientists thought that superconductors would work only at very low temperatures. Materials that are superconducting at room temperature would be a technological breakthrough.

The first superconductors were discovered in 1911 when certain metals were cooled to below -270 °C. In 1986, two European scientists, Georg Bednorz and Alex Müller, reported they had made a superconductor at 30 °C higher (-240 °C). Scientists round the world turned their attention to superconductors. Teams of scientists in different countries raced against each other to find superconductors which would work at higher temperatures. By March 1987, the temperature limit had been raised to -179 °C. By then the political leaders of the major industrial nations realised how important superconductors could become. The race continues because the benefits would be immense. The importance of Bednorz and Müller's work in starting the race was recognised when they received the 1987 Nobel Prize for physics.

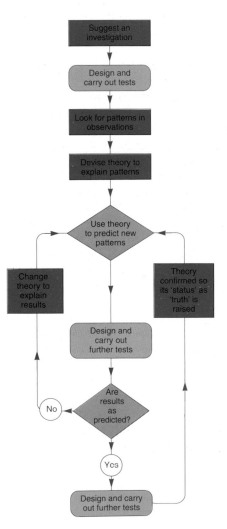

Figure B ● A good theory

Models and theories

In 1987, scientists in laboratories throughout the world worked round the clock to try to make new superconductors. Other scientists were busy with computers trying to work out why the new materials were superconducting. They made models of how the atoms of the new materials might be fixed together. They tested and changed the models using their computers until they found a model that explained what was observed in the laboratories. Understanding how and why new superconductors work is important since it can give vital clues to better superconductors.

A good theory or model is one which gives predictions that turn out to be correct. The predictions are tested in the laboratory. If they do not pass the test, the theory or model must be altered or even thrown out. The more tests a model or theory passes, the more 'faith' scientists have in it. However, complete faith is never possible. Someone, somewhere might one day make a discovery which can't be explained by the theory.

Beyond the frontiers

Who knows where a new discovery can lead? When Michael Faraday discovered how to generate electricity, he was asked 'What use is electricity, Mr Faraday?' He is reported to have replied 'What use is a new baby?' No one knows what might become of a new invention or a new discovery. More than 150 years later, life without electricity for most of us is unthinkable. Michael Faraday would have been truly astonished.

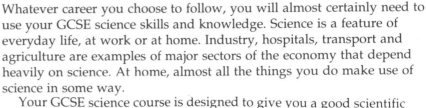

Whatever career you choose to follow, you will almost certainly need to use your GCSE science skills and knowledge. Science is a feature of everyday life, at work or at home. Industry, hospitals, transport and agriculture are examples of major sectors of the economy that depend heavily on science. At home, almost all the things you do make use of science in some way.

Your GCSE science course is designed to give you a good scientific background so that you can, if you wish, carry on with science studies after GCSE.

To work in science requires specific personal qualities in addition to academic qualifications. Scientists are very creative and imaginative people and the work of an individual scientist can bring huge benefits to everyone. For example, Alexander Fleming's discovery of penicillin has saved countless lives. But do not be misled into thinking that the life of a scientist is one of continual discoveries. Scientists have to be very patient and methodical to discover anything; they have to be good at working together and at communicating their ideas to each other and to other people. The qualifications needed to become a scientist are outlined on p. xii; the qualities needed to become a scientist are just the same as you need in your GCSE course – enthusiasm, hard work, imagination, awareness and concern.

What jobs are done by scientists? In industry, scientists design, develop and test new products. For example, scientists in the glass industry are developing amazingly clear glass for use as **optical fibres** in communication links. In medicine, scientists are continually finding applications for scientific discoveries; for example, medical scientists have developed a high-power ultrasonic transmitter for destroying kidney stones, thus avoiding a surgical operation. These are just two examples of the work of scientists. You will find scientists at work in research laboratories, industrial laboratories, forensic laboratories, hospitals, schools, on field trips, expeditions, radio, TV and lots of other places. Scientists have to be very versatile as science is a very wide and varied field.

The skills and knowledge you gain through studying science will enable you to gain the benefits of new technologies. Ask your parents what aspects of life have **not** changed since they were children – they may be stuck for an answer! New technologies force the pace of change and if you do not learn to use them, you will not share their benefits. Studying science encourages you to develop an open mind and to seek new approaches. That is why a wide range of careers involve further studies in science.

For many careers further studies in science are essential, for example, medicine, dentistry, pharmacy, engineering and computing. For other careers such as law, business studies, administration, the armed forces and retail management, studying science after GCSE will be helpful. Thus by continuing your science studies after GCSE, you keep many career options open, which is important when you are considering your choice of career.

The next steps after GCSE

Read this section carefully, bearing in mind that your working life will probably be about forty years. If you want your life ahead to be interesting; if you want to make your own decisions about what you do; if you want to make the most of your talents; then you should continue your studies. After GCSE, you can continue in full-time education at school or at college, or you can train in a job through part-time study. If you take a job without training, you will soon find that your friends who stayed on have much better prospects.

Most students aiming for a career in science or technology continue full-time study for two years, taking either GCE A- and AS-levels or a GNVQ course in Science. Successful completion of a suitable combination of these courses can then lead to a degree course at a university or a college of higher education or, alternatively, straight into employment.

The A-level route to higher education requires successful completion of a two-year full-time course, consisting of three or more A-level subjects. Students taking A-level Chemistry frequently choose their other subjects from Biology, Physics, Mathematics, Geography, Geology and Home Economics. Students with an interest in taking further courses in chemistry, biochemistry or chemical engineering are well advised to choose Mathematics. Students who plan further studies in chemistry or chemical engineering should choose Physics.

The GNVQ science route is also a two-year full-time course at advanced level, leading to a qualification equivalent to double award A-level Science. All students study some biology, chemistry and physics and can then specialise. The course is assessed continuously through tests and laboratory work. Students who do not have grade CC in science to enter advanced level GNVQ can take the one-year GNVQ science intermediate course as a preparation for the advanced course.

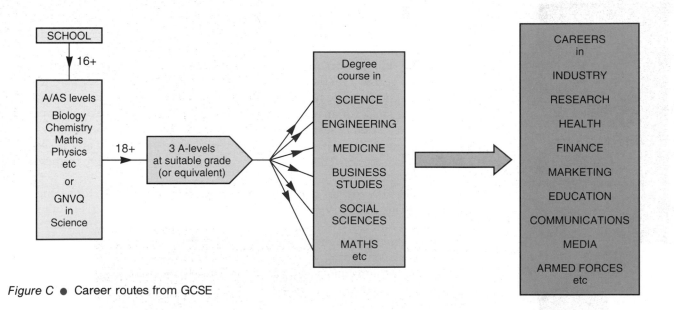

Figure C ● Career routes from GCSE

THEME A
Planet Earth

What is the Earth like 5000 km below the surface?
Why does South America have volcanoes whereas northern Europe has none?
Why does the west coast of North America suffer from earthquakes while the east coast has none?
On which atlas would you find Pangaea, Gondwanaland and Laurasia?
Why do geologists believe that South America and Africa were once joined together?
How can geologists tell the age of the Earth's surface?
You will find the answers to these questions in this theme.

TOPIC 1 *PLANET EARTH*

 1.1 **The structure of the Earth**

FIRST THOUGHTS

What is Earth like deep down, many kilometres below the surface? It sounds an impossible question to answer, but scientists have found methods of investigating the deep structure of the Earth.

The study of the Earth is called **geology**, and a person who works in this branch of science is called a **geologist**. The research work of geologists has enabled them to construct a model of Earth's structure (see Figure 1.1A).

How was Earth formed? A molten mass cooled down over millions of years. Dense materials sank deeper into the centre to form a core of dense molten rock. Less dense material remained on the surface to form a **crust** of solid rock (50 km thick). Gaseous matter outside the crust is the **atmosphere**. Earth's atmosphere is chiefly oxygen and nitrogen.

As Earth cooled, water vapour condensed to form rivers, lakes and oceans on the surface of Earth. No other planet has oceans and lakes, though Mars has some water vapour and polar ice caps.

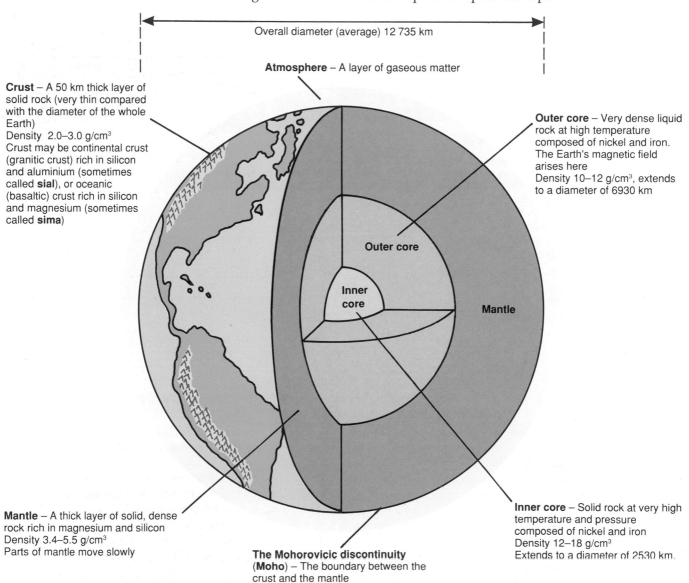

Overall diameter (average) 12 735 km

Atmosphere – A layer of gaseous matter

Crust – A 50 km thick layer of solid rock (very thin compared with the diameter of the whole Earth)
Density 2.0–3.0 g/cm³
Crust may be continental crust (granitic crust) rich in silicon and aluminium (sometimes called **sial**), or oceanic (basaltic) crust rich in silicon and magnesium (sometimes called **sima**)

Outer core – Very dense liquid rock at high temperature composed of nickel and iron. The Earth's magnetic field arises here
Density 10–12 g/cm³, extends to a diameter of 6930 km

Outer core

Inner core

Mantle

Mantle – A thick layer of solid, dense rock rich in magnesium and silicon
Density 3.4–5.5 g/cm³
Parts of mantle move slowly

The Mohorovicic discontinuity (Moho) – The boundary between the crust and the mantle

Inner core – Solid rock at very high temperature and pressure composed of nickel and iron
Density 12–18 g/cm³
Extends to a diameter of 2530 km.

Figure 1.1A ● The structure of the Earth

Volcanoes
(videodisc)

Use this disc to look at a model of the Earth's structure. You will find information on the interior of the Earth and the nature of the crust. Do you think the model is a good one? Can you think of evidence to support the model?

Earth's crust

Earth's crust is composed of rocks and **soils**. Soils have been formed by the breakdown of rocks and vegetation. The crust is divided into continental and oceanic crust. Earth's overall diameter is 12 735 km.

Continental crust
- Forms continents and their shelves
- Up to 70 km thick in mountain ranges
- Density ~ 2.7 g/cm^3
- **Age: up to 3700 million years**
- Same composition as granite rock
- Often called granitic crust
- Rich in silicon and aluminium
- The deeper parts of continental crust are of a denser material similar to oceanic crust.

Oceanic crust
- Beneath deep sea floors
- Average thickness 6 km
- Density ~ 3.0 g/cm^3
- **Age: up to 220 million years**
- Same composition as basalt rock
- Often called basaltic crust
- Rich in silicon and magnesium
- Material similar to oceanic crust is thought to lie beneath the continents.

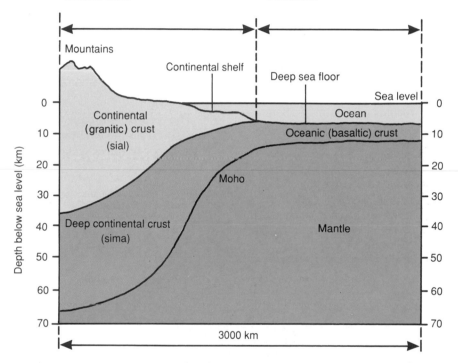

Figure 1.1B ● Continental crust and oceanic crust

SUMMARY

The Earth has a layered structure: inner core, outer core, mantle, crust and atmosphere. The crust is composed of oceanic (basaltic) crust beneath the ocean floors and continental (granitic) crust, which forms the Earth's land masses. The boundary between the crust and the mantle is called the Moho. The study of the Earth is called geology.

FIRST THOUGHTS

We cannot obtain information about the deep structure of Earth by mining or by drilling. The world's deepest mine is 3.5 km in depth, and the radius of Earth is over 6000 km. Evidence about the interior of Earth comes from the study of earthquakes and volcanoes.

1.2 ● The layered structure of the Earth

Earthquakes

The study of earthquakes is called **seismology**. About 500 000 earthquakes occur every year. Only about 1000 of these are strong enough to cause damage, and only a few are serious. An earthquake occurs when forces inside Earth become strong enough to fracture large masses of rock and make them move. The energy which is released travels through the Earth as a series of shock waves. Earthquakes are limited to the rigid part of the crust. They cannot occur in the molten part of the mantle. Most earthquakes are generated within 600 km of Earth's surface. The point where an earthquake originates is called the

Volcanoes
(videodisc)

Use this disc to find out about how movements inside the Earth can lead to earthquakes and volcanoes. There are case studies containing film of real volcanoes. Examine evidence of the devastation caused by volcanic eruptions in the past.

Can you design and construct a model seismometer? Make it a team effort. Test your model by seeing whether it records waves when you shake the table beneath it.

focus. The nearest point on Earth's surface directly above it is the **epicentre**. Shock waves are felt most strongly at the epicentre and then spread out from it. Earthquake shocks are recorded by an instrument called a **seismometer** (see Figure 1.2A).

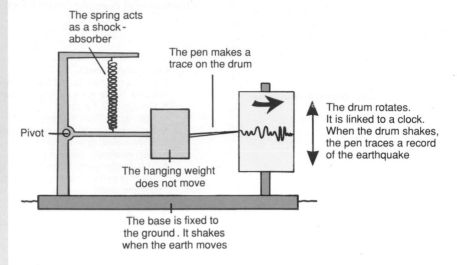

Figure 1.2A ● A seismometer

The recording is called a **seismogram**. The energy of the earthquake is measured on the **Richter scale**. Each point on the scale means an increase by a factor of ten: a scale of 5 is ten times as powerful as a scale of 4.

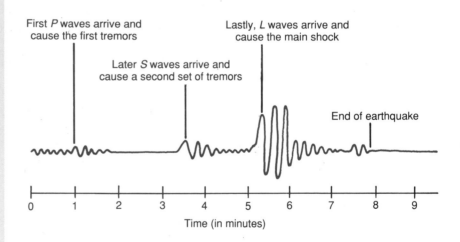

Figure 1.2B ● A seismogram

The seismogram in Figure 1.2B records three types of shock waves. P waves – **primary waves**, S waves – **secondary waves** and L waves – **long waves**. Figure 1.2C shows the difference between them.

The pattern of waves received by a seismometer depends on what the waves have passed through inside the Earth. Waves are either reflected (bounced back) or refracted (bent) when they travel from one type of material into another.

The positions of boundaries, e.g. the Moho, and the thicknesses and densities of the zones have all been worked out from the way that earthquake shock waves have been affected by passing through them. S waves do not pass through the outer core at all. Since it is known that S waves do not travel through liquids, this is evidence that the outer core is in the liquid state. This shows that its temperature must be very high.

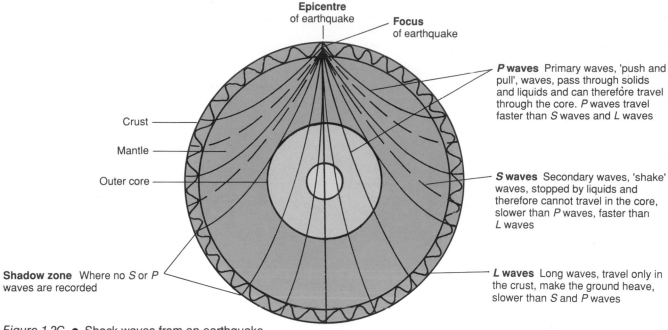

Epicentre of earthquake

Focus of earthquake

P waves Primary waves, 'push and pull', waves, pass through solids and liquids and can therefore travel through the core. *P* waves travel faster than *S* waves and *L* waves

Crust

Mantle

Outer core

S waves Secondary waves, 'shake' waves, stopped by liquids and therefore cannot travel in the core, slower than *P* waves, faster than *L* waves

Shadow zone Where no *S* or *P* waves are recorded

L waves Long waves, travel only in the crust, make the ground heave, slower than *S* and *P* waves

Figure 1.2C ● Shock waves from an earthquake

Volcanoes

The lava erupted by volcanoes gives information about the crust and upper mantle, where lava is produced, but not about deeper layers. Plotting on a map the places where volcanoes have occurred gives information about the regions of Earth where heat is being generated and causing volcanic activity.

Meteorites

Meteorites reach Earth from space. They are pieces of rock and dust which have been attracted towards Earth by Earth's gravity. Most meteorites burn up when they reach Earth's atmosphere, but some fall to Earth's surface. Geologists believe that meteorites may be samples of planetary material dating from the time of formation of the solar system.

Magnetism

Earth's magnetic field is evidence for the presence of iron in the core.

SUMMARY

Evidence for the structure of the Earth comes from:
• the patterns of shock waves produced by earthquakes
• material erupted by volcanoes
• the positions of earthquakes and volcanoes on the map
• meteorites
• the Earth's magnetic field.

CHECKPOINT

❶ The overall density of Earth is 5.5 g/cm³. The rocks in the Earth's crust have densities of 2.5 to 3.0 g/cm³. How can you explain the difference between these values and the much higher density of the whole Earth?

❷ Take a piece of string 3 m long. Imagine that this length represents the 3000 million years that have passed since the first living things appeared on Earth. Mark on the string the length that represents the 2 million years since the human race appeared.

1.3 **Earthquakes and volcanoes**

FIRST THOUGHTS

Why do many parts of Earth experience earthquakes and volcanoes? Why do some parts of Earth have neither? Geologists have found answers to this puzzle.

Earthquakes and volcanoes occur in certain parts of Earth's crust but not in others. Geologists speculated for many years on reasons for the difference. Some patterns emerge from Figure 1.3A.

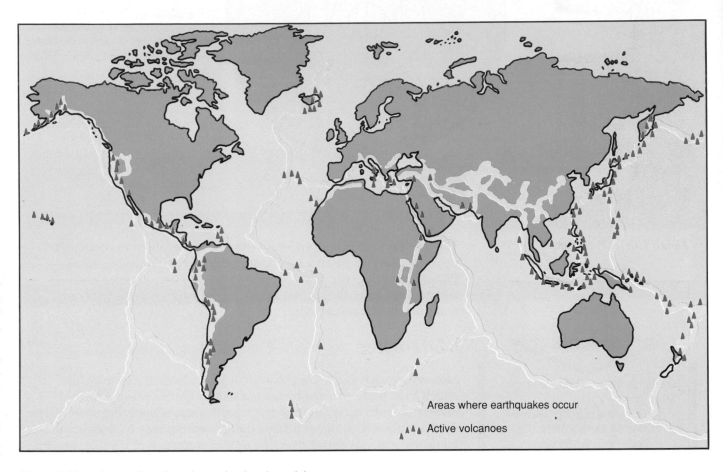

Areas where earthquakes occur

▲▲▲ Active volcanoes

Figure 1.3A ● Areas of earthquake and volcanic activity

- Earthquakes and volcanoes occur in belts of activity. The belts are hundreds of kilometres wide and thousands of kilometres long. In places, belts join up.
- On land, belts occur along chains of high mountains like the Alps.
- Beneath the sea, belts pass through the centres of oceans and through chains of volcanic islands like the Philippines.
- Surveys of the mid-ocean belts show that the sea floor rises to form chains of huge mountains beneath the sea. These mountain chains are called **oceanic ridges**. The mid-Atlantic ridge rises above sea-level to form Iceland.
- Surveys show that belts which pass through chains of islands are close to deep **oceanic trenches** on the sea floor. The same is true of mountain ranges near the edges of continents. A trench in the Pacific called the Peru–Chile trench runs parallel to the Andes.

SUMMARY

Earthquakes and volcanoes occur in belts of activity. These belts run:
- along mountain ranges
- along oceanic ridges (mountains on the sea floor)
- along oceanic trenches (deep channels in the sea floor)

1.4  Plate tectonics

Is Earth's crust really composed of separate moving pieces? It sounds amazing, but this is one of the newest scientific theories.

Geologists believe that the outer layer of Earth is made up of separate pieces called **plates**. Each plate is a piece of **lithosphere** (crust and uppermost layer of mantle) of 80–120 km in thickness. Movements in the mantle beneath make the plates move very slowly, a few centimetres per year. As a result, plates sometimes rub against each other. If stress builds up to a large extent, the plates may bend. When they spring back into shape, the ground shakes violently: there is an **earthquake**. There has to be a source of energy to produce the movement of plates. Many geologists believe that it is the heat given out when radioactive elements decay (see Figure 1.4L).

Figure 1.4A ● The result of an earthquake in Mexico City in 1986

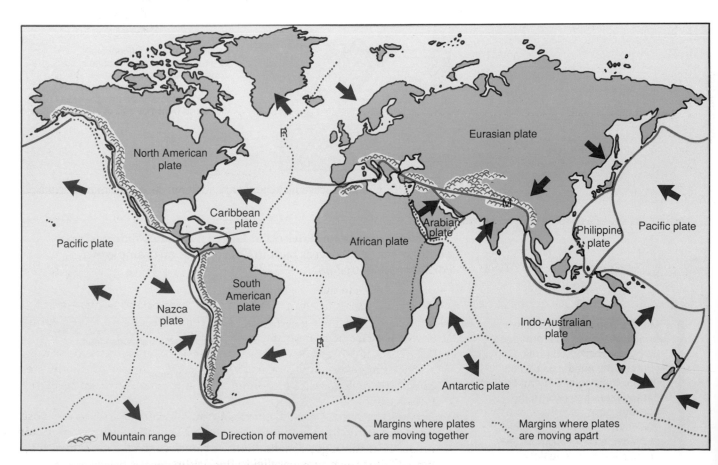

Figure 1.4B ● The plates which make up Earth's crust

Boundaries

Boundaries between plates may be **constructive**, **destructive** or **conservative**.

Constructive boundaries occur where the plates are moving apart (see Figure 1.4C). Evidence for plate movement was found in the 1960s when the volcanoes along oceanic ridges were studied. Lava erupts from these volcanoes and cools to form new oceanic crust along the edges of the plates on either side. The plates move away from the ridge and the width of the ocean floor is increased. The process is called **sea-floor spreading**. An example is the mid-Atlantic trench.

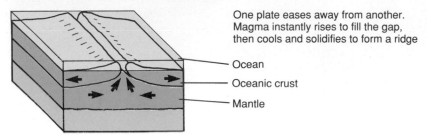

One plate eases away from another. Magma instantly rises to fill the gap, then cools and solidifies to form a ridge

— Ocean
— Oceanic crust
— Mantle

Figure 1.4C ● A constructive plate boundary. Material is added at the plate boundary and the plates move apart

Destructive boundaries occur where the plates are in collision (see Figure 1.4D). Oceanic trenches are regions where plates meet as they move together. When they meet, the edge of one plate is forced to slide beneath the other and move down into the mantle. This process is called **subduction**. The descending plate edge melts to become part of the mantle. Some oceans are shrinking in size as subduction occurs.

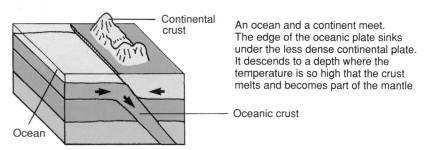

Continental crust

An ocean and a continent meet. The edge of the oceanic plate sinks under the less dense continental plate. It descends to a depth where the temperature is so high that the crust melts and becomes part of the mantle

— Oceanic crust

Ocean

Figure 1.4D ● A destructive plate boundary. Material descends from the plate boundary

Conservative boundaries occur where two plates slide past one another. The San Andreas fault in California is an example of a conservative boundary.

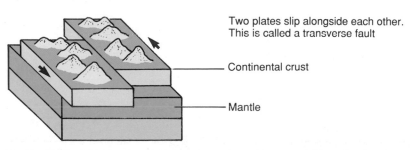

Two plates slip alongside each other. This is called a transverse fault

— Continental crust

— Mantle

Figure 1.4E ● A conservative plate boundary. No material is gained or lost

Volcanoes
(videodisc)

Use this disc to compare world maps of volcanic activity with the pattern of 'plates' in the Earth's crust. Look at the films showing activity at the plate margins. Do you think that plate movement could affect the landscape?

The conveyor belt

As material is subducted at an oceanic trench and added at an oceanic ridge, the net effect is to convey material from one edge of a plate to another. Since the mass of material taken away at oceanic trenches is equal to the mass of material added at oceanic ridges, the plate remains the same size.

This movement of plates has been called a **conveyor belt**. Continents ride the conveyor belt beneath them: as the plates move, the continents on them move. This has been happening for thousands of millions of years. Although the rate is slow, only a few centimetres a year, the continents have already travelled thousands of kilometres. About 300 million years ago, northern Europe was near the equator, and tropical forests grew there. These later decayed to form coal deposits.

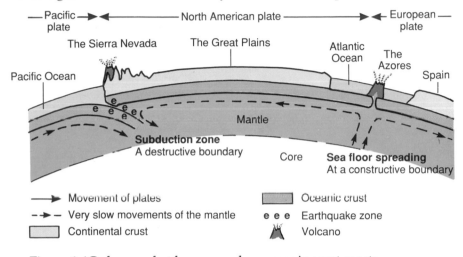

Figure 1.4F ● Movement of plates on the 'conveyor belt'

Figure 1.4G shows what happens when oceanic crust meets continental crust at a destructive plate boundary.

The oceanic plate sinks below the less dense continental plate. Some of the sediment on the surface of the oceanic plate is scraped off. It piles up on the landward side

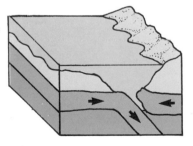

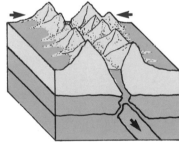

The sediment is compressed by the continental plate. It folds to form a mountain range. The Andes, on the west coast of South America, formed in this way

SUMMARY

Earth's crust and the upper part of the mantle are together called the lithosphere. Geologists believe that the lithosphere consists of separate plates and that these plates are moving slowly. At oceanic ridges, the plates are moving apart. At oceanic trenches, the plates are moving together. When plates meet, continental crust is pushed up to form mountain ranges. Plate movement gives rise to earthquakes and volcanoes.

Figure 1.4G ● Collision between an oceanic plate and a continental plate: formation of a mountain range. (When two continents collide the sediments from both are squeezed up to form mountains.)

The theory of plate tectonics

You will remember the theory that Earth originated from a cloud of hot gas which condensed to form a ball of hot liquid. The densest materials sank into the **core** of the liquid while the less dense materials rose to the surface. There they cooled and solidified to form a **crust**. Until the twentieth century, scientists considered that Earth went on cooling until the present day. They thought that as the interior cooled and contracted, the crust folded to fit it and as a result mountain ranges were formed. Scientists compared the folding crust with the wrinkling of the skin of an apple as the inner part of the fruit dries out and shrinks.

There were some mysteries which this view of Earth could not explain. For example, it is possible in the UK to find rocks and fossils which can only be formed in desert conditions. Mountains occur in ranges, not as isolated peaks. Earthquakes and volcanoes are collected in belts of activity and not scattered over Earth's surface.

Geologists noticed the similarity between the east coastline of South America and the west coastline of Africa (see Figure 1.4H). The coastlines look as though the continents could have been joined together in a previous era.

Figure 1.4H ● The 'jigsaw' fit of South America and Africa

If Africa and South America were once joined, how did the Atlantic come to be formed? In the eighteenth and nineteenth centuries, it was widely believed that a vast flood, such as Noah's flood in the Old Testament, had forced the continents apart. *And how did the Pacific ocean originate?* The nineteenth century belief was that a huge section of continental land mass was gouged out to form the moon and the hole that was left became the Pacific.

It was a German meteorologist (weather scientist) called Alfred Wegener who, in 1915, promoted the theory that the continents bordering the Atlantic were at one time joined together and had subsequently drifted apart. Wegener's ideas were not new, but he amassed more evidence in support of the theory of **continental drift** than previous workers had done.

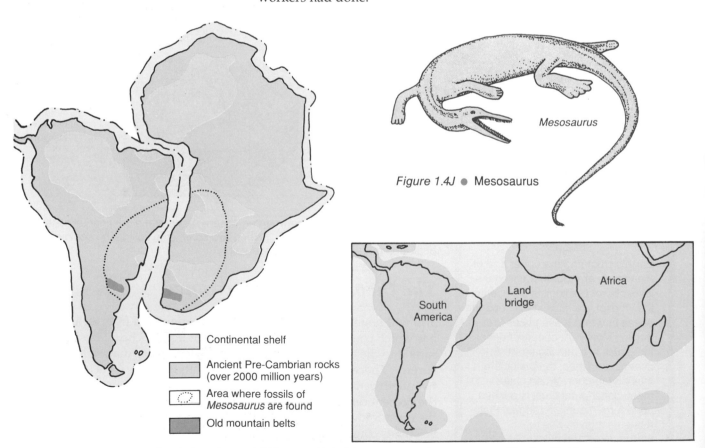

Figure 1.4J ● Mesosaurus

Continental shelf

Ancient Pre-Cambrian rocks (over 2000 million years)

Area where fossils of *Mesosaurus* are found

Old mountain belts

Figure 1.4I ● Evidence that South America and Africa were joined

Figure 1.4K ● The land bridge theory

Wegener described the matching up of geology, fossils and plant and animal populations on both sides of the Atlantic. He said, 'It is just as if we were to refit the torn pieces of a newspaper by matching the edges and check whether the lines of print run smoothly across. If they do, there is nothing left but to conclude that the pieces were in fact joined in this way.'

How can continents drift? Wegener put forward the theory that continental land masses float on a fluid, denser crust beneath them. According to his theory, continents can move horizontally, provided that there are forces acting on them, forces which last for geological eras. Wegener was unable to come up with a convincing idea to explain the force which must have driven continents apart.

The theory of continental drift is not restricted to South America and Africa. It is believed that long ago all the southern continents were joined together as one land mass, called **Gondwanaland** and the northern continents were joined to form a supercontinent called **Laurasia**. It is believed that at a still earlier time all the present-day continents were part of a single land mass, known as **Pangaea**, which began to break up and spread about 200 million years ago (see Figure 1.4O).

Wegener's theory was derided in 1915, but is accepted today. *Why did it take so long?* People's ideas were dominated by the belief that Earth was cooling, shrinking and folding. Some physicists considered that Earth's crust was too rigid for sideways movements of the kind Wegener was describing. A turning point came when geologists came to realise the magnitude of the heat produced by Earth's **radioactive materials**. It provides for all volcanic activity with plenty of heat to spare. *What happens to the excess heat?* In 1931, Arthur Holmes, a British geologist, put forward a suggestion that the excess heat was discharged by convection currents and that continental drift was powered by such currents. It took 30 years for Holmes' ideas to be widely accepted by earth scientists.

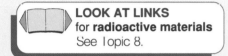

LOOK AT LINKS
for **radioactive materials**
See Topic 8.

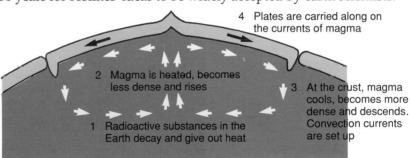

4 Plates are carried along on the currents of magma

2 Magma is heated, becomes less dense and rises

3 At the crust, magma cools, becomes more dense and descends. Convection currents are set up

1 Radioactive substances in the Earth decay and give out heat

Figure 1.4L ● Convection currents of magma

Why did the theory of continental drift eventually meet with success? Some of the evidence is summarised below.

● *Evidence of sea-floor spreading*

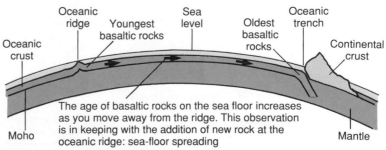

Oceanic crust
Oceanic ridge
Youngest basaltic rocks
Sea level
Oldest basaltic rocks
Oceanic trench
Continental crust
Moho
The age of basaltic rocks on the sea floor increases as you move away from the ridge. This observation is in keeping with the addition of new rock at the oceanic ridge: sea-floor spreading
Mantle

Figure 1.4M ● Sea-floor spreading

RESOURCE ACTIVITY PACK

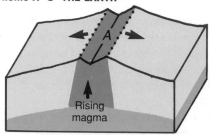

Band A The basalt which solidifies as the magma cools is magnetised normally with the North Pole as we know it

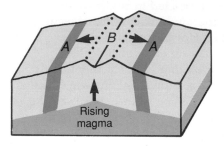

Band B The sea-floor has spread. The Earth's magnetic field has reversed. The basalt which now solidifies shows reversed magnetism. As band *B* solidifies, it pushes the two sides of band *A* apart

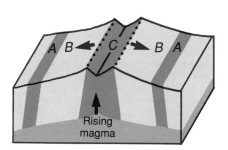

Band C The polarity of the Earth's magnetic field has changed again. Normally magnetised basalt solidifies at the centre of the ridge

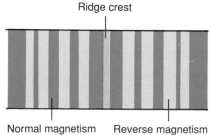

Note the symmetrical pattern of bands of normally magnetised basalt and reversely magnetised basalt on either side of the crest of the oceanic ridge. How could this symmetrical pattern have arisen without sea-floor spreading?

Figure 1.4N ● Pattern of magnetic stripes at an oceanic ridge

The youngest basalts on the ocean floor are at the oceanic ridges. As you move away from the ridges, the basalts get progressively older. The oldest parts of the oceanic crust are in the oceanic trenches and at the borders of the continents. No oceanic basalt more than 220 million years old has been found. Continental crust, in contrast, is over 1000 million years old.

Every few thousand years Earth's magnetism reverses its polarity. That is to say that over certain periods of time in the past the compass needle would have pointed to the South Pole instead of the North Pole. Oceanic floor basalts include minerals which contain iron and are therefore magnetic. As they crystallise, particles of iron line up with the magnetic field which is operating at the time.

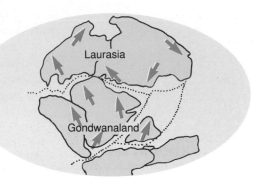

180 million years ago
The original land mass, Pangaea, had split into two major parts. Gondwanaland had started to break up

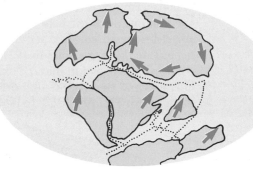

135 million years ago
Gondwanaland and Laurasia drifted northwards. The North Atlantic and Indian Oceans widened. The South Atlantic rift lengthened

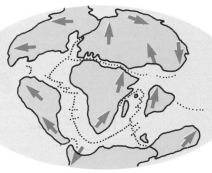

65 million years ago
South America had separated from Africa. Australia and Antarctica were still combined. The Mediterranean Sea had appeared. India was moving towards Asia

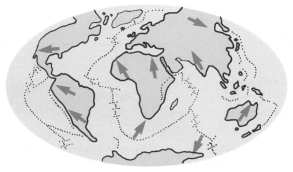

Today
South America has connected with North America. Australia has separated from Antarctica. India has collided with Asia

Figure 1.4O ● The formation of the continents

SUMMARY

According to the theory of sea-floor spreading, new oceanic crust is being formed along oceanic ridges. The new crust pushes older crust away from the ridges in mid-ocean towards the continents. In this way the sea floor is constantly spreading.

Sea-floor spreading is very uneven. As different lengths of a ridge spread by different amounts, cracks appear between them. These cracks are called transform faults. It is possible to calculate the rate of sea-floor spreading from magnetic data. It comes to 2–10 cm per year for different parts of ridges in different oceans. Figure 2.4O summarises the way continents have drifted apart due to sea-floor spreading.

This model of the Earth, in which rigid slabs of the crust jostle with one another on the surface of a sphere, is called **plate tectonics** (tectonics = construction). From the theory, it has been possible to predict accurately where earthquakes will occur. Unfortunately, earth scientists cannot yet predict the time when a future earthquake will occur.

CHECKPOINT

❶ Refer to Figure 1.4B on p 7.
   (a) Name a plate that is surrounded on all sides by subduction zones.
   (b) Name the mountain range, M.
   (c) The line R—R is the mid-Atlantic ridge. Lava erupts along this ridge. What type of lava is it? Explain how eruptions arise from the movement of plates.
   (d) The San Andreas fault passes through California. What is happening to the plates along this fault?
   (e) Why does the UK experience few earthquakes and no serious quakes?
   (f) Rocks on the Isle of Skye in Scotland show that volcanoes erupted there about 50 million years ago. Explain how this could have happened.

❷ Refer to the figure below.
   (a) Name the features A, B and C.
   (b) What type of rock has formed the mountain range, A?
   (c) State the direction of movement in (i) Plate P, (ii) Plate Q and (iii) Plate R.
   (d) Name the zone labelled E. Say what part this zone plays in plate movement.
   (e) Explain what is happening at D.
   (f) Explain how the mountain range A has been formed.

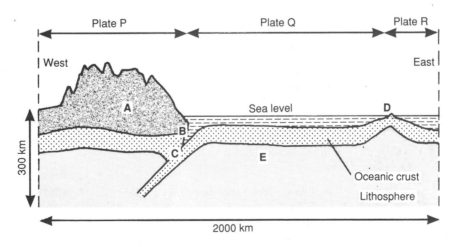

❸ The North Atlantic is spreading at a rate of about 5 cm/year.
   (a) How far will it spread during your lifetime (say 80 years)?
   (b) How tall are you?
   (c) How does your answer to (a) compare with your answer to (b)?

❹ Outline two theories which explain how the Pacific Ocean was formed: the plate tectonic theory and a historical theory.

❺ What are the three different types of boundary between plates?

⑥ Refer to the figure below.
   (a) How old is the ocean floor basalt at (i) the western edge of the ocean, (ii) 500 km east of the centre of the ocean and (iii) at the centre of the ocean?
   (b) How do the measurements shown in the graph agree with the theory of sea-floor spreading?

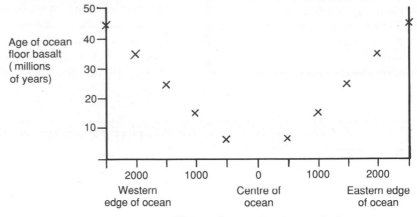

Distance from centre of ocean (km)

⑦ Imagine that you are a journalist who has just attended the 1926 conference at which Wegener put forward his theory of plate tectonics. Write an account for your magazine, *Science Weekly*, describing how other scientists reacted to Wegener's theory.

TOPIC 2 ## *ROCKS*

2.1 **Types of rock**

There are three main types of rock: igneous, sedimentary and metamorphic. In this section, you can find out which is which.

Igneous rocks

Sometimes enough heat is generated in the crust and upper mantle to melt rocks. The molten rock is called **magma**. Once formed, magma tends to rise. If it reaches Earth's surface, it is called **lava**. When cracks appear in Earth's crust, magma is forced out from the mantle on to the surface of Earth. It erupts as a **volcano**, a shower of burning liquid, smoke and dust.

Igneous rocks formed when the lava erupted from a volcano cools are:

- **basalt**, from free-flowing mobile lava,
- **rhyolite**, from slow-moving lava,
- **pumice**, from a foam of lava and volcanic gases.

Types of rock erupted by a volcano are:

- **agglomerate**, the largest rock fragments which settle close to the vent,
- **volcanic ash**, finer fragments of rock,
- **tuff**, compacted volcanic ash,
- **dust**, which may be carried over great distances by the wind. Sometimes, dust rises high into the atmosphere and affects the weather. This is what happened at Mount St Helens in the USA in 1980.

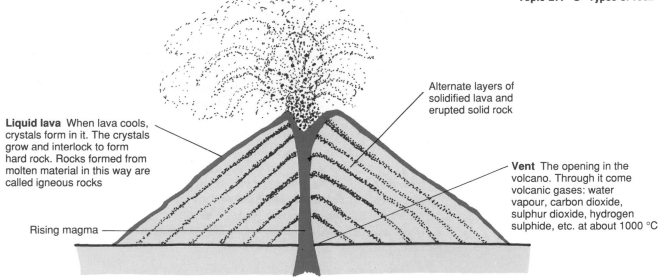

Liquid lava When lava cools, crystals form in it. The crystals grow and interlock to form hard rock. Rocks formed from molten material in this way are called igneous rocks

Alternate layers of solidified lava and erupted solid rock

Vent The opening in the volcano. Through it come volcanic gases: water vapour, carbon dioxide, sulphur dioxide, hydrogen sulphide, etc. at about 1000 °C

Rising magma

If lava solidifies in the vent, gas pressure builds up and there is likely to be a violent eruption. If this happens, lava and rock are forced out of the vent in a jet of volcanic gas. The mixture can travel rapidly down the side of a volcano causing death and destruction in its path.

Figure 2.1A ● A volcano

Figure 2.1B ● Lava rolling down the slopes of Mount Etna, one of the most active volcanoes.

Figure 2.1C ● Mount St. Helens during its second eruption. Dust and ash was spread for many miles around and stayed in the atmosphere for months

Figure 2.1D ● In 79ᴀᴅ, the city of Pompei was destroyed when a mixture of lava, rock and ash travelled quickly down the side of the volcano Vesuvias

Figure 2.1E ● The Giant's causeway in Northern Ireland is made from basalt, solidified lava

There are 540 active volcanoes, including 80 on the sea bed. In addition to **active** volcanoes (which have erupted in the past 80 years) there are **dormant** (resting) and **extinct** (dead) volcanoes. Rocks found in the Lake District and in North Wales prove that volcanoes were erupting there 450–500 million years ago, showing that Britain must have been in contact with plate margins in the past.

Sedimentary rocks

The formation of sedimentary rocks begins when solid particles settle out of a liquid or an air stream to form a **sediment**. The solid material comes from older rocks or from living organisms. All rocks exposed on Earth's surface are worn away by **weathering** and by **erosion**. The material that is worn away is transported by gravity, wind, ice, rivers and seas. The transported material may be fragments of rock, pebbles and grains of sand, or it may be dissolved in water. Eventually the transported material is deposited as a **bed** (layer) of sediment. It may be deposited on a sea bed, on a sea shore or in a desert. The beds of sediment are slowly compacted (pressed together) as other material is deposited above. Eventually, after millions of years, the pieces of sediment become joined together into a sedimentary rock. This process is called **lithification**. Examples of sedimentary rocks are:

- **limestone**, formed from the shells of dead animals,
- **coal**, formed from the remains of dead plants,
- **sandstone**, compacted grains of sand.

Metamorphic rocks

Igneous and sedimentary rocks can be changed by high temperature or high pressure into harder rocks. The new rocks are called **metamorphic rocks** (from the Greek for 'change of shape'). Examples of metamorphic rocks are:

- **marble**, formed when limestone is close to hot igneous rocks,
- **slate**, formed from clay, mud and shale at high pressure,
- **metaquartzite**, from metamorphism of sandstone.

The composition of the Earth's crust is: igneous rocks 65%, sedimentary rocks 8% and metamorphic rocks 27%. The differences are summarised in Table 2.1.

Table 2.1 ● Types of rock

	Igneous	Sedimentary	Metamorphic
Type of grain	Crystalline	Fragmental: grains do not usually interlock (They do in some limestones)	Crystalline
Direction of grain	Grains usually not lined up	Grains usually not lined up	Grains usually lined up
Mode of formation	Crystallisation of magma	Deposition of particles	Recrystallisation of other rocks
Fossil remains	Absent	May be present	Absent
Appearance when broken	Shiny	Usually dull	Shiny
Ease of breaking	Hard, not easily split, may crumble if weathered	May be soft and crumble, but some are hard to break	Hard, but may split in layers, may crumble if weathered
Examples	Basalt, granite, rhyolite, pumice	Limestone, clay, sandstone, mudstone	Marble, hornfels, slate, schist

The rock cycle

Only igneous rocks are formed from new material brought into the crust. The original crust of Earth must have been made entirely from igneous rocks. The slow processes by which metamorphic and sedimentary rocks are formed from igneous rock and also converted back into igneous rock is called the rock cycle (see Figure 2.1G).

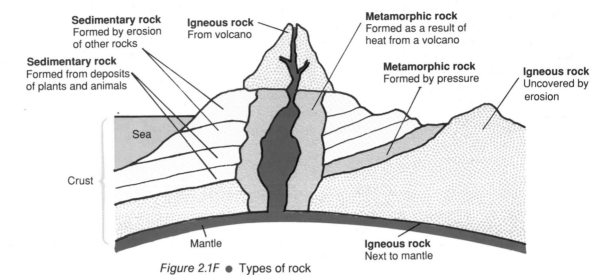

Figure 2.1F ● Types of rock

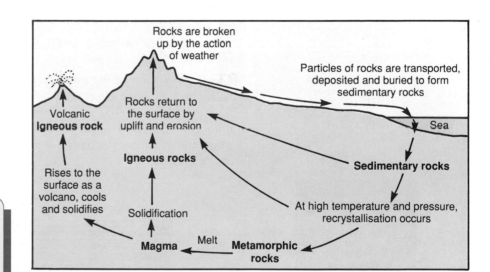

Figure 2.1G ● The rock cycle

SUMMARY

When volcanoes erupt, they emit volcanic gases, lava, ash and pieces of solid rock. Lava cools and solidifies to form igneous rock. The solid rock settles as agglomerate. Ash is compacted to form tuff. Rocks are weathered into smaller particles. These particles are deposited as a sediment, which becomes compacted to form sedimentary rock.
Igneous and sedimentary rocks may be changed by high temperature and pressure into metamorphic rock. The slow processes by which rock material is recycled are called the rock cycle.

CHECKPOINT

❶ Copy and complete this passage.
   When volcanoes erupt, molten rock called _____ streams out of the Earth. It solidifies to form _____ rocks, e.g. _____ . When deposits of solid materials are compressed to form rocks, _____ rocks are formed, e.g. _____ . The action of heat and pressure can turn _____ rocks and _____ rocks into _____ rocks.

2.2   Deformation of rocks

The deformation of rocks is caused by forces acting within the crust. It results in the formation of **folds**, **faults**, **cleavage** and **joints**. The extent of deformation produced by a force depends on the type of rock: brittle rocks may fracture to produce a fault, while soft rocks may crumple to produce a fold.

Folding

When rocks are compressed (squeezed) they may become folded. Sedimentary rocks are often folded. The beds of rock which have been laid down are no longer horizontal; they are folded to bulge upwards (an **anticline**) or downwards (a **syncline**).

(a) A fold in limestone rock

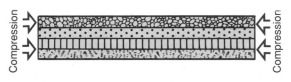

Figure 2.2A ● Folding in sedimentary rock

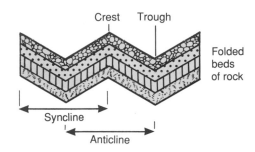

The formation of a fold

Cleavage

Cleavage is the splitting into thin sheets of rock under pressure. Slate is easily broken along **cleavage planes**. Slate is formed by the metamorphism of shales. Clay minerals and flaky minerals, such as mica, are recrystallised to lie perpendicular to the direction of maximum stress. The slate which results has a weakness in one plane along which it can be easily broken.

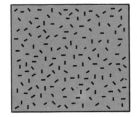

A mass of clay minerals, non-aligned

Figure 2.2B ● The formation of slate

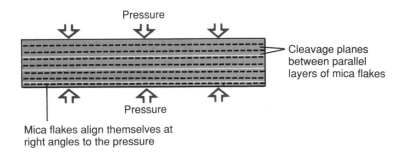

Faults

A powerful force can break rocks and cause faults (breaks) in the rock. A break allows rocks on one side of a fault to move against rocks on the other side of the fault. As long as forces keep acting, rocks will continue to keep moving against each other along the line of the fault. Since there is friction between the edges of rocks, the movement takes place in jerks. Each of these jerks may cause an earthquake.

Figure 2.2C ● (a) A vertical fault

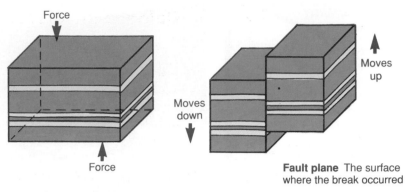

(b) A fault developing

Fault plane The surface where the break occurred

In Figure 2.2C the faults are vertical displacements. Horizontal displacements occur at faults where two of the Earth's plates slide past each other. The San Andreas fault which lies beneath California in the USA is of this type. There were severe earthquakes in California in 1838, 1906 and 1989.

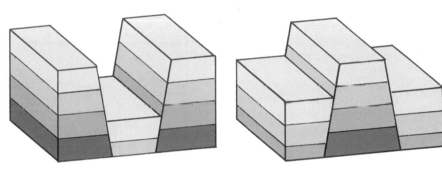

A rift valley. The subsidence of rock between faults creates a valley

A horst. A block of rock is left upstanding after rocks on both sides subside

Figure 2.2D ● (a) A rift valley. The valley has subsided between faults. It may become the course of a river or the site of a lake.
(b) A horst left standing when rocks on both sides subside

Joints

A fracture may occur without the rocks on either side moving relative to one another. Such a break cannot be called a fault because there is no displacement so it is called a **joint**. Joints in igneous rock may be caused by cooling and shrinking. Figure 2.2F shows joints in the limestone pavement at Malham Cove. Joints in sedimentary rocks may be caused by loss of water. Joints make a rock permeable to water. They provide weaknesses which may be affected by weathering.

 IT'S A FACT

Ever since 'the trembler' of 1906, California has been waiting for 'the big one'. The earthquake of 1906 registered 8.3 on the Richter scale and flattened three-quarters of San Francisco leaving 700 dead. Scientists predicted that there would be another earthquake. They predicted the site and the size of the future earthquake but could not predict the date. In 1989 an earthquake, measuring 6.9 on the Richter scale hit San Francisco killing 300 people. This time the damage was limited because engineers had designed buildings to withstand an earthquake. Scientists predict that another earthquake, of magnitude 8, is still to come.

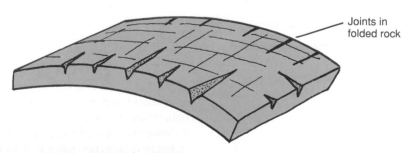

Joints in folded rock

Figure 2.2E ● Joints in folded rock

Sedimentary rock is laid down in horizontal layers. These may be deformed by:
- folding, with the formation of synclines and anticlines
- cleavage, splitting into thin sheets
- faulting, developing breaks which allow slabs of rock to slide against one another
- jointing, developing breaks without any movement of rock.

Figure 2.2F ● Limestone pavement at Malham Cove

CHECKPOINT

❶ The figure shows a section through some layers of rock.
  (a) Explain the statement: *Limestone, shale and sandstone are* **sedimentary** *rocks.*
  (b) What type of rock is granite?
  (c) Explain how the granite could have pushed through the layers of sedimentary rock.
  (d) Explain the statement: *The layers of sedimentary rock in regions B and C have been* **metamorphosed**.
  (e) Why have the sedimentary rocks at *A* and *D* not been metamorphosed?

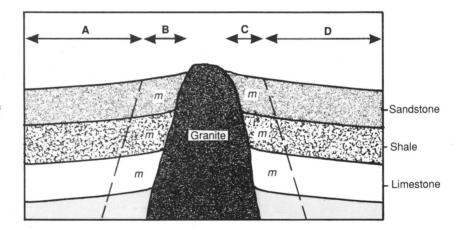

FIRST THOUGHTS

2.3 The forces which shape landscapes

What forces shaped the varied landscape that we see around us? What forces pushed up the mountain ranges, smoothed the plains and carved out the valleys?

Rocks are continually being broken down into smaller particles by forces in the environment. These processes are called **weathering**. Weathering may be brought about by:
- **physical forces**, especially in deserts and high mountains,
- **chemical reactions**, especially in warm, wet climates.

Rain

Water is an important weathering agent. Water expands when it freezes. If water enters a crack in a rock and then freezes, it will force the crack to open wider. When the ice thaws, water will penetrate further into the rock. After cycles of freeze and thaw, pieces of rock will break off.

Water reacts with some minerals, like mica. The reaction produces tiny particles which are easily transported away and deposited as a sediment of mud or clay.

Rivers and streams

LOOK AT LINKS
for the **water cycle**
See Topic 16.1.

Rivers and streams carry water back to the oceans as part of the **water cycle**. A fast-flowing stream can carry a lot of particles in suspension (see Figure 2.3A). A very fast stream can push sand and pebbles along with it.

Running water causes erosion. The bed load and the suspension load rub against the bed and sides of the river channel. In addition, there are chemical reactions between water and rocks.

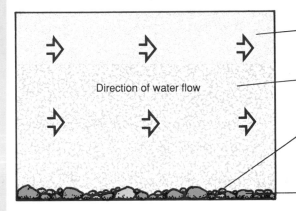

Solution load Material is dissolved in the water

Suspension load Fine particles are suspended in the flowing water

Bed load Small particles may be carried some distance, then dropped, then picked up again

Large stones and boulders may be rolled along the river bed by the force of the moving water

Direction of water flow

Figure 2.3A ● The load carried by a stream

SUMMARY

Some of the weathering processes are:
• expanding ice breaks up rock
• water breaks up some minerals, e.g. mica, and dissolves others
• running water brings about erosion.
    Sediments are deposited when rivers lose speed: on the inside curve of river bends and where rivers flow into lakes and seas.

● **Deposition**

Rivers deposit the sand and gravel which they carry. The deposits form when rivers lose speed:
• on the inside curves of river beds,
• when a river flows into the sea or a lake.

TRY THIS

Accelerated weathering

1 Find a small screw-top bottle. Fill it completely with water, and screw on the cap. Place the bottle inside a plastic bag, and tie the bag. Place the bottle in a freezer, and leave it overnight.
    What happens? What has this to do with weathering?
2 Take 4 pieces of sandstone. Soak two pieces in water, and leave the other two dry. Place all four pieces in a freezer, and treat them as shown in the table.

Piece	Wet or dry	Treatment
1	Dry	Leave in the freezer for two weeks.
2	Dry	Take out every day for two hours; then replace.
3	Wet	Leave in the freezer for two weeks.
4	Wet	Take out every day, allow to thaw and replace.

What do you observe? What conclusions about weathering can you draw from your observation?

Underground water

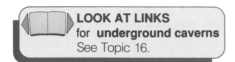

LOOK AT LINKS
for **underground caverns**
See Topic 16.

Water can pass through certain kinds of rocks. It can seep through joints in beds of sedimentary rock. Such rocks, like limestone, are described as **permeable**. Other rocks, like sandstone, have tiny spaces between their mineral grains which allow water to enter. These rocks are **porous**.

Water held in rocks below the surface is called **ground water**. Some of it finds a way back to the surface to become a **spring**. Some of it remains as an underground reserve. About one third of the UK water supply is drawn from ground water.

Rain water causes weathering of rocks. This happens also below ground. Rain water permeates the ground and dissolves some minerals; gypsum is dissolved by rain. The acid in rain water reacts with limestone to form soluble compounds. Small openings in the rock become wider and in time form large underground passages which carry underground rivers and streams. When you visit **underground caves** you walk along dried-up river beds.

SUMMARY

Underground water reacts with limestone and other rocks to form soluble compounds. In time, underground caverns are formed.

The sea

Erosion by the sea is illustrated in Figure 2.3B The cave has been gouged out by waves at a weak point in the cliff, e.g. a joint. The arch has been formed by two caves meeting back to back. The sea stack has been left where the top of an arch has fallen away.

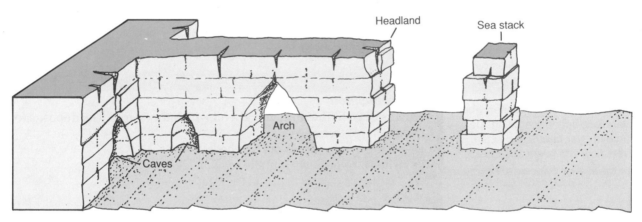

Figure 2.3B ● Landscaping by the sea

Glaciers

Figure 2.3C ● Glacier

In some regions the temperature is low enough for snow to exist all the year round. As layers of snow build up, the lower levels become compressed into a mass of solid ice. When the ice begins to move under the influence of gravity, a **glacier** develops. Glaciers move slowly downhill, usually at less than 1 m/day. As glacial ice moves over a land surface, it wears down rocks by:
- **plucking**, freezing round pieces of rock and carrying them along,
- **grinding**, wearing down the rocks over which it is moving by means of the sharp rocks which become attached to the bottom of the glacier.

When a glacier melts, all the material which it carries is deposited.

Wind

In dry desert regions, wind is a landscaping agent, for example in shaping sand dunes. Moisture holds particles of sand and soil together and makes it much more difficult for the wind to remove them. In moist regions, plants grow, and their roots bind the soil, making it much more difficult to erode.

SUMMARY

Forces which mould the landscape are:
• Rain, causing the formation of cracks in rock when it freezes
• Rivers and streams, transporting and depositing rock fragments
• Underground water, dissolving soluble minerals, e.g. gypsum, and reacting with limestone and other rocks to form soluble substances
• The sea, eroding rocks
• Glaciers, plucking rocks from the landscape and grinding land surfaces
• Wind, eroding sand and soil, especially in dry regions.

Figure 2.3D ● Sand dunes

CHECKPOINT

❶ Which of the following are needed for the formation of a glacier?
(a) mountains (b) steep-sided valleys (c) heavy snowfall (d) heavy rainfall
(e) low temperatures

❷ Which of the following are examples of (a) erosion and (b) weathering?
(i) Waves breaking against a cliff.
(ii) Rocks splitting after a cold winter.
(iii) Soil carried by the wind.
(iv) Sand carried along the bed of a river.
(v) The surface of a rock cracking after repeated heating and cooling.

❸ What kind of rock (sedimentary, igneous or metamorphic) is formed under each of the following conditions?
(a) Fragments of rock are formed by the action of frost and fall to the foot of a mountain.
(b) Particles of clay come out of suspension in still water.
(c) Dead plants sink to the bottom of a swamp.
(d) Shells and shell fragments are rolled along a sea floor.

2.4 ● Soil

As a result of weathering, rocks break up into small particles. These particles become part of the soil. Soil also contains water and **humus**: matter that was formed by the decay of dead plants and animals. Soils differ in the type of rock from which they were formed, the size of the particles, the amounts of water, salts and humus which they contain and the pH. The way in which the size of particles affects the properties of soils is shown in Figure 2.4A and Table 2.2. Most soils are mixtures of particles of different sizes and have properties in between those of sandy soils and clay soils. Such a mixture is called **loam**.

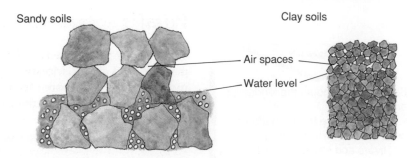

Figure 2.4A ● Clay soils and sandy soils

LOOK AT LINKS
for **humus**
See *KS: Biology*, Topic 5.1.

Table 2.2 ● The properties of sandy soils and clay soils

	Sandy soils	Clay soils
Particle size	Large, >0.2 mm	Small, <0.002 mm
Air spaces	Large	Small
Drainage	Rapid, leaving a dry soil	Slow, leaving a wet soil
Temperature	Fluctuates, tending to be higher	More consistent, tending to be lower
Cultivation	Easy to dig and plough because they are dry and loose	Difficult to dig and plough because they are wet and sticky
For plant growth . . .	Plants may suffer from lack of water. Minerals may be leached (washed) from the soil by rain	Plant roots may lack oxygen if the soil becomes waterlogged. The mineral content is high since minerals tend to stick to clay particles.

SUMMARY

Soil contains small particles of rock, water, humus and salts. Sandy soils have larger particles than clay soils. Loams are mixtures of sandy soils and clay soils. Humus improves all soils.

Humus improves soils by the following means:

- It improves the texture of clay soils by helping the particles to stick together to form crumbs.
- It improves the texture of sandy soils by increasing the soil's ability to hold water.
- Humus reduces the leaching of minerals.
- By absorbing water, humus makes soil more fertile. This is especially important in sandy soils.
- Humus provides food for detritus feeders, e.g. woodlice and earthworms, which in turn fertilise the soil.
- Its high water content enables humus to absorb heat and warm the soil.

2.5 The geological time scale

Geologists can date the different layers of rocks which they unearth. Inspecting the fossils which they find helps, and the radioactivity of the rock tells a story.

Figure 2.5A shows the **geological column**. It divides Earth's history (4600 million years) into **eras**. Each era is divided into **periods**. Human life evolved in the Quaternary period. When geologists describe a rock as being of the Silurian age, they mean that the rock was formed between 435 and 395 million years ago. It was during the Pre-Cambrian period that the Earth's crust solidified, oceans and atmospheres developed, and the first living organisms appeared.

Fossils

Geologists are able to say what period a rock dates from by examining the fossils that the rock contains. Fossils are the preserved remains of or marks made by dead plants and animals (see Figure 2.5B). If a rock contains the imprints of the shells of creatures known to have been living 300 million years ago, the rock may be Carboniferous.

Figure 2.5B ● Carboniferous fossils

Relative dating

Relative dating does not give the age of rocks but enables you to classify them (arrange them) in order of age. If one sedimentary rock lies above another, it is very likely that the upper rock is younger than the lower one (although folding of the rocks can reverse the order). Fragments of rock included in another rock must be older than the rock that surrounds them.

Dating from radioactivity

Some elements are **radioactive**. They have unstable atoms which split up (decay) to form atoms of stable elements. Imagine that a rock contains the radioactive element A, which decays very slowly to form element B. Then the ratio of B to A in the rock increases as the years go by. A measurement of the ratio of B to A will give the length of time for which A has been decaying, that is, the age of the rock. It is by radioactive dating that the age of Earth has been established as 4600 million years.

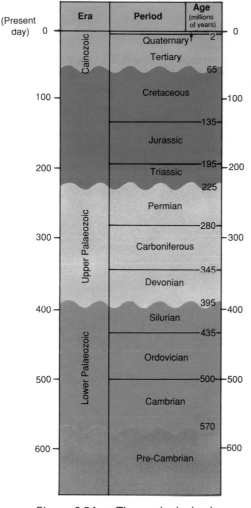

Figure 2.5A ● The geological column

CHECKPOINT

❶ A geologist made a sketch of the beds of rock in a quarry (see the figure below). He tabulated the fossils which he found in the four layers of rock.

Layer	Fossils					
	A	B	C	D	E	F
1	✓		✓	✓		
2	✓			✓	✓	✓
3	✓	✓		✓	✓	
4		✓		✓	✓	

Soil
1 Limestone
2 Shale
3 Limestone
4 Shale
Floor of quarry

(a) Which fossil is found in all 4 layers?
(b) Which fossil is found only in the youngest limestone?
(c) Which fossil can be used to give the age of layer 2? Explain your answer.

? THEME QUESTIONS

1 A river carries material along with it, as large particles, as small particles and in solution.
   (a) How does the river transport large particles, e.g. pebbles?
   (b) How does the river transport fine particles, e.g. clay?
   (c) When the river flows into a lake, what happens to (i) the large particles and (ii) the fine particles?
   (d) Explain how (c) leads to the formation of sedimentary rocks.

2 The diagram shows part of the rock cycle.

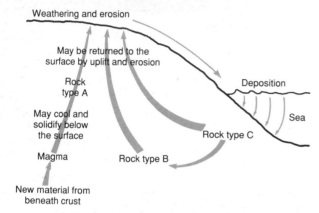

   (a) What names are given to the **three types** of rock shown in the diagram?
   (b) Suggest **two** features you would expect sedimentary rocks to have.
   (c) Suggest **two** conditions necessary to convert sedimentary rocks to metamorphic rocks.
   (d) Suggest **two** methods by which rocks at the surface might be weathered.
   (e) The diagram below shows the type and structure of the rocks and the underground water level in a place where crude oil (petroleum) and natural gas (methane) are likely to be trapped. Copy the diagram and write the letter O on the diagram where you would expect crude oil to be and the letter G where you would expect natural gas to be. Show the oil level.

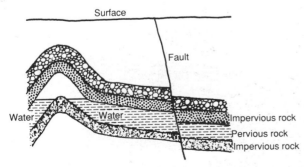

   (f) The type and structure of the rocks underground can affect our lives in many ways. The following table shows two ways in which this can happen. Copy the table and add **three** more ways.

Cause	Effect
Local stone used as building material	Many local buildings have a similar appearance
Underground deposits of crude oil (petroleum)	Chemical industries have been set up in the area

(SEG)

3 The following diagrams represent changes, at the same place on Earth, over millions of years to the present day.

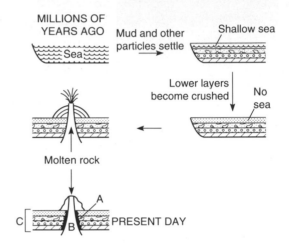

Study the diagrams and answer the following questions.
   (a) Name the three types of rock labelled on the "PRESENT DAY" diagram. Choose names from this list:
   igneous, metamorphic, sedimentary

Rock area	Type of rock
A	
B	
C	

   (b) State, using the appropriate letter, the rock area in which the chance of finding the remains of early plants and animals (fossils) is
   (i) most likely
   (ii) not at all possible.
   (c) Explain your answer to (b)(ii).

(WJEC)

4 A certain soil contains particles of clay and particles of sand.
   (a) State one advantage and one disadvantage which clay particles give to the soil.
   (b) State one advantage and one disadvantage which sand particles give to the soil.
   (c) State two other things which the soil needs to become good for growing plants.

THEME B
Matter and Particles

Earth and all the other planets in the universe are made of matter. How many different kinds of matter are there? They are countless: far too many for one book to mention, but in Theme B we start to answer the question. What does matter do: how does it behave? The study of how matter behaves is called chemistry. Can one kind of matter change into a different kind of matter? The answer to the question is 'yes, it can': provided that it receives energy.

MATTER

FIRST THOUGHTS **3.1** **The states of matter**

What is a state of matter? How do states of matter differ? These are questions for you to explore in this topic.

Figure 3.1A ● Ice skaters

Everything you see around you is made of matter. The skaters, the skates, the ice and all the other things in Figure 3.1A are different kinds of matter.

The skaters are overcoming friction as they glide across the frozen pond. *How do they manage this?* They change one kind of matter, ice, into another kind of matter, water. A thin layer of water forms between a skate and the ice, and this reduces friction and enables the skater to glide across the pond. The water beneath the skate refreezes as the skater moves on. *How are skaters able to melt ice? Why does it refreeze behind them?* Read on to find out.

The different kinds of matter are **solid** and **liquid** and **gaseous** matter. These are called the **states of matter**. Table 3.1 summarises the differences between the three chief states of matter. The symbols (s) for solid, (l) for liquid and (g) for gas are called **state symbols**. Later you will use another state symbol, (aq), which means 'in aqueous (water) solution'.

Table 3.1 ● States of matter

State	Description
Solid (s)	Has a fixed volume and a definite shape. The shape is usually difficult to change.
Liquid (l)	Has a fixed volume. Flows easily; changes its shape to fit the shape of its container.
Gas (g)	Has neither a fixed volume nor a fixed shape; changes its volume and shape to fit the size and shape of its container. Flows easily; liquids and gases are called **fluids**. Gases are much less dense than solids and liquids.

SUMMARY

The three chief states of matter are:
* solid (s): fixed volume and shape,
* liquid (l): fixed volume; shape changes,
* gas (g): neither volume nor shape is fixed.
Liquids and gases are fluids.

3.2 Pure substances

Most of the solids, liquids and gases which you see around you are mixtures of substances.
- Rock salt, the impure salt which is spread on roads in winter, is a mixture of salt and sand and other substances.
- Crude oil is a mixture of petrol, paraffin, diesel fuel, lubricating oil and other liquids.
- Air is a mixture of gases.

Some substances, however, consist of one substance only. Such substances are **pure substances**. For example, from the mixture of substances in rock salt chemists can obtain pure salt which is 100% salt.

Figure 3.2A ● Rock salt and pure salt

SUMMARY

A pure substance is a single substance.

3.3 Density

Table 3.1 tells you that gases are much less **dense** than solids and liquids. What does **dense** mean? What is **density**? The two lengths of car bumper shown in Figure 3.3A have the same volume. You can see that they do not have the same mass. The steel bumper is heavier than the plastic bumper. This is because steel is a more **dense** material than the plastic; steel has a higher **density** than the plastic has.

$$\text{Density} = \frac{\text{Mass}}{\text{Volume}}$$

The unit of density is kg/m^3 or g/cm^3. The density values of some common substances are shown in Table 3.2.

Table 3.2 ● Density

Substance	Density (g/cm³)
Air	1.2×10^{-3}
Aluminium	2.7
Copper	8.92
Ethanol	0.789
Gold	19.3
Hydrogen	8.33×10^{-5}
Iron	7.86
Lead	11.3
Methane	6.67×10^{-4}
Oxygen	1.33×10^{-3}
Silver	10.5
Water	1.00

Figure 3.3A ● Two objects with the same volume

You can see that the gases are much less dense than any of the solid or liquid substances.

SUMMARY

Density = $\dfrac{\text{Mass}}{\text{Volume}}$

Mass = Volume × Density

Volume = $\dfrac{\text{Mass}}{\text{Density}}$

The density triangle:

To find the quantity you want, cover up that letter.
The other letters in the triangle show you the formula.

CHECKPOINT

❶ A worker in an aluminium plant taps off 300 cm³ of the molten metal. It weighs 810 g. What is the density of aluminium?

❷ An object has a volume of 2500 cm³ and a density of 3.00 g/cm³. What is its mass?

❸ Mercury is a liquid metal with a density of 13.6 g/cm³. What is the mass of 200 cm³ of mercury?

❹ 50.0 cm³ of metal A weigh 43.0 g
52.0 cm³ of metal B weigh 225 g
Calculate the density of each metal. Say whether they will float or sink in water.

3.4   Change of state

Matter can change from one state into another. Some changes of state are summarised in Figure 3.4A.

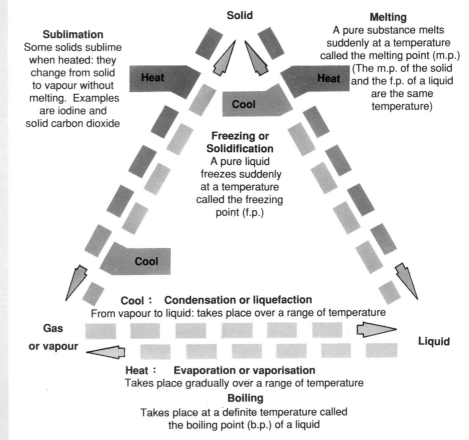

Figure 3.4A ● Changes of state

Sometimes gases are described as **vapours**. A liquid evaporates to form a vapour. A gas is called a vapour when it is cool enough to be liquefied.

Vapour $\xrightarrow{\text{Either cool or compress without cooling}}$ Liquid

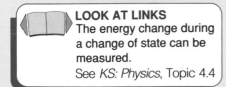

LOOK AT LINKS
The energy change during a change of state can be measured.
See *KS: Physics*, Topic 4.4

SUMMARY

Matter can change from one state into another.
The changes of state are:
• melting
• freezing or solidification
• evaporation or vaporisation
• condensation or liquefaction
• sublimation

The hotter a liquid is, the faster it evaporates. At a certain temperature, it becomes hot enough for vapour to form in the body of the liquid and not just at the surface. Bubbles of vapour appear inside the liquid. When this happens, the liquid is boiling, and the temperature is the boiling point of the liquid.

CHECKPOINT

❶ Copy the diagram below. Fill in the names of the changes of state. (Some of them have two names.)

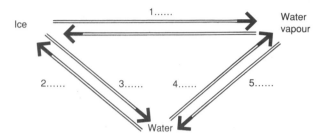

❷ Give the scientific name for each of these changes.
(a) A puddle of water gradually disappears.
(b) A mist appears on your glasses.
(c) A mothball gradually disappears from your wardrobe.
(d) The change that happens when margarine is heated.

 FIRST THOUGHTS

3.5 Finding melting points and boiling points

You can identify substances by finding their melting points and boiling points. Accuracy is essential, as you will see in this section

● Finding the melting point of a solid
Figure 3.5A shows an apparatus which can be used to find the melting point of a solid. To get an accurate result:
• first note the temperature at which the solid melts,
• allow the liquid which has been formed to cool and note the temperature at which it freezes.

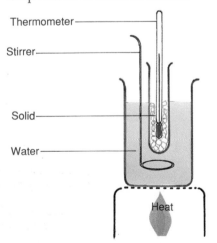

First, find the m.p. of the solid. Heat the water in the beaker. Stir. Watch the thermometer. When the solid melts, note the temperature. Stop heating

Now find the f.p. of the liquid. Let the liquid cool. watch the thermometer. When the liquid begins to freeze, the temperature stops falling. It stays the same until all the liquid has solidified

Figure 3.5A ● Finding the melting point of a solid (for solids which melt above 100 °C, a liquid other than water must be used)

SCIENCE AT WORK
Advanced laboratories use a temperature sensor and a microcomputer to find melting points.

The apparatus shown in Figure 3.5A will work between 20 °C and 100 °C. For solids with melting points above 100 °C, a liquid with a higher boiling point than water must be used. For liquids which freeze below room temperature, a liquid with a lower freezing point than water must be used. A mixture of ice and salt can be used down to –18 °C (see Figure 3.5B).

Stir. Watch the thermometer. The temperature falls and then remains constant at freezing point of the liquid while the liquid freezes

— Thermometer
— Stirrer
— Water
— Ice-salt mixture (freezes below 0 °C)

Figure 3.5B ● Finding the freezing point of water

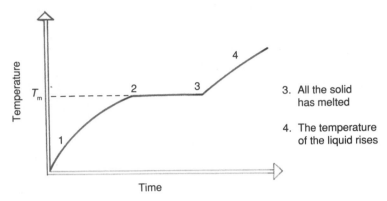

1. The temperature of the solid rises.
2. The solid starts to melt. The temperature remains steady at T_m (the m.p. of the solid) as long as there is any solid left

3. All the solid has melted

4. The temperature of the liquid rises

Figure 3.5C ● A graph of temperature against time when a pure solid melts

SUMMARY

The apparatus shown in Figure 3.5A or 3.5B can be used to find the melting point of a solid and the freezing point of a liquid.
- The temperature of a pure solid stays constant while it is melting.
- The temperature of a pure liquid stays constant while it is freezing.

If the solid is a pure substance, a graph of temperature against time as the solid melts will look like Figure 3.5C.

If you have a pure solid and you do not know what it is, you can use its melting point to find out. Chemists have drawn up lists of pure substances with their melting points. You find the melting point of the unknown solid and compare it with the listed melting points. *Which of the solids could be X?*

Solid	Melting point (°C)
Unknown solid X	116
Benzamide	132
Butanamide	116
Ethanamide	82

Figure 3.5D ● A truck spreading salt on an icy road in winter

SUMMARY

The melting point can be used to identify an unknown pure solid. The presence of an impurity lowers the melting point

If a solid is not pure, the melting point will be low, and the impure solid will melt gradually over a range of temperatures. Look at Figure 3.5D and explain why the ice on the road melts.

● *Finding the boiling point of a liquid*

Figure 3.5E shows an apparatus which can be used to find the boiling point of a non-flammable liquid. For a flammable liquid, a distillation apparatus, e.g. Figure 11.6A, must be used.

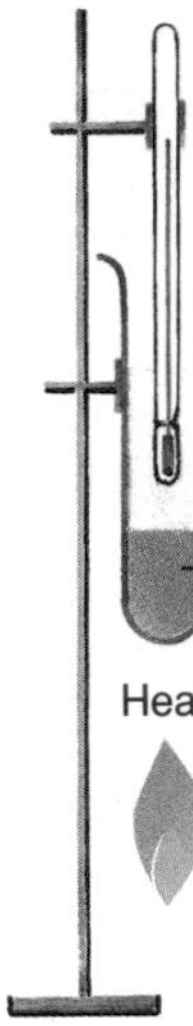

When the liquid boils, note the temperature shown by the thermometer. This is the boiling point of the liquid

Test tube

Liquid

Heat

Figure 3.5E ● Finding the boiling point of a non-flammable liquid

1. The temperature of the liquid rises
2. The liquid starts to boil. The temperature stays constant at the boiling point until all the liquid has vaporised

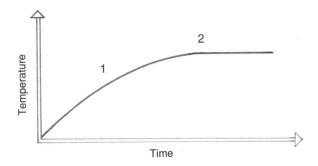

Figure 3.5F ● A graph of temperature against time when a pure liquid is heated

SUMMARY

The apparatus in Figure 3.5E can be used to find the boiling point of a liquid. The temperature of a pure liquid stays constant while it is boiling. Dissolving a solid in a liquid raises the boiling point.

While a pure liquid is boiling, the temperature remains steady at its boiling point. All the heat going into the liquid is used to vaporise the liquid and not to raise its temperature. Figure 3.5F shows a graph of temperature against time when a pure liquid is heated.

A mixture of liquids, such as crude oil, boils over a range of temperature. If a solid is dissolved in a pure liquid, it raises the boiling point.

Figure 3.5G ● Why will he have difficulty in brewing a strong cup of tea?

The boiling point of a liquid depends on the surrounding pressure. If the surrounding pressure falls, the boiling point falls. The boiling point of water on a high mountain is lower than 100 °C. An increase in the surrounding pressure raises the boiling point (see Figure 3.5H).

1. The lid is tightly fastened to the pan

2. A rubber sealing ring prevents steam escaping

3. The pressure of the steam builds up. The b.p. of water rises to about 120 °C. Food cooks more quickly than at 100 °C

4. The control valve. If the pressure of steam becomes too high, it lifts the weight. Some steam escapes, and the weight falls back into position

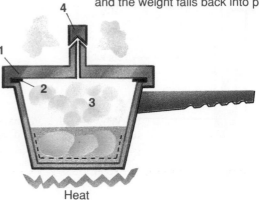

Heat

Figure 3.5H ● How a pressure cooker works

Boiling points are stated at standard pressure (atmospheric pressure at sea level). Boiling points can be used to identify pure liquids. You take the boiling point of the liquid you want to identify; then you look through a list of boiling points of known liquids and find one which matches. *Which of the liquids could be the unknown substance X?*

No two substances have both the same boiling point and the same melting point.

Liquid	Boiling point (°C)
Substance X	111
Benzene	80
Methylbenzene	111
Naphthalene	218

SUMMARY

The boiling point of a liquid is stated at standard pressure.
- At lower pressures, the boiling point is lower.
- At higher pressure, the boiling point is higher.

CHECKPOINT

❶ A pupil heated a beaker full of ice (and a little cold water) with a Bunsen burner. She recorded the temperature of the ice at intervals until the contents of the beaker had turned into boiling water. The table shows the results which the pupil recorded.

Time (*minutes*)	0	2	4	6	8	10	12	14	16
Temperature (°C)	0	0	0	26	51	76	100	100	100

(a) On graph paper, plot the temperature (on the vertical axis) against time (on the horizontal axis).
(b) On your graph, mark the m.p. of ice and the b.p of water.
(c) What happens to the temperature while the ice is melting?
(d) What happens to the temperature while the water is boiling?
(e) The Bunsen burner gives out heat at a steady rate. Explain what happens to the heat energy (i) when the beaker contains a mixture of ice and water at 0 °C, (ii) when the beaker contains water at 100 °C and (iii) when the beaker contains water at 50 °C.

❷ Bacteria are killed by a temperature of 120 °C. One way of sterilising medical instruments is to heat them in an autoclave (a sort of pressure cooker: see the figure opposite). The table shows the effect of pressure on the boiling point of water.

Boiling point of water (°C)	Pressure (kPa)
80	47
90	68
100	101
110	140
120	195
130	273

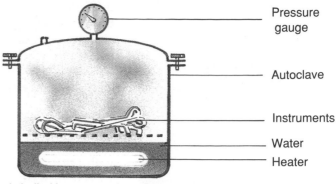

Pressure gauge

Autoclave

Instruments

Water

Heater

(a) Why are the instruments not simply boiled in a covered pan?

(b) What pressure must the autoclave reach to sterilise the instruments?

(c) What is the value of standard pressure in kPa (kilopascals)?

FIRST THOUGHTS

3.6 Properties and uses of materials

Types of matter from which things are made are called **materials**. Different materials are used for different jobs. The reason is that their different **properties** (characteristics) make them useful for different purposes. You will see a list of properties in the margin. Some of them have been mentioned earlier in this topic. Conduction of heat and electricity will be covered in later topics. Now let us look at the rest.

Properties of materials
- Hardness
- Toughness
- Strength
- Flexibility
- Elasticity
- Solubility
- Density (Topic 3.3)
- Melting point (Topic 3.5)
- Boiling point (Topic 3.5)
- Conduction of heat (*KS: Physics*, Topics 6.5 and 6.6)
- Conduction of electricity (Topic 9.1)

Hardness

It is difficult to change the shape of a hard material. A hard material will dent or scratch a softer material. A hard material will withstand impact without changing. Table 3.3 shows the relative hardness of some materials on a 1–10 scale.

Table 3.3 ● Relative hardness of materials

Material	Relative hardness	Uses
Diamond	10.0	Jewellery, cutting tools
Silicon carbide	9.7	Abrasives
Tungsten carbide	8.5	Drills
Steel	7–5	Machinery, vehicles, buildings
Sand	7.0	Abrasives, e.g. sandpaper
Glass	5.5	Cut glass can be made by cutting glass with harder materials.
Nickel	5.5	Used in coins; hard-wearing
Concrete	5–4	Building material
Wood	3–1	Construction furniture
Tin	1.5	Plating steel food cans

LOOK AT LINKS
for **diamond**
To find out what makes diamond so hard, see Topic 5.9; for metals see Topic 19.

Toughness and brittleness

Construction workers on building sites wear 'hard hats' to protect themselves from falling objects. A hard hat is designed to absorb the energy of an impact. The hat material is **tough**, that is, it is difficult to break, although it may be dented by the impact. In comparison, a brick is

RESOURCE ACTIVITY PACK

Figure 3.6A ● Hardness

difficult to dent and will shatter if dropped onto a concrete floor. The brick is **brittle**. Glass is another brittle material. These materials cannot absorb the energy of a large force without cracking. If a still larger force is applied the cracks get bigger and the materials shatter.

Composite materials

To make brittle materials tougher you have to try to stop them cracking. Mixing a brittle material with a material made of fibres, e.g. glass fibre or paper, will often do this. The fibres are able to absorb the energy of a force and the brittle material does not crack. Plaster is a brittle material. Plasterboard is much tougher. It is made by coating a sheet of plaster with paper fibres. It is a **composite material**. Glass-reinforced plastic (GRP) is a mixture of glass fibre and a plastic resin.

Figure 3.6C ● A GRP canoe

Paper
Made of fibres

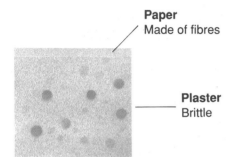

Plaster
Brittle

Figure 3.6B ● Plasterboard

Figure 3.6D ● Corrugated plastic roof

Concrete has great compressive strength, but it can crack if it is stretched. For construction purposes, **reinforced concrete** is used. Running through this composite material are steel rods which act like the glass fibres in GRP.

The shape of a piece of material alters its strength. Corrugated cardboard is used for packaging. Corrugated iron sheet and corrugated plastic sheet are used for roofing.

SUMMARY

Materials may be:
- hard – resistant to impact, difficult to scratch,
- tough – difficult to break, will 'give' before breaking,
- brittle – will break without 'giving',
- composite – made of more than one substance.

Strength

A strong material is difficult to break by applying force. The force may be a stretching force (e.g. a pull on a rope), or a squeeze (e.g. a vice tightening round a piece of wood), or a blow (e.g. a hammer blow on a lump of stone). A material which is hard to break by stretching has good **tensile strength**; a material which is hard to break by crushing has good **compressive strength**. The tensile strength of a material depends on its cross-sectional area.

Flexibility

LOOK AT LINKS
There is further discussion of materials in Topic 30.3.

While a material is pulled it is being stretched: it is under **tension**. While a material is squashed it is being compressed: it is under **compression**. When a material is bent, one side of the material is being stretched while the opposite side is being compressed. A material which is easy to bend without breaking has both tensile strength and compressive strength. It is **flexible**.

Figure 3.6E ● It's flexible

Elasticity

You can change the shape of a material by applying enough force. When you stop applying the force, some materials retain their new shapes; these are **plastic** materials. Other materials return to their old shape when you stop applying the force; these are **elastic materials**.

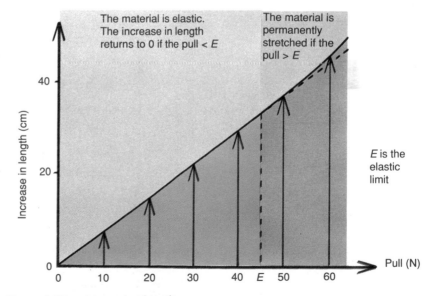

Figure 3.6F ● Increasing length

SUMMARY

Materials may have tensile strength (resistance to stretching), compressive strength (resistance to pressure), flexibility (both tensile and compressive strength) or elasticity (the ability to return to their original shape after being stretched).

When you pull an elastic material, it stretches – increases in length. At first, when you double the pull, you double the increase in length. As the pull increases, however, you reach a point where the material no longer returns to its original shape. This pull is the **elastic limit** of the material. Increasing the pull still more eventually makes the material break (see Figure 3.6F).

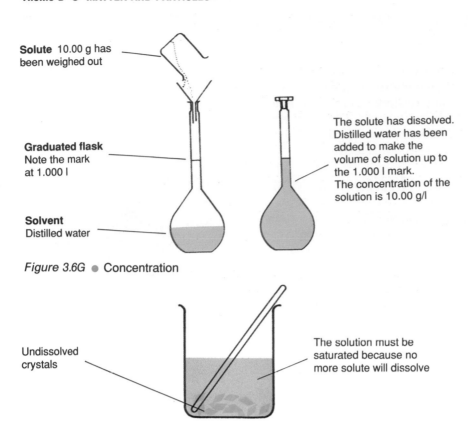

Solute 10.00 g has been weighed out

Graduated flask
Note the mark at 1.000 l

Solvent
Distilled water

The solute has dissolved. Distilled water has been added to make the volume of solution up to the 1.000 l mark. The concentration of the solution is 10.00 g/l

Figure 3.6G ● Concentration

Undissolved crystals

The solution must be saturated because no more solute will dissolve

Figure 3.6H ● A saturated solution

Solubility

A solution consists of a **solvent** and a **solute**. The solute, which may be a solid or a liquid or a gas, dissolves in the solvent. Water is the most common solvent, but there are many others such as ethanol (alcohol) and trichloroethane (trichlor). A **concentrated** solution contains a high proportion of solute; a **dilute** solution contains a small proportion of solute. The **concentration** of a solution is the mass of solute dissolved in a certain volume, say one litre, of the solution (see Figure 3.7G).

A solution that contains as much solute as it is possible to dissolve at that temperature is a **saturated** solution (Figure 3.7H).

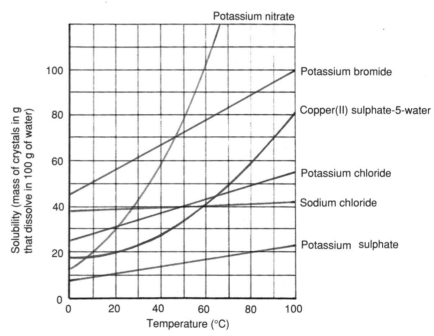

Figure 3.6I ● Some solubility curves

The concentration of solute in a saturated solution is the **solubility** of the solute at that temperature. Solubility is stated as the mass in grams of the solute that will saturate 100 grams of solvent at a certain temperature. A graph of solubility against temperature is called a **solubility curve**.

As you can see from the Figure 3.6I the solubilities of most solutes increase with temperature. When a saturated solution is cooled, it can hold less solute at the lower temperature and some solute comes out of solution: it crystallises. Gases, on the other hand, are less soluble at higher temperatures.

SUMMARY

Solubility = the mass of solute that will saturate 100 g of solvent at a stated temperature.

❶ (a) Name two materials than can be used to drill through steel.
  (b) Why can a steel blade slice through tin?
  (c) Name a material that is used to make wood smooth.
  (d) Why are diamond-tipped saws used to slice through concrete?
  (e) Name two materials which can be used to cut glass.
  (f) Why is nickel a better coinage material than tin?

❷ (a) Name two tough materials. Say what they are used for.
  (b) Name two brittle materials. Say what can be done to make them tougher.

❸ Use the information in the table below to answer this question.

Cross-sectional area (mm^2)	1	2	4	6	8	10
Breaking force (N)	0.5	1	2	3	4	5

  (a) On a piece of graph paper, plot the breaking force (on the vertical axis) against the cross-sectional area (on the horizontal axis).
  (b) From the shape of the graph, say how the breaking force alters when the cross-sectional area (i) doubles, (ii) increases by a factor of 10.
  (c) Write a sentence saying how the tensile strength of the material depends on the cross-sectional area.

❹ Name a material to fit each of the following descriptions:
  hard, soft, strong, flexible, tough, brittle, composite.

❺ Do the following materials need good tensile strength or good compressive strength?
  a tent rope, a tow bar, a stone wall, an anchor chain, a concrete paving stone, a building brick, a rope ladder.

❻ Classify the following materials as either plastic or elastic:
  plasticine, pottery clay, a rubber band, a balloon, 'potty putty'.

❼ Taran needs to make a solution of some large crystals. Which of the following suggestions would help her?
  (a) Crush the crystals before adding them to water.
  (b) Add the crystals to water one at a time.
  (c) Warm the water.
  (d) Stir the mixture of water and crystals.
  (e) Add the water dropwise to the crystals.

❽ Refer to Figure 3.6l.
  (a) What mass of potassium bromide will dissolve in 100 g of water at
    (i) 20 °C and (ii) 100 °C?
  (b) What will happen when the solution in (a) is cooled from 100 °C to 20 °C?
  (c) What mass of water is needed to dissolve 40 g of potassium sulphate at 80 °C?
  (d) When 1.00 kg of water saturated with copper(II) sulphate-5-water is cooled from 70 °C to 20 °C, what mass of solid crystallises?
  (e) A 100 g mass of water is saturated with sodium chloride and potassium chloride at 100 °C. When the solution is cooled to 20 °C, what mass of (i) sodium chloride and (ii) potassium chloride crystallises?

❾ On graph paper, plot a solubility curve for potassium chlorate(V) from the data given below.

Solubility (g/100 g water)	8	11	14	18	24	31	39	50
Temperature (°C)	20	30	40	50	60	70	80	90

One kilogram of water is saturated with potassium chlorate(V) at 90 °C and then cooled to 40 °C. What mass of crystals will separate from solution?

TOPIC 4 PARTICLES

4.1 The atomic theory

FIRST THOUGHTS

What exactly is an atom, and why did the atomic theory take nearly 2000 years to catch on? Combine your imagination and your experimental skills in this section to find out why.

The idea that matter consists of tiny particles is very, very old. It was first put forward by the Greek thinker Democritus in 500 BC. For centuries the theory met with little success. People were not prepared to believe in particles which they could not see. The theory was revived by a British chemist called John Dalton in 1808. Dalton called the particles **atoms** from the Greek word for 'cannot be split'. According to Dalton's **atomic theory**, all forms of matter consist of atoms.

The atomic theory explained many observations which had puzzled scientists. Why are some substances solid, some liquid and others gaseous? When you heat them, why do solids melt and liquids change into gases? Why are gases so easy to compress? How can gases diffuse so easily? In this topic, you will see how the atomic theory provides answers to these questions and many others.

4.2 Elements and compounds

LOOK AT LINKS
for **elements** and **compounds**
This account of elements and compounds continues in Topic 5.

There are two kinds of pure substances: **elements** and **compounds**. An element is a simple substance which cannot be split up into simpler substances. Iron is an element. Whatever you do with iron, you cannot split it up and obtain simpler substances from it. All you can do is to build up more complex substances from it. You can make it combine with other elements. You can make iron combine with the element sulphur to form iron sulphide. Iron sulphide is made of two elements chemically combined: it is a compound.

The smallest particle of an element is an **atom**. In some elements, atoms do not exist on their own: they join up to form groups of atoms called **molecules**. Figure 4.2A shows models of the molecules of some elements.

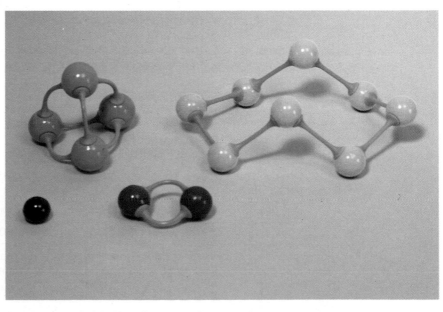

Figure 4.2A ● Models of molecules (helium, He; oxygen, O_2; phosphorus, P_4; sulphur, S_8)

LOOK AT LINKS
for **compounds**
The differences between
mixtures and compounds
are described in
Topic 5.10.

LOOK AT LINKS
for **chemical bonds**
See Topic 10.

LOOK AT LINKS
for **ions**
See Topic 9.

SUMMARY

Compounds are pure substances
that contain two or more elements
chemically combined. Many
compounds are made up of
molecules; others are made up of
ions. All gases consist of
molecules.

A compound is a pure substance that contains two or more elements. The elements are not just mixed together: they are chemically combined. Many compounds consist of molecules, groups of atoms which are joined together by **chemical bonds**. All gases, whether they are elements or compounds, consist of molecules. Figure 4.2B shows models of molecules of some compounds.

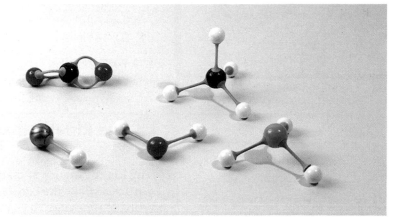

Figure 4.2B ● Models of molecules of some compounds (carbon dioxide, CO_2; methane, CH_4; ammonia, NH_3; water, H_2O; hydrogen chloride, HCl)

Some compounds do not consist of molecules; they are made up of electrically charged particles called **ions**. The word particle can be used for an atom, a molecule and an ion. Gaseous compounds and compounds which are liquid at room temperature consist of molecules.

4.3 How big are atoms?

IT

Molecules
(program)

Use this program to study the way
that particles behave in a container
according to the kinetic theory of
matter.
• Change the temperature. What
  happens to the particles?
• Change another of the variables.
  What happens?
Remember: this is a model of the
way particles behave – it is a
computer simulation.

Hydrogen atoms are the smallest. One million hydrogen atoms in a row would stretch across one gain of sand. Five million million hydrogen atoms would fit on a pinhead. A hydrogen atom weighs 1.7×10^{-24} g; the heaviest atoms weigh 5×10^{-22} g.

How big is a molecule?

You can get an idea of the size of a molecule by trying the experiment shown in Figure 4.3A. This experiment uses olive oil, but a drop of detergent will also work. You can try this experiment at home.

1   Fill a clean tea tray
    with clean water

2   Sprinkle fine talcum
    powder on the surface

3   Dip a fine piece of wire into
    the olive oil. Lift out a tiny
    drop of oil. Aim to get a drop
    about 0.5 mm in diameter

4   Dip the wire into the water.
    The droplet of oil spreads
    out and pushes back the
    talcum powder. Measure
    as well as you can the area
    of the patch of olive oil

Figure 4.3A ● Estimating the size of a molecule

Sample results

Diameter of drop = 0.5 mm
Volume of drop = $(0.5\ \text{mm})^3$ = 0.125 mm³
Area of patch = $(25\ \text{cm})^2$ = $(250\ \text{mm})^2$ = 6.25×10^4 mm²
Volume of patch = area x depth (d)
  0.125 mm³ = 6.25×10^4 mm² x d

$$d = \frac{0.125\ \text{mm}^3}{6.25 \times 10^4\ \text{mm}^2}$$

$d = 2 \times 10^{-6}$ mm

The layer is only 2×10^{-6} mm deep (two millonths of a millimetre). We assume that it is one molecule thick.

SUMMARY

Atoms are tiny! A pinhead would hold 5×10^{12} hydrogen atoms. Olive oil molecules are 2×10^{-6} mm in diameter.

4.4 ● The kinetic theory of matter

FIRST THOUGHTS

Particles in motion: what does this idea explain? The difference between solids, liquids and gases for a start, and the beauty of crystalline solids.

The **kinetic theory of matter** states that matter is made up of small particles which are constantly in motion. (Kinetic comes from the Greek word for 'moving'.) The higher the temperature, the faster they move. In a solid, the particles are close together and attract one another strongly. In a liquid the particles are further apart and the forces of attraction are weaker than in a solid. Most of a gas is space, and the particles shoot through the space at high speed. There are almost no forces of attraction between the particles in a gas.

Scientists have been able to explain many things with the aid of the kinetic theory.

Solid, liquid and gaseous states

The differences between the solid, liquid and gaseous states of matter can be explained on the basis of the kinetic theory.

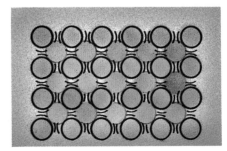

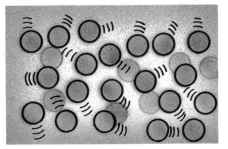

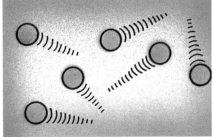

A solid is made up of particles arranged in a regular 3-dimensional structure. There are strong forces of attraction between the particles. Although the particles can vibrate, they cannot move out of their positions in the structure.

When a solid is heated, the particles gain energy and vibrate more and more vigorously. Eventually they may break away from the solid structure and become free to move around. When this happens, the solid has turned into a liquid: it has melted.

In a liquid the particles are free to move around. A liquid therefore flows easily and has no fixed shape. There are still forces of attraction between the particles.

When a liquid is heated, some of the particles gain enough energy to break away from the other particles. The particles which escape from the body of the liquid become a gas.

In a gas, the particles are far apart. There are almost no forces of attraction between them. The particles move about at high speed. Because the particles are so far apart, a gas occupies a very much larger volume than the same mass of liquid.

The molecules collide with the container. These collisions are responsible for the pressure which a gas exerts on its container.

Figure 4.4A ● The arrangement of particles in a solid, a liquid and a gas

TRY THIS

Growing crystals
Prepare a saturated solution of alum (aluminium potassium sulphate). Put a few drops on to a microscope slide. Watch through the microscope while crystals form. Draw the shape of the crystals. Other solutions you can try are copper(II) sulphate, sodium chloride, lead(II) iodide, sodium ethanoate, potassium manganate(VII), chrome alum (chromium potassium sulphate).

Crystals

Crystals are a very beautiful form of solid matter. A crystal is a piece of solid which has a regular shape and smooth faces (surfaces) which reflect light (see Figure 4.4B). Different salts have differently shaped crystals. *Why are many solids crystalline?* Viewing a crystal with an electron microscope, scientists can actually see individual particles arranged in a regular pattern (see Figure 4.4C). It is this regular pattern of particles which gives the crystal a regular shape.

Figure 4.4B ● Crystals of copper(II) sulphate

LOOK AT LINKS
for X-rays
X-ray photographs played a big part in working out the structure of DNA.
See *KS: Biology*, Topic 10.5.

X-rays can be used to work out the way in which the particles in a crystal are arranged. Figure 4.4D shows the effect of passing a beam of X-rays through a crystal on to a photographic film. X-rays blacken photographic film. The pattern of dots on the film shows that the particles in the crystal must be arranged in a regular way. From the pattern of dots, scientists can work out the arrangement of particles in the crystals.

SUMMARY

The kinetic theory of matter can explain the differences between the solid, liquid and gaseous states, and also how matter can change state. X-ray photographs show that crystals consist of a regular arrangement of particles.

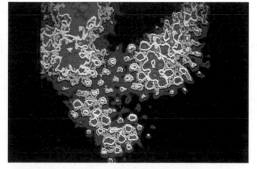

Figure 4.4C ● An electron microscope picture of uranyl acetate: each spot represents a single uranium atom

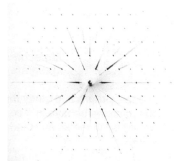

Figure 4.4D ● X-ray pattern from crystals of the metal palladium

How do solids dissolve, how do liquids vaporise and how do gases diffuse? Imagine particles in motion, and you will be able to explain all these changes.

Dissolving

Crystals of many substances dissolve in water. You can explain how this happens if you imagine particles splitting off from the crystal and spreading out through the water. See Figure 4.4E.

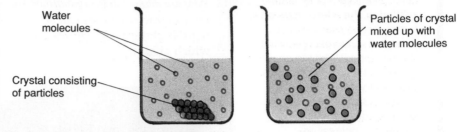

Water molecules

Particles of crystal mixed up with water molecules

Crystal consisting of particles

Figure 4.4E ● A coloured crystal dissolving

Diffusion

What evidence have we that the particles of a gas are moving? The diffusion of gases can be explained. **Diffusion** is the way gases spread out to occupy all the space available to them. Figure 4.4F shows what happens when a jar of the dense green gas, chlorine, is put underneath a jar of air.

On the theory that gases consist of fast-moving particles, it is easy to explain how diffusion happens. Moving molecules of air and chlorine spread themselves between the two gas jars.

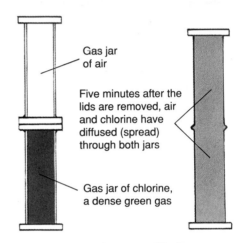

Gas jar of air

Five minutes after the lids are removed, air and chlorine have diffused (spread) through both jars

Gas jar of chlorine, a dense green gas

Figure 4.4F ● Gaseous diffusion

SUMMARY

Gases diffuse: they spread out to occupy all the space available to them. The kinetic theory of matter explains gaseous diffusion.

This section describes Brownian motion. You will see how neatly the kinetic theory can explain it.

Brownian motion

Figure 4.4G shows a smoke cell and the erratic path followed by a particle of smoke.

Molecules
(program)

Use this program to study the way that particles behave in a container according to the kinetic theory of matter.

• Change the temperature. What happens to the particles?

• Change another of the variables. What happens?

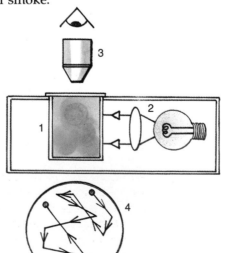

1 A small glass cell is filled with smoke

2 Light is shone through the cell

3 The smoke is viewed through a microscope

4 You see the smoke particles constantly moving and changing direction. The path taken by one smoke particle will look something like this

Figure 4.4G ● A smoke cell

Brownian motion in a liquid
In 1785, Robert Brown was using a microscope to observe pollen grains floating on water. He was amazed to see that the pollen grains were constantly moving about and changing direction. It was as if they had a life of their own.

Brown could not explain what he saw. You have the kinetic theory of matter to help you. Can you explain, with the aid of a diagram, what was making the pollen grains move?

We call this kind of motion **Brownian motion** after the botanist, Robert Brown, who first observed it. Figure 4.4H shows the explanation of Brownian motion.

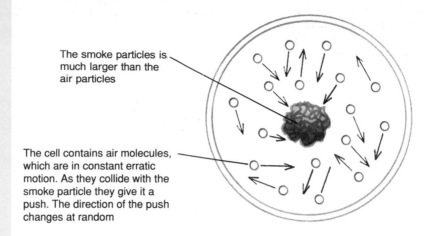

The smoke particles is much larger than the air particles

The cell contains air molecules, which are in constant erratic motion. As they collide with the smoke particle they give it a push. The direction of the push changes at random

Figure 4.4H ● Brownian motion

SUMMARY

Brownian motion puzzled scientists until the kinetic theory of matter offered an explanation.

Evaporation

When a liquid evaporates, it becomes cooler (see Figure 4.4I).

FIRST THOUGHTS

Why can you cool a hot cup of tea by blowing on it? Read this section to see if your idea is correct.

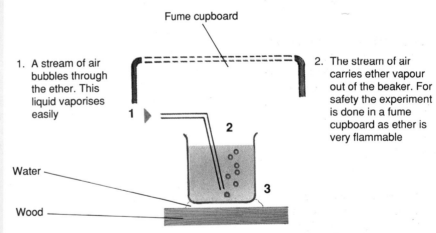

Fume cupboard

1. A stream of air bubbles through the ether. This liquid vaporises easily

2. The stream of air carries ether vapour out of the beaker. For safety the experiment is done in a fume cupboard as ether is very flammable

Water

Wood

3. As ether evaporates, it takes heat from its surroundings. The water between the beaker and the wood freezes

Figure 4.4I ● The cooling effect produced when a liquid evaporates

LOOK AT LINKS
The kinetic theory explains the gas laws.
See *KS: Physics*, Topic 6.2.

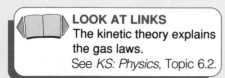
RESOURCE -ACTIVITY- PACK

The kinetic theory can explain this cooling effect. Attractive forces exist between the molecules in a liquid (see Figure 4.4J). Molecules with more energy than average can break away from the attraction of other molecules and escape from the liquid. After the high energy molecules have escaped, the average energy of the molecules which remain is lower than before: the liquid has become cooler.

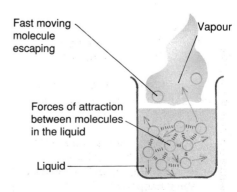

Fast moving molecule escaping

Vapour

Forces of attraction between molecules in the liquid

Liquid

Figure 4.4J ● Evaporation

SUMMARY

When a liquid evaporates, it takes heat from its surroundings. You can speed up evaporation by heating the liquid and by blowing air over it.

What happens when you raise the temperature? More molecules have enough energy to break away from the other molecules in the liquid. The rate of evaporation increases.

What happens if you pass a stream of dry air across the surface of the liquid? The dry air carries vapour away. The particles in the vapour are prevented from re-entering the liquid, that is, condensing. The liquid therefore evaporates more quickly.

CHECKPOINT

❶ Explain the following statements in terms of the kinetic theory.

(a) Water freezes when it is cooled sufficiently.
(b) Heating a liquid makes it evaporate more quickly.
(c) Heating a solid makes it melt.

❷ The solid X does not melt until it is heated to a very high temperature. What can you deduce about the forces which exist between particles of X?

❸ Of the five substances listed in the table below which is/are (a) solid (b) liquid (c) gaseous (d) unlikely to exist?

Substance	Distance between particles	Arrangement of particles	Movement of particles
A	Close together	Regular	Move in straight lines
B	Far apart	Regular	Random
C	Close together	Random	Random
D	Far apart	Random	Move in straight lines
E	Close together	Regular	Vibrate a little

❹ Supply words to fill in the blanks in this passage.

A solid has a fixed _____ and a fixed _____. A liquid has a fixed _____, but a liquid can change its _____ to fit its container. A gas has neither a fixed _____ nor a fixed _____. Liquids and gases flow easily; they are called _____. There are forces of attraction between particles. In a solid, these forces are _____, in a liquid they are _____ and in a gas they are _____.

❺ Imagine that you are one of the millions of particles in a crystal. Describe from your point of view as a particle what happens when your crystal is heated until it melts.

❻ Which of the two beakers in the figure opposite represents (a) evaporation, (b) boiling? Explain your answers.

❼ When a stink bomb is left off in one corner of a room, it can soon be smelt everywhere. Why?

❽ Beaker A and dish B contain the same volume of the same liquid. Will the liquid evaporate faster in A or B? Explain your answer.

❾ Why can you cool a cup of hot tea by blowing on it?

❿ A doctor dabs some ethanol (alcohol) on your arm before giving you an injection. The ethanol makes your arm feel cold. How does it do this?

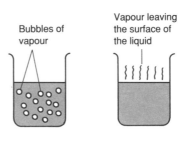

Bubbles of vapour

Vapour leaving the surface of the liquid

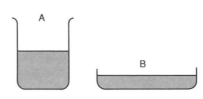

A

B

4.5 🔥 Gases

Gases have much lower densities than solids and liquids. The reason is that most of a gas is space. The particles are far apart, and move through the space at high speed. From time to time they collide with the walls of the container and with other particles. Gases are much more compressible than solids and liquids:

* Increase the pressure ➜ the volume decreases.
* Decrease the pressure ➜ the volume increases.

The kinetic theory explains this by saying that there is so much space between the particles that it is easy for the particles to move closer together when the gas is compressed.

Gases exert pressure because their particles are colliding with the walls of the container. If the volume of gas is decreased, the particles will hit the walls more often and the pressure will increase. The pressure of a given mass of gas changes with temperature:

* increase the temperature, while keeping the volume constant ➜ the pressure increases.
* increase the temperature while keeping the pressure constant ➜ the gas expands.

The kinetic theory explains these observations because, as the temperature rises, the particles have more energy, move faster and collide more frequently with the walls of the container.

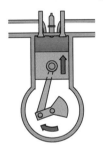

One cylinder of a four-stroke car engine. During the compression stroke, the piston moves up the cylinder and decreases the volume of the mixture of petrol vapour and air; therefore the pressure increases.

(a) The volume decreases therefore the pressure increases

A cylinder of gas under pressure. When the valve is opened, the pressure decreases. The gas expands out of the cylinder.

(b) The gas expands as the pressure decreases

When a balloon full of gas is heated, the pressure of gas increases until the balloon bursts.

(c) The pressure increases as the temperature rises

A lump of bread dough contains air and carbon dioxide. When the dough is heated in an oven, the volume of gas increases and the dough 'rises'.

(d) The gas expands as the temperature rises

Figure 4.5A ● Factors which determine the volume of a gas

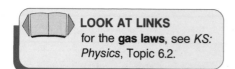

LOOK AT LINKS

for the **gas laws**, see *KS: Physics*, Topic 6.2.

The gas laws

A French scientist called Jacques Charles made measurements on the exact way in which the volume of a gas changes with temperature. He obtained the sort of results which are shown in Figure 4.5B.

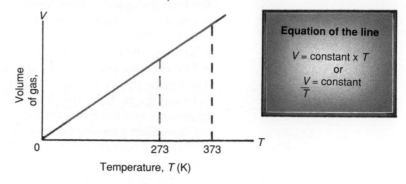

Figure 4.5B ● Charles' law

By extrapolating (continuing) the graph, you can see that at a temperature of $-273\,°C$, the volume of gas should become zero. All gases give the same value, and the temperature $-273°$ is called absolute zero. In fact, all gases liquefy before they reach this low temperature. The temperature scale called the Kelvin scale takes $-273\,°C$ as its zero. On this scale, temperatures are called absolute temperatures and are measured in kelvins (K). The relationship between the two scales is:

Temperature (K) = Temperature (°C) + 273

$$0\,K = -273\,°C$$

$$273\,K = 0\,°C$$

Charles' Law (1660) states that the volume of a gas is directly proportional to its absolute temperature:

Volume = constant × Absolute temperature (kelvin)
(provided the mass and pressure of gas remain constant).

An Irish scientist called Robert Boyle studied the effect of pressure on the volume of a gas. His results are summarised in **Boyle's Law** (1662) which states:

Pressure × Volume = constant

$$pV = \text{constant}$$

(provided the mass of gas and the temperature remain constant).

A third gas law is the **Pressure Law**. This states:

Pressure = constant × Absolute temperature

$$p = \text{constant} × T(K)$$

(provided the mass and volume of gas remain constant).

● The combined gas law

The three gas laws can be combined into a single equation. This is:

$$\frac{\text{Pressure} × \text{Volume}}{\text{Absolute temperature}} = \text{constant}$$

$$\frac{pV}{T} = \text{constant}$$

(for a fixed mass of gas).

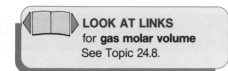

LOOK AT LINKS
for **gas molar volume**
See Topic 24.8.

Stating gas volume

It would not be very informative to state the volume of a gas without stating the temperature and pressure at which the volume is measured. It is the custom to quote gas volumes either at standard temperature and pressure (s.t.p., 0 °C and 1 atm) or at room temperature and pressure (r.t.p., 20 °C and 1 atm).

CHECKPOINT

❶ Explain why people are advised to let some of the air out of their car tyres when driving from the UK to a very hot country.

❷ A weather balloon is released into the atmosphere, where it will rise to a height of 10 km. It is only partially inflated at ground level. Why should it not be filled completely?

❸ A gas syringe holds 100 cm³ of nitrogen at a pressure of 1 atm. The gas is allowed to expand into a 1 l flask. What can you say about the pressure of gas in the 1 l flask?

? THEME QUESTIONS

1 A solid X melts at 58 °C. Which of the following substances could be X?
A Chloroethanoic acid, m.p. 63 °C
B Diphenylamine, m.p. 53 °C
C Ethanamide, m.p. 82 °C
D Trichloroethanoic acid, m.p. 58 °C

2 An impure sample of a solid Y melts at 77 °C. Which of these solids could be Y?
E Dibromobenzene, m.p. 87 °C
F 1, 4-Dinitrobenzene, m.p. 72 °C
G 3-Nitrophenol, m.p. 97 °C
H Propanamide, m.p. 81 °C

3 An impure sample of liquid Z boils at 180 °C. Which of these liquids could be Z?
I Benzoic acid, b.p. 249 °C
J Butanoic acid, b.p. 164 °C
K Hexanoic acid, b.p. 205 °C
L Methanoic acid, b.p. 101 °C

4 Which of the substances listed below are (a) solid (b) liquid (c) gaseous at room temperature?

Pure substance	A	B	C	D	E	F
Melting point (°C)	8	–92	41	63	–111	–30
Boiling point (°C)	101	–21	182	189	11	172

5 The graph shows how the temperature of a substance rises as it is heated. At A, the substance is a solid.
(a) Say what happens:
  (i) between A and B,   (iv) between D and E,
  (ii) between B and C,   (v) between E and F.
  (iii) between C and D,
(b) Name the temperatures T_1 and T_2.

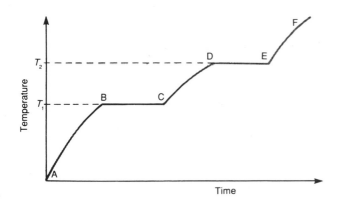

6 The table gives the solubility of potassium nitrate at various temperatures.

Temperature (°C)	0	10	20	40	60
Solubility (g per 100 g of water)	13	21	32	64	110

(a) On graph paper, plot solubility (on the vertical axis) against temperature (on the horizontal axis). Draw a smooth curve through the points. Use your graph to answer (b) and (c).
(b) What is the solubility of potassium nitrate at 30 °C?
(c) At what temperature is the solubility of potassium nitrate 85 g/100 g?
(d) A 100 g mass of water is saturated with potassium nitrate at 60 °C and cooled at 30 °C. What happens?

7 The diagram shows what happens when you breathe.

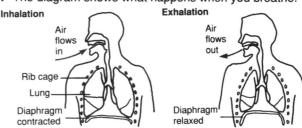

The diaphragm contracts and flattens, so increasing the volume of the chest cavity. What does this do to the air pressure in the cavity? Why does it cause air to flow into the lungs?

The diaphragm relaxes and pushes up into the chest cavity. What does this do to the air pressure in the cavity? Why does it cause air to flow out of the lungs?

8 A king in medieval times asked his goldsmith to make him a new crown, nothing fancy, just plain with no jewels. He gave the goldsmith 2.00 kg of gold. When the crown arrived, the king had it weighed. It was exactly 2.00 kg in mass. He tried it on. The crown did not look exactly the right colour, and the king wondered whether the goldsmith had kept back some of the gold and alloyed the rest with a cheaper metal to make up the mass. What could the king do to find out whether the crown was pure gold? Describe the measurements which he would have to make.

9 (a) In which state of matter is water in (i) the sea (ii) icicles and (iii) steam?
(b) How does the kinetic theory explain the difference between the three states?
(c) In the Great Salt Lake in the USA, a person can float very easily.
(i) Explain why the high concentration of salt makes it easier to float.
(ii) Describe an experiment which you could do to find out what mass of salt is present in 100 cm³ of the lake water.

THEME C
The Atom

'Matter is composed of atoms.'
This is what John Dalton said in 1808. What a simple statement this appears to be! Yet the complex developments that followed from this statement fill the whole of physics and chemistry.

What are atoms?

How many different kinds are there?

How do atoms differ from even smaller particles?

You will find some of the answers to these questions in Theme C.

ELEMENTS AND COMPOUNDS

5.1 ⚛ Silicon

FIRST THOUGHTS

Elements; they're *elementary*: they are simple substances, and there are 106 of them – all different. You have already met many useful elements, like silicon and gold.

LOOK AT LINKS
for **silicon chips**
See *KS: Physics*, Topic 23

The tiny chip in the mighty micro

Before 1950, a computer was a massive combination of circuits and valves which took up a whole room. Nowadays, a microcomputer the size of a typewriter can do the same job as the old-style computer. Microcomputers can be fitted in aeroplanes and spacecraft. The size and weight of the old-style computers made this impossible. Many people own a personal computer, a PC, to streamline jobs such as budgeting their expenses. The change in computer size has been brought about by the use of **silicon chips**. Figure 5.1A shows an electronic circuit built on to the surface of a silicon chip. Such circuits are very reliable because they are less affected by age, moisture and vibration than the old-style circuits.

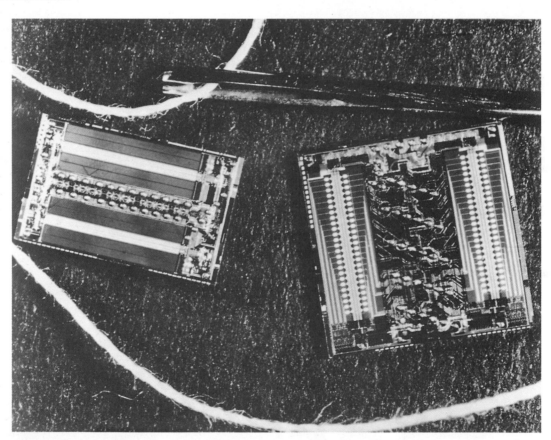

Figure 5.1A ● An integrated circuit built on the surface of a silicon chip

IT

Chemdata
(program)

Use this computer program to explore the properties of different elements. What patterns can you discover by using this program?

Silicon is an **element**. An element is a pure substance which cannot be split up into simpler substances. Elements are classified as metallic elements and non-metallic elements. Silicon is a non-metallic element. Most elements are either electrical conductors (substances which allow an electric current to flow through them) or electrical insulators (substances which do not allow an electric current to pass through them). Silicon is an unusual element in being a **semiconductor**: its behaviour is between that of a conductor and that of an insulator.

Figure 5.1B ● An electron diffraction micrograph of silicon

The first step in making a silicon chip is to slice large crystals of silicon into wafers. Then tiny areas of the wafer are treated with other elements which make silicon become an electrical conductor. The result is the creation of a thousand electronic circuits on each wafer. A chip 0.5 to 1.0 cm across contains thousands of tiny electrical switches called **transistors**. They allow on–off electric signals, which are the basis of computers, to occur at very high speeds.

● *Number crunching*

Very difficult and long calculations can be done on a computer. Computers operate so rapidly that they can solve in minutes problems that would take weeks to compute by hand. This is why computers are used to obtain fast and accurate weather forecasts. Earth scientists use computers to record the readings of their **seismometers** all over the world. The use of computers to process information is called **information technology**.

● *Tedious jobs*

Keeping track of records is a job for a computer. Debiting and crediting accounts, keeping a list of the stock in a shop or factory and such jobs are handled easily by a computer because a computer can repeat the same procedure over and over without error.

● *Planning ahead*

Using computers can help businesses to plan ahead. They can try out various plans and see how each will affect profits. In a similar way, computers can be used to try out various approaches to environmental problems to see how each approach will affect the animal and plant populations.

● *Data-logging*

Computers are used in science for recording and storing measurements. A sensor which measures pH, temperature, humidity, etc., can be connected to a microcomputer. The readings are recorded and displayed as a table or as a chart or as a graph. This can be useful for taking readings over a long period of time, e.g. monitoring water pollution, monitoring weather conditions, recording the temperature in the core of a nuclear reactor, recording the light emitted by a distant star.

LOOK AT LINKS
for **weather forecasting**
You will find more about forecasting the weather
See *KS: Physics*, Topic 3.

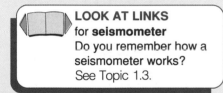

LOOK AT LINKS
for **seismometer**
Do you remember how a seismometer works?
See Topic 1.3.

IT'S A FACT

Plans for the car of the future include a radar set and a computer to tell the driver how far ahead the next car is. It will make fast driving much safer.

SCIENCE AT WORK

Jobs which once used to be done laboriously by hand are now done in a fraction of the time by computer. Computers have taken the drudgery out of a lot of jobs. They have also contributed to safety in aeroplanes and in industrial plants. The use of microcomputers has created jobs in the manufacture of computers (hardware) and the writing of computer programs which tell the computer what to do (software). The new jobs are technical jobs. Skilled people are needed to fill them.

Figure 5.1C ● Using microcomputers

Computer programs are used as sources of information on a multitude of topics; such programs are called **data bases**. In forensic science, materials found at the scene of a crime are analysed. This information can be fed into a computer to become part of a data base which could help the police to solve a similar crime.

● *Word processing*

Computers with a word processor program are used in place of typewriters. Some programs enable the user to do 'desktop publishing'. This is used to present reports for business and scientific purposes in a neat, clear, easy-to-read format.

● *Your science lessons*

You can use microcomputers in your science lessons for various purposes:
- extracting information from data bases or making your own database,
- using a spreadsheet to display results from an experiment in a table or graph,
- using computer programs to test your ideas,
- recording readings from sensors and displaying and analysing the results.

SUMMARY

The element silicon is a semiconductor. Silicon is used to make transistors (devices which allow current to flow in one direction only). Circuits built on to silicon chips are the basis of the microcomputer industry.

CHECKPOINT

❶ Can you think of jobs which are done by microcomputers in;
(a) shops (b) the car industry (c) schools (d) homes?

How do you and your family benefit from the use of micros in shops and the car industry? Do you benefit from the use of micros in any other way?

❷ (a) What kinds of jobs are lost through the introduction of computers?
(b) What kinds of jobs are created by the spread of computers?
(c) Can the people who lose their jobs as a result of (a) take the jobs created in (b)? If your answer is *yes*, explain why. If your answer is *no*, explain what you think could be done to solve the problem.

❸ Would space travel have been possible before the age of the microcomputer? Explain your answer.

5.2 Gold

The prospector's dream element

Gold has always held a great fascination for the human race. Gold occurs **native**, that is, as the free element, not combined with other elements. For thousands of years, people have been able to collect gold dust from river beds and melt the dust particles together to form lumps of gold. We can still see some of the objects which the goldsmiths made thousands of years ago because gold never tarnishes. For centuries, people used gold coins.

Figure 5.2A ● Panning for gold

Figure 5.2B ● Gold jewellery

Gold is a metallic element. The shine and the ability to be worked into different shapes are characteristic of metals. Unlike gold, most metals tarnish in air. Gold is not attacked by air or water or any of the other chemicals in the environment. It is used in electrical circuits when it is essential that the circuits do not corrode. For example, in microcomputers gold wires connect silicon chips to external circuits. Spacecraft use gold connections in their electrical circuits.

SUMMARY

Gold is a metallic element. It conducts electricity and is easily worked. Unlike many other metals, gold never becomes tarnished. Gold is used for jewellery and in electrical circuits which must not corrode.

CHECKPOINT

❶ Why has gold always been a favourite metal for jewellers to work with?

❷ What use is made of gold in modern technology?

❸ Dentists use gold to fill teeth. Why is gold a suitable metal for this job?

5.3 Copper and bronze

More metallic elements and some alloys:
- Copper, bronze and the Bronze Age
- Iron, steel and the Industrial Revolution

IT'S A FACT

What happened in 1886? Three thousand miles of copper cable were laid under the Atlantic. What for? This was the start of the transatlantic telegraph system.

A step up from the Stone Age

Thousands of years ago, Stone Age humans found lumps of copper embedded in rocks. Attracted by the colour and shine of copper, they hammered it into bracelets and necklaces. Then they started to use copper to make arrowheads, spears, knives and cooking pans. They found that copper tools did not break like the stone tools they were used to. Copper knives could be ground to a sharper edge than stone, and copper dishes did not crack as pottery bowls did. The shine, the ability to be worked into different shapes and the ability to conduct heat are typical of metals. Copper is a metallic element.

Stone Age people also discovered the alloy of copper and tin called **bronze** (a mixture of metals is called an alloy). Bronze is harder than copper or tin and can be ground to a sharper edge. Bronze weapons and tools made such a difference to the way people lived that they gave their name to the Bronze Age. Thanks to the new tools hunting and farming no longer occupied all the time of all the members of the community. Some people were able to spend time on painting, making pottery and building homes. The arrival of the Bronze Age was the beginning of civilisation.

Copper is still an important metal. Copper is a good electrical conductor – a typical metal. It is easily drawn into wire. Copper wire is used in electrical circuits. Wires, cables, overhead power lines, switches and windings in electrical motors are made of copper. Half the world's production of copper (total 8 million tonnes a year) is used by electrical industries.

Figure 5.3A ● Bronze in use

CHECKPOINT

❶ What advantages did copper tools have over tools made of (a) stone and (b) gold?

❷ Explain why the discovery of bronze was so important that it gave its name to the Bronze Age.

❸ What is the most important use of copper today?

5.4 Iron

Our most important metal

Our ancestors discovered how to extract iron from iron-bearing rocks over 3000 years ago. They used iron to make hammers, axes and knives, which did not break like stone tools and did not bend like bronze tools. Iron tools and weapons made such a difference to the way the human race lived that they gave their name to the Iron Age.

Many centuries later, iron and steel made the Industrial Revolution possible. The various types of **steel** are alloys of iron with carbon and other elements. Iron and steel are hard and strong: they can be hammered into flat blades and ground to a cutting edge. Our way of life in the twentieth century depends on machines made of iron and steel, buildings constructed on a framework of steel girders, and cars, lorries, trains, railways and ships made of steel.

Iron is a typical metallic element. An exceptional characteristic of iron is that it can be magnetised.

> **LOOK AT LINKS**
> for **iron** and **steel**
> You can read about the chemistry of iron and steel in Topic 19.10.

> **SUMMARY**
>
> Iron is a metallic element. Steel is an alloy. Being hard and strong, iron and steel are used for the manufacture of tools, machinery, motor vehicles, trains and ships.

Figure 5.4A ● A steel furnace

5.5 Carbon

> **FIRST THOUGHTS**
>
> Some non-metallic elements:
> • Can brilliant diamond and greasy graphite be the same element?
> • Can the killer, chlorine, save lives?

Diamond and graphite

Why are diamonds so often used in engagement rings? The sparkle of diamonds comes from their ability to reflect light. The 'fire' of diamonds arises from their ability to split light into flashes of colour. The hardness of diamonds means that they can be worn without becoming scratched. Diamonds are 'for ever'. Diamond is the hardest naturally occurring substance. The only thing that can scratch a diamond is another diamond.

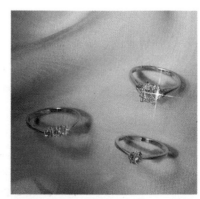

Figure 5.5A ● Diamonds

LOOK AT LINKS

Biologists who want to look at thin sections of plant and animal material with an electron microscope use diamond knives for cutting thin sections
See *KS: Biology*, Topic 9.1.

IT'S A FACT

Nineteenth century chemists thought graphite was a source of riches. All they had to do was to find the right conditions and graphite would change into diamond. It was more difficult than they expected. Not until the 1950s was a method found. Now 20 tonnes of manufactured diamonds are produced every year.

SUMMARY

Diamonds are beautiful jewels. Diamond is the hardest naturally occurring material. Small diamonds are used in industry for cutting, grinding and drilling.

Graphite is a slightly shiny grey solid, which conducts electricity. Graphite is used as a lubricant and as an electrical conductor. Diamond and graphite are allotropes (pure forms) of the element, carbon.

The hardness of diamond finds it many uses in industry. It is able to cut through metals, ceramics, glass, stone and concrete. Diamond-tipped saws slice silicon wafers from large crystals of silicon. As well as for cutting, diamonds are used for grinding, sharpening, etching and polishing. Oil prospectors would not be able to drill through hard rock without the help of drills studded with small diamonds (see Figure 5.5B). A 20 cm bit may be studded with 60 g of small diamonds.

Figure 5.5B ● Oil rig workers using a diamond studded drill

Diamond is a strange material, prized for its beauty and its usefulness. It is one form of the non-metallic element carbon.

Graphite is a shiny dark grey solid. It is so soft that it rubs off on your fingers and on paper. Pencil 'leads' contain graphite mixed with clay. Graphite is also used as a lubricant, in cars for example. Graphite is a second form of the element carbon. It is unusual in that it is the only non-metallic element which conducts electricity (see Topic 9.1).

Diamond and graphite are both pure forms of carbon. They are called **allotropes** of carbon. The existence of two or more crystalline forms of an element is called **allotropy**. Small diamonds (micron size, 10^{-6} m) can be made by heating graphite to a high temperature (1300 °C) under high pressure (60 000 atmospheres) for a few minutes. You can read more about the difference between the allotropes in Topic 5.8.

CHECKPOINT

❶ How are small industrial diamonds made?

❷ If the process for making industrial diamonds is carried on for a week, large gem-sized diamonds can be obtained. These large diamonds are dearer than natural diamonds. Why do you think this is so?

❸ What uses are made of diamonds? What characteristics of diamond make it suitable for the uses you mention?

❹ What is graphite used for? Why is graphite a suitable material for the uses you mention?

5.6 Chlorine

IT'S A FACT

The man who thought up the plan of using chlorine in warfare was the German chemist, Fritz Haber. Before the war, Haber won the Nobel prize for his discovery of a method of manufacturing ammonia. Haber said, 'A man belongs to the world in time of peace but to his country in times of war'. Haber's wife was so distressed by his involvement in the war that in 1916 she committed suicide.

The British scientist, Michael Faraday, was asked to develop poison warfare during the Crimean War, but he refused.

IT'S A FACT

In 1831, an epidemic of cholera hit London, and 50 000 people died. The cholera germs had been carried in the drinking water. That can't happen now that the water supply is disinfected with chlorine.

 LOOK AT LINKS
for chlorine and safe drinking water
See Topic 16.3
for oxidising reactions
See Topics 21.5, 21.6.

The life-saver

Chlorine is a killer. Its most infamous use was in the First World War when the German Army released cylinders of chlorine gas. The poisonous green cloud was driven by the wind into the trenches occupied by the British and French forces. Thousands of soldiers died, either choking on the gas or being shot as they retreated.

Chlorine is now used to kill germs. It is the bactericide which the water industry uses to make sure that our water supply is safe to drink. Chlorine has saved more lives than any other chemical. Before chlorine was used in the disinfection of the water supply, deaths from water-borne diseases, such as cholera and dysentery, were common. These diseases are still common in parts of the world which do not have safe water to drink.

Chlorine is a non-metallic element. Many household bleaches and disinfectants contain chlorine.

Figure 5.6A ● Bleach and disinfectant

5.7 Elements

FIRST THOUGHTS

You will have noticed that in science there is a need to classify things: to sort them into groups of similar members. Elements can be classified as metallic and non-metallic elements, with some exceptions.

Gold, copper, and iron are typical **metallic elements**. Like all metallic elements, they conduct electricity. Many of the metallic substances we use are not elements; they are **alloys**. Steel, brass, bronze, gunmetal, solder and many others are alloys. An alloy is a combination of two or more metallic elements and sometimes non-metallic elements also. Silicon, carbon and chlorine are **non-metallic elements**. Some of their characteristics are **typical** of non-metallic elements, but some are **atypical** (not typical). Diamond is atypical in being shiny; most non-metallic elements are dull. Graphite is the only non-metallic element that conducts electricity. Silicon is one of the few semiconductors of electricity.

There are 92 elements found on Earth. A further 14 elements have been made by scientists. Table 5.1 summarises the characteristics of metallic elements and non-metallic elements. You will see that there are many differences. Metallic and non-metallic elements also differ in their **chemical reactions**.

Table 5.1 ● Characteristics of metallic and non-metallic elements

Metallic elements	Non-metallic elements
Solids, except for mercury which is a liquid.	Solids and gases, except for bromine which is a liquid.
Hard and dense.	Most of the solid elements are softer than metals, but diamond is very hard.
A smooth metallic surface is shiny, but many metals tarnish in air, e.g. iron rusts.	Most are dull, but diamond is brilliant.
The shape can be changed by hammering: they are **malleable**. They can be pulled out into wire form: they are **ductile**.	Many solid non-metallic elements break easily when you try to change their shape. Diamond is the exception in being hard and strong.
Conduct heat, although highly polished surfaces reflect heat.	Poor thermal conductors.
Good electrical conductors.	Poor electrical conductors, except for graphite. Some are semi-conductors, e.g. silicon.
Make a pleasing sound when struck: are **sonorous**.	Are not **sonorous**.

LOOK AT LINKS
for **chemical reactions**
For oxides of metals and non-metals, see Topics 12.2 and 15.6. For the reactions of metals, see Topic 19.3.

SUMMARY

Table 5.1 lists the differences between metallic and non-metallic elements. Alloys are combinations of metallic elements and sometimes non-metallic elements also.

5.8 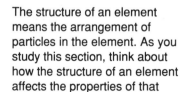 The structures of some elements

FIRST THOUGHTS

The structure of an element means the arrangement of particles in the element. As you study this section, think about how the structure of an element affects the properties of that element.

LOOK AT LINKS
for **molecules**
See Topic 4.1.

Can you name an element which fits each of these descriptions?
(a) a solid metallic element
(b) a liquid metallic element
(c) a solid non-metallic element
(d) a gaseous non-metallic element
(e) a hard solid element
(f) a soft solid element
(g) a shiny element
(h) a dull element

How do these differences arise? The reason lies in the different arrangements of atoms in the different elements. In many elements the atoms are bonded together in groups called **molecules**.

Oxygen, chlorine and many non-metallic elements consist of individual molecules. There are strong bonds between the atoms in the molecules, but between molecules there is only a very weak attraction. The molecules move about independently, and these elements are gaseous.

Sulphur is a yellow solid. There are two forms of sulphur, which form differently shaped crystals. The crystals of **rhombic sulphur** are octahedral; those of **monoclinic sulphur** are needle-shaped. Rhombic and monoclinic sulphur are **allotropes** of sulphur: the only difference between them is the shape of their crystals (see Figure 5.8A). The reason why the crystals are shaped differently is that the sulphur molecules are packed into different arrangements in the allotropes.

LOOK AT LINKS
for the **metallic bond**
Why are metals so strong?
See Topic 19.2.

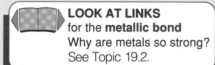

DID YOU KNOW?
In 1985, new allotropes of carbon were discovered. They consist of molecules containing 30–70 carbon atoms. The allotropes are known as **fullerenes**. The structure of a fullerene molecule is a closed cage of carbon atoms. The most symmetrical structure is buckminsterfullerene, C_{60}, which is a perfect sphere of carbon atoms. It is named after the geodesic domes designed by the architect Buckminster Fuller. The atoms are bonded together in 20 hexagons and 12 pentagons, which fit together like those on the surface of a football. Fullerenes have been nicknamed 'bucky balls'.

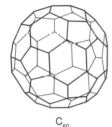

C_{60}

Scientists believe they will find important applications as semiconductors, superconductors, lubricants, catalysts and in batteries.

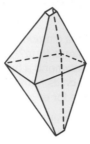

Figure 5.8A ● Allotropes of sulphur (x20)

Figure 5.8B shows a model of diamond. The structure is described as **macromolecular** or **giant molecular**. You can see that the carbon atoms form a regular arrangement. A crystal of diamond contains millions of carbon atoms arranged in this way. Every carbon atom is joined by chemical bonds to four other carbon atoms. It is very difficult to break this structure. The macromolecular structure is the source of diamond's hardness.

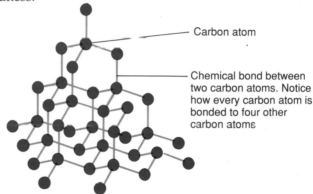

Carbon atom

Chemical bond between two carbon atoms. Notice how every carbon atom is bonded to four other carbon atoms

Figure 5.8B ● The arrangement of carbon atoms in diamonds

Figure 5.8C shows a model of graphite. Like diamond, graphite is a crystalline solid with a macromolecular structure. Graphite has a **layer structure**. Within each layer, the carbon atoms are joined by chemical bonds. Between layers, there are only weak forces of attraction. These weak forces allow one layer to slide over the next layer. This is why graphite is soft and rubs off on your fingers. The structure of graphite enables it to be used as a lubricant and in pencil 'leads' to mark paper.

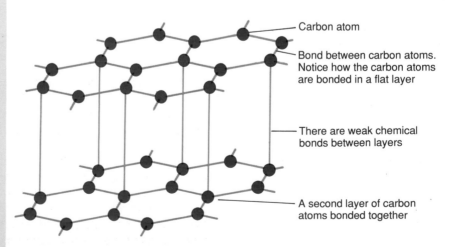

Carbon atom

Bond between carbon atoms. Notice how the carbon atoms are bonded in a flat layer

There are weak chemical bonds between layers

A second layer of carbon atoms bonded together

Figure 5.8C ● The arrangement of carbon atoms in graphite

SUMMARY

The carbon atoms in diamond are joined by chemical bonds to form a giant molecular structure.

CHECKPOINT

❶ (a) What kind of atoms are there in a diamond?
(b) What kind of structure do the atoms form?
(c) Why does this structure make diamond a hard substance?
(d) What uses of diamond depend on its hardness?
(e) Why are diamonds often chosen for engagement rings?

❷ (a) Explain how its structure makes graphite less hard than diamond.
(b) Why are diamond and graphite called allotropes?
(c) What is graphite used for?

❸ Say how the allotropes differ and how they resemble one another.

FIRST THOUGHTS

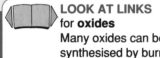

5.9   Compounds

> Most of the substances you see around you are not elements: they are compounds. Compounds can be made from elements by chemical reactions.

LOOK AT LINKS
for **oxides**
Many oxides can be synthesised by burning elements in oxygen
See Topic 15.6.

The Verey distress rocket contains the metallic element magnesium. When it is heated – when the fuse is lit – magnesium burns in the oxygen of the air. It burns with a brilliant white flame. A white powder is formed. This powder is the compound, magnesium oxide. A change which results in the formation of a new substance is called a **chemical reaction**. A chemical reaction has taken place between magnesium and oxygen. The elements have combined to form a compound.

Magnesium + Oxygen   →   Magnesium oxide
Element + Element   →   Compound

A compound is a pure substance which contains two or more elements chemically combined. A compound of oxygen and one other element is called an **oxide**. Making a compound from its elements is called **synthesis**.

Chemical reactions can synthesise compounds, and chemical reactions can also decompose (split up) compounds. Some compounds can be decomposed into their elements by heat. An example is silver oxide (see Figure 5.9B). The chemical reaction that takes place is

Silver oxide   _{Heat} →   Silver + Oxygen
Compound   Elements

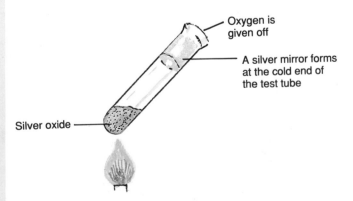

Figure 5.9A ● The thermal decomposition of silver oxide

Splitting up a compound by heat is called **thermal decomposition**.

Some compounds can be decomposed into their elements by the passage of a direct electric current. An example is sodium chloride (common salt) which is a compound of sodium and chlorine. A compound of chlorine with one other element is called a **chloride**.

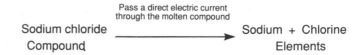

The chemical reaction that occurs when a compound is split up by means of electricity is called **electrolysis**.

Water is a compound. It can be electrolysed to give the elements hydrogen and oxygen (see Figure 5.9B).

Water is a compound of hydrogen and oxygen. You could call it hydrogen oxide. It is possible to make water by a chemical reaction between hydrogen and oxygen.

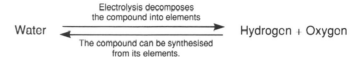

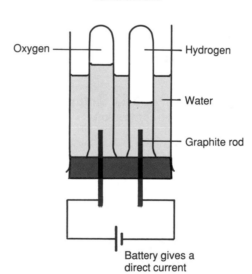

Figure 5.9B ● The decomposition of water by electrolysis

SUMMARY

Elements combine to form compounds. Some compounds can be split up by heat in thermal decomposition. Some compounds are decomposed by electrolysis (the passage of a direct electric current through the molten compound or a solution of the compound). Molten common salt can be electrolysed to give the elements sodium and chlorine. Water can be electrolysed to give the elements hydrogen and oxygen.

CHECKPOINT

❶ In a chemical change, a new substance is formed. In a physical change, no new substance is formed. Say which of the following changes are chemical changes.
(a) Evaporating water.　　(d) Cracking an egg.
(b) Electrolysing water.　　(e) Boiling an egg.
(c) Melting wax.

❷ Which of the following brings about a chemical change?
(a) Heating magnesium.
(b) Heating silver oxide.
(c) Heating sodium chloride.
(d) Passing a direct electric current through copper wire.
(e) Passing a direct electric current through molten salt.

5.10 Mixtures and compounds

Try to explain the difference between a mixture and a compound. Then read this section, and see whether you are right.

There are a number of differences between a compound and a mixture of elements. A mixture can contain its components in any proportions. A compound has a **fixed composition**. It always contains the same elements in the same percentages by mass. You can mix together iron filings and powdered sulphur in any proportions, from 1% iron and 99% sulphur to 99% iron and 1% sulphur. When a chemical reaction takes place between iron and sulphur, you get a compound called iron(II) sulphide. It always contains 64% iron and 36% sulphur by mass. Table 5.2 summarises the differences between mixtures and compounds.

Table 5.2 ● Differences between mixtures and compounds

Mixtures	Compounds
A mixture can be separated into its parts by methods such as distillation and dissolving.	A chemical reaction is needed to split a compound into simpler compounds or into its elements.
No chemical change takes place when a mixture is made.	When a compound is made, a chemical reaction takes place, and often heat is given out or taken in.
A mixture behaves in the same way as its components.	A compound does not have the characteristics of its elements. It has a new set of characteristics.
A mixture can contain its components in any proportions.	A compound always contains its elements in fixed proportions by mass; for example, calcium carbonate (marble) always contains 40% calcium, 12% carbon and 48% oxygen by mass.

SUMMARY

A compound is a pure substance which consists of two or more elements chemically combined. The components of a mixture are not chemically combined. Table 5.2 lists the differences between mixtures and compounds.

CHECKPOINT

❶ Group the following into mixtures, compounds and elements:
rain water, sea water, common salt, gold dust, aluminium oxide, ink, silicon, air.

❷ Name an element which can be used for each of the following uses:
surgical knife, pencil 'lead', wedding ring, saucepan, crowbar, thermometer, plumbing, electrical wiring, microcomputer circuit, disinfecting swimming pools, fireworks, artists' sketching material.

❸ What are the differences between a mixture of iron and sulphur and the compound iron sulphide?

❹ (a) What happens when you connect a piece of copper wire across the terminals of a battery? What happens when you disconnect the wire from the battery? Is the copper wire the same as before or has it changed?
(b) What happens when you hold a piece of copper in a Bunsen flame for a minute and then switch off the Bunsen? Is the copper wire the same or different?
(c) What type of change or changes occur in (a) and (b)?

SYMBOLS; FORMULAS; EQUATIONS

TOPIC 6

6.1 Symbols

FIRST THOUGHTS

Molly had a little dog
Her dog don't bark no more
'Cos what she thought was H_2O
Was really H_2SO_4
Anon
This rhyme shows the importance of getting your formulas right!

For every element there is a **symbol**. For example, the symbol for sulphur is S. The letter S stands for one atom of sulphur. Sometimes, two letters are needed. The letters Si stand for one atom of silicon: the symbol for silicon is Si. The symbol of an element is a letter or two letters which stand for one atom of the element. In some cases the letters are taken from the Latin name of the element, Ag from *argentum* (silver) and Pb from *plumbum* (lead) are examples. Table 6.1 gives a short list of symbols. There is a complete list at the end of the book.

Table 6.1 ● The symbols of some common elements

Element	Symbol	Element	Symbol	Element	Symbol
Aluminium	Al	Gold	Au	Oxygen	O
Barium	Ba	Hydrogen	H	Phosphorus	P
Bromine	Br	Iodine	I	Potassium	K
Calcium	Ca	Iron	Fe	Silver	Ag
Carbon	C	Lead	Pb	Sodium	Na
Chlorine	Cl	Magnesium	Mg	Sulphur	S
Copper	Cu	Mercury	Hg	Tin	Sn
Fluorine	F	Nitrogen	N	Zinc	Zn

SUMMARY

The symbol of an element is a letter or two letters which stand for one atom of the element.

6.2 Formulas

Figure 6.2A ● A model of one molecule of carbon dioxide

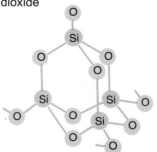

Figure 6.2B ● The structure of silicon(IV) oxide

For every compound there is a formula. The formula of a compound contains the symbols of the elements present and some numbers. The numbers show the ratio in which atoms are present. The compound carbon dioxide consists of molecules. Each molecule contains one atom of carbon and two atoms of oxygen. The formula of the compound is CO_2. The 2 below the line multiplies the O in front of it. To show three molecules of carbon dioxide you write $3CO_2$.

Sand is impure silicon dioxide (also called silicon(IV) oxide). It consists of macromolecules, which contain millions of atoms. There are twice as many oxygen atoms as silicon atoms in the macromolecule. The formula of silicon dioxide is therefore SiO_2.

The formulas of some of the compounds mentioned in this chapter are:
- Water, H_2O (two H atoms and one O atom; the 2 multiplies the H in front of it).
- Sodium chloride, NaCl (one Na; one Cl).
- Silver oxide, Ag_2O (two Ag: one O).
- Iron(II) sulphide, FeS (one Fe; one S).

The formula for aluminium oxide is Al_2O_3. This tells you that the compound contains two aluminium atoms for every three oxygen atoms. The numbers below the line multiply the symbols immediately in front of them.

The formula for calcium hydroxide is $Ca(OH)_2$. The 2 multiplies the symbols in the brackets. There are 2 oxygen atoms, 2 hydrogen atoms and 1 calcium atom. To write $4Ca(OH)_2$ means that the whole of the formula is multiplied by 4. It means 4Ca, 8O and 8H atoms. Table 6.2 lists the formulas of some common compounds.

Analysis
(program)

Use the program to practise identifying a range of unknown elements and compounds.

Table 6.2 ● The formulas of some common compounds

Compound	Formula
Water	H_2O
Carbon monoxide	CO
Carbon dioxide	CO_2
Sulphur dioxide	SO_2
Hydrogen chloride	HCl
Hydrochloric acid	HCl(aq)
Sulphuric acid	$H_2SO_4(aq)$
Nitric acid	$HNO_3(aq)$
Sodium hydroxide	NaOH
Sodium chloride	NaCl
Sodium sulphate	Na_2SO_4
Sodium nitrate	$NaNO_3$
Sodium carbonate	Na_2CO_3
Sodium hydrogencarbonate	$NaHCO_3$
Calcium oxide	CaO
Calcium hydroxide	$Ca(OH)_2$
Calcium chloride	$CaCl_2$
Calcium sulphate	$CaSO_4$
Calcium carbonate	$CaCO_3$
Calcium hydrogencarbonate	$Ca(HCO_3)_2$
Copper(II) oxide	CuO
Copper(II) sulphate	$CuSO_4$
Aluminium chloride	$AlCl_3$
Aluminium oxide	Al_2O_3
Ammonia	NH_3
Ammonium chloride	NH_4Cl
Ammonium sulphate	$(NH_4)_2SO_4$

SUMMARY

The formula of a compound is a set of symbols and numbers. The symbols show which elements are present in the compound. The numbers give the ratio in which the atoms of different elements are present.

6.3 Valency

Some atoms can form only one chemical bond. They can therefore combine with only one other atom. Elements with such atoms are said to have a **valency** of one. Hydrogen has a valency of one, and chlorine has a valency of one.

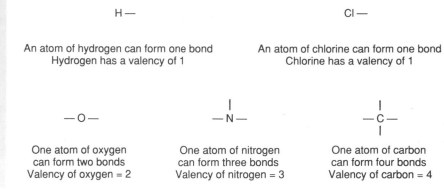

H—

An atom of hydrogen can form one bond
Hydrogen has a valency of 1

Cl—

An atom of chlorine can form one bond
Chlorine has a valency of 1

—O—

One atom of oxygen
can form two bonds
Valency of oxygen = 2

—N—

One atom of nitrogen
can form three bonds
Valency of nitrogen = 3

—C—

One atom of carbon
can form four bonds
Valency of carbon = 4

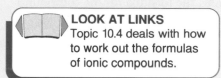

LOOK AT LINKS
Topic 10.4 deals with how to work out the formulas of ionic compounds.

The formula of a compound depends on the valencies of the elements in the compound. When hydrogen combines with chlorine, oxygen, nitrogen and carbon, the compounds formed have the formulas:

$$H - Cl \qquad H - O - H \qquad H - \overset{\displaystyle H}{\underset{}{N}} - H \qquad H - \overset{\displaystyle H}{\underset{\displaystyle H}{C}} - H$$

These are the compounds hydrogen chloride (HCl), water (H_2O), ammonia (NH_3), and methane (CH_4). They have different formulas because of the valencies of the elements Cl, O, N and C are different.

Figure 6.3A ● Models of HCl, H_2O, NH_3 and CH_4

SUMMARY

The formula of a compound depends on the valencies of the elements in the compound.

Some elements have more than one valency. Sulphur, for example, forms compounds in which it has a valency of 2, e.g. H_2S, compounds in which it has a valency of 4, e.g. SCl_4, and compounds in which it has a valency of 6, e.g. SF_6.

6.4 Equations

An equation tells what happens in a chemical reaction. A word equation gives the names of the reactants and the products. A chemical equation gives their symbols and formulas.

In a chemical reaction, the starting materials, the **reactants**, are changed into new substances, the **products**. The atoms present in the reactants are not changed in any way, but the bonds between the atoms change. Chemical bonds are broken, and new chemical bonds are made. The atoms enter into new arrangements as the products are formed. Symbols and formulas give us a nice way of showing what happens in a chemical reaction. We call this way of describing a chemical reaction a **chemical equation**.

Example 1   Copper and sulphur combine to form copper sulphide. Writing a **word equation** for the reaction

Copper + Sulphur   →   Copper sulphide

The arrow stands for **form**.

Writing the symbols for the elements and the formula for the compound gives the chemical equation

$Cu + S$ → CuS

IT   *Balancing Equations* (program)

Practise the art of balancing chemical equations by using this program. You will need to supply the correct formulas for some elements and compunds in the equation.

Adding the state symbols

$Cu(s) + S(s)$ → $CuS(s)$

Why is this called an equation? The two sides are equal. On the left hand side, we have one atom of copper and one atom of sulphur; on the right hand side, we have one atom of copper and one atom of sulphur combined as copper sulphide. The atoms on the left hand side and the atoms on the right hand side are the same in kind and in number.

Example 2   Calcium carbonate decomposes when heated to give calcium oxide and carbon dioxide. The word equation is

Calcium carbonate   →   Calcium oxide + Carbon dioxide

The chemical equation is

$CaCO_3(s)$ → $CaO(s) + CO_2(g)$

Example 3   Carbon burns in oxygen to form the gas carbon dioxide. The word equation is

Carbon + Oxygen   →   Carbon dioxide

We must use the formula O_2 for oxygen because oxygen consists of molecules which contain two oxygen atoms. The chemical equation is

$C(s) + O_2(g)$ → $CO_2(g)$

Example 4 Magnesium burns in oxygen to form the solid magnesium oxide.

$$\text{Magnesium} + \text{Oxygen} \rightarrow \text{Magnesium oxide}$$

$$Mg(s) + O_2(g) \rightarrow MgO(s)$$

There is something wrong here! The two sides are not equal. The left hand side has two atoms of oxygen; the right hand side has only one. Multiplying MgO by 2 on the right hand side should fix it

$$Mg(s) + O_2(g) \rightarrow 2MgO(s)$$

Now there are two oxygen atoms on both sides, but there are two magnesium atoms on the right hand side and only one on the left hand side. Multiply Mg by 2

$$2Mg(s) + O_2(g) \rightarrow 2MgO(s)$$

The equation is now **a balanced chemical equation**. Check up. On the left hand side, number of Mg atoms = 2; number of O atoms = 2. On the right hand side, number of Mg atoms = 2; number of O atoms = 2. The equation is balanced.

SUMMARY

How to write a balanced chemical equation;
• Write the word equation.
• Put in the symbols of the elements and the formulas of the compounds.
• Add the state symbols.
• Balance the equation. Do this by multiplying symbols or formulas. **Never** change a formula.
• Check again;

| no. of atoms of each element on LHS | = | no. of atoms of each element on RHS |

CHECKPOINT

❶ Refer to the table of elements at the end of the book, (i.e. the Periodic Table). Write down the names and symbols of the elements with atomic numbers 8, 10, 20, 24, 38, 47, 50, 80, 82. Say what each of these elements is used for. (The term 'atomic number' will be explained in Topic 7.)

❷ Give meanings of the state symbols (s), (l), (g), (aq). (See Topic 3.1 if you need to revise.)

❸ Try writing balanced chemical equations for the following reactions.
(a) Zinc and sulphur combine to form zinc sulphide.
(b) Copper reacts with oxygen to form copper(II) oxide.
(c) Sulphur and oxygen form sulphur dioxide.
(d) Magnesium carbonate decomposes to form magnesium oxide and carbon dioxide.
(e) Hydrogen and copper(II) oxide form copper and water.
(f) Carbon and carbon dioxide react to form carbon monoxide.
(g) Magnesium reacts with sulphuric acid to form hydrogen and magnesium sulphate.
(h) Calcium reacts with water to form hydrogen and a solution of calcium hydroxide.
(i) Zinc reacts with steam to form hydrogen and zinc oxide.
(j) Aluminium and chlorine react to form aluminium chloride.

❹ Write balanced chemical equations for the following reactions.
(a) Zinc and sulphur combine to form zinc sulphide, ZnS
(b) Copper and chlorine combine to form copper(II) chloride, $CuCl_2$
(c) Sulphur burns in oxygen to form sulphur dioxide, SO_2
(d) Magnesium carbonate decomposes to form magnesium oxide and carbon dioxide
(e) Calcium burns in oxygen to form calcium oxide.

❺ How many atoms are present in the following?
(a) $CaCl_2$ (b) $3CaCl_2$ (c) $Cu(OH)_2$ (d) $5Cu(OH)_2$ (e) H_2SO_4 (f) $3H_2SO_2$
(g) $2NaNO_3$ (h) $3Cu(NO_3)_2$

TOPIC 7 ● INSIDE THE ATOM

7.1 Becquerel's key

> FIRST THOUGHTS

This topic will give you a glimpse of the fascinating story of the discovery of radioactivity.

In 1896, a French physicist, Henri Becquerel, left some wrapped photographic plates in a drawer. When he developed the plates, he found the image of a key. The plates were 'fogged' (partly exposed). The areas of the plates which had not been exposed were in the shape of a key. Looking in the drawer, Becquerel found a key and a packet containing some uranium compounds. He did some further tests before coming to a strange conclusion. He argued that some unknown rays, of a type never met before, were coming from the uranium compounds. The mysterious rays passed through the wrapper and fogged the photographic plates. Where the key lay over the plates, the rays could not penetrate, and the image formed on the plates.

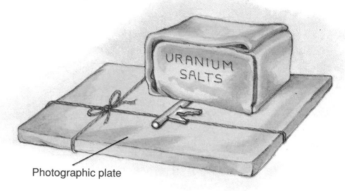

Photographic plate

Figure 7.1A ● Becquerel's key

A young research worker called Marie Curie took up the problem in 1898. She found that this strange effect happened with all uranium compounds. It depended only on the amount of uranium present in the compound and not on which compound she used. Madame Curie realised that this ability to give off rays must belong to the 'atoms' of uranium. It must be a completely new type of change, different from the chemical reactions of uranium salts. This was a revolutionary new idea. Marie Curie called the ability of uranium atoms to give off rays **radioactivity**.

Marie Curie's husband, Pierre, joined in her research into this brand new branch of science. Together, they discovered two new radioactive elements. They called one **polonium**, after Madame Curie's native country, Poland. They called the second **radium**, meaning 'giver of rays'. Its salts glowed in the dark.

Many scientists puzzled over the question of why the atoms of these elements, uranium, polonium and radium, give off the rays which Marie Curie named radioactivity. The person who came up with an explanation was the British physicist, Lord Rutherford. In 1902 he suggested that radioactivity is caused by atoms splitting up. This was another revolutionary idea. The word 'atom' comes from the Greek word for 'cannot be divided'. When the British chemist John Dalton put forward his Atomic Theory in 1808, he said that atoms cannot be created or destroyed or split. Lord Rutherford's idea was proved by experiment to be correct. We know now that many elements have atoms which are unstable and split up into smaller atoms.

> **Who's behind the science**

The Curies worked for four years in a cold, ill-equipped shed at the University of Paris. From a tonne of ore from the uranium mine, Madame Curie extracted a tenth of a gram of uranium. The Curies published their work in research papers and exchanged information with leading scientists in Europe. A year later, they were awarded the Nobel prize, the highest prize for scientific achievement. Pierre Curie died in a road accident. Marie Curie went on with their work and won a second Nobel prize. She died in middle-age from leukaemia, a disease of the blood cells. This was caused by the radioactive materials she worked with.

7.2 ● Protons, neutrons and electrons

Protons, neutrons and electrons

FIRST THOUGHTS

Atoms are made up of even smaller particles: protons, neutrons and electrons. As you read this section try to visualise how these particles are arranged inside the atom.

The work of Marie and Pierre Curie, Rutherford and other scientists showed that atoms are made up of smaller particles. These **subatomic particles** differ in mass and in electrical charge. They are called **protons**, **neutrons** and **electrons** (see Table 7.1).

Table 7.1 ● Sub-atomic particles

Particle	Mass (in atomic mass units)	Charge
Proton	1	$+e$
Neutron	1	0
Electron	0.0005	$-e$

Protons and neutrons both have the same mass. We call this mass one **atomic mass unit**, one u ($1.000\ u = 1.67 \times 10^{-27}$ kg). The mass of an atom depends on the number of protons and neutrons it contains. The electrons in an atom contribute very little to its mass. The number of protons and neutrons together is called the **mass number**.

Electrons carry a fixed quantity of negative electric charge. This quantity is usually written as $-e$. A proton carries a fixed charge equal and opposite to that of the electron. The charge on a proton can be written as $+e$. Neutrons are uncharged particles. Whole atoms are uncharged because the number of electrons in an atom is the same as the number of protons. The number of protons (which is also the number of electrons) is called either the **atomic number** or the **proton number**. You can see that

Number of neutrons – Mass number – Atomic (proton) number

For example, an atom of potassium has a mass of 39 u and an atomic (proton) number of 19. The number of electrons is 19, the same as the number of protons. The number of neutrons in the atom is

$$39 - 19 = 20$$

Who's behind the science

Rutherford had an easier life in science than the Curies. He arrived in Britain from New Zealand in 1895. By the age of 28, he was a professor. He worked in the universities of Montreal in Canada, Manchester and Cambridge. He was knighted in 1914 and made Lord Rutherford of Nelson in 1931. His co-worker Otto Hahn (see p. 92) described him as 'a very jolly man'. Many stories were told about Rutherford. He would whistle 'Onward, Christian soldiers' when the research work was going well and 'Fight the good fight' when difficulties had to be overcome.

Relative atomic mass

The lightest of atoms is an atom of hydrogen. It consists of one proton and one electron. Chemists compared the masses of other atoms with that of a hydrogen atom. They use **relative atomic mass**. The relative atomic mass, A_r, of calcium is 40. This means that one calcium atom is 40 times as heavy as one atom of hydrogen.

$$\text{Relative atomic mass of an element} = \frac{\text{Mass of one atom of the element}}{\text{Mass of one atom of hydrogen}}$$

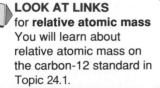

LOOK AT LINKS
for **relative atomic mass**
You will learn about relative atomic mass on the carbon-12 standard in Topic 24.1.

CHECKPOINT

❶ Some relative atomic masses are:

$A_r(H) = 1$, $A_r(He) = 4$, $A_r(C) = 12$, $A_r(O) = 16$, $A_r(Ca) = 40$.

Copy and complete the following sentences.

(a) A calcium atom is _____ times as heavy as an atom of hydrogen.
(b) A calcium atom is _____ times as heavy as an atom of helium.
(c) One carbon atom has the same mass as _____ helium atoms.
(d) _____ helium atoms have the same mass as two oxygen atoms.
(e) Two calcium atoms have the same mass as _____ oxygen atoms.

❷ Element E has atomic number 9 and mass number 19. Say how many protons, neutrons and electrons are present in one atom of E.

❸ State (i) the atomic number and (ii) the mass number of:
(a) an atom with 17 protons and 18 neutrons,
(b) an atom with 27 protons and 32 neutrons,
(c) an atom with 50 protons and 69 neutrons.

7.3 ⚛ The arrangement of particles in the atom

IT'S A FACT

If the nucleus of an atom was the size of a cricket ball, the nearest electron would be in the stands.

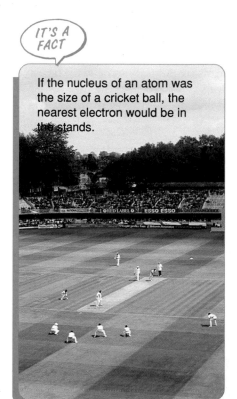

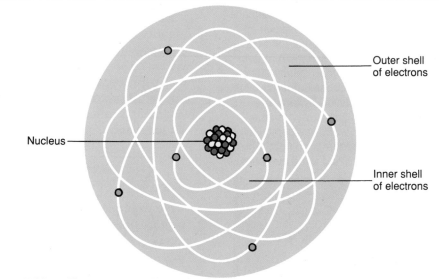

Outer shell of electrons

Nucleus

Inner shell of electrons

Figure 7.3A ● The structure of an atom

Lord Rutherford showed, in 1914, that most of the volume of an atom is space. Only protons and electrons were known in 1914; the neutron had not yet been discovered. Rutherford pictured the massive particles, the protons, occupying a tiny volume in the centre of the atom. Rutherford called this the **nucleus**. We now know that the nucleus contains neutrons as well as protons. The electrons occupy the space outside the nucleus. The nucleus is minute in volume compared with the volume of the atom.

The electrons of an atom are in constant motion. They move round and round the nucleus in paths called **orbits**. The electrons in orbits close to the nucleus have less energy than electrons in orbits distant from the nucleus.

7.4 ⚛ How are the electrons arranged?

Figure 7.4A illustrates the electrons of an atom in their orbits. The orbits are grouped together in **shells**. A shell is a group of orbits with similar energy. The shells distant from the nucleus have more energy than those close to the nucleus. Each shell can hold up to a certain number of electrons. In any atom, the maximum number of electrons in the outermost group of orbits is eight.

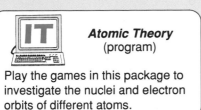

Atomic Theory (program)

Play the games in this package to investigate the nuclei and electron orbits of different atoms.

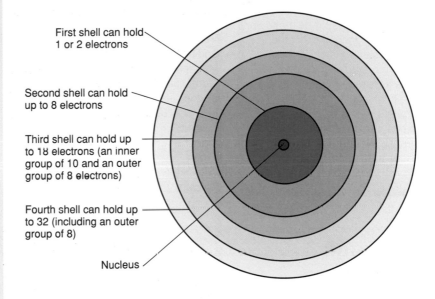

First shell can hold 1 or 2 electrons

Second shell can hold up to 8 electrons

Third shell can hold up to 18 electrons (an inner group of 10 and an outer group of 8 electrons)

Fourth shell can hold up to 32 (including an outer group of 8)

Nucleus

Figure 7.4A ● Shells of electron orbits in an atom

The atomic number of an element tells you the number of electrons in an atom of the element. The electrons fill the innermost orbits in the atom first. An atom of oxygen has 8 electrons. Two electrons enter the first shell, which is then full. The other 6 electrons go into the second shell (see Figure 7.4B).

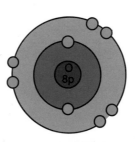

Figure 7.4B ● The arrangement of electrons in the oxygen atom

An atom of sodium has atomic number 11. The first shell is filled by 2 electrons, the second shell is filled by 8 electrons, and 1 electron occupies the third shell. The arrangement of electrons can be written as (2.8.1). It is called the **electron configuration** of sodium. Table 7.2 gives the electron configurations of the first 20 elements.

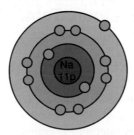

Figure 7.4C ● The electron configuration of sodium

Element	Symbol	Atomic (proton) number	Number of electrons in ...				Electron configuration
			1st shell	2nd shell	3rd shell	4th shell	
Hydrogen	H	1	1				1
Helium	He	2	2				2
Lithium	Li	3	2	1			2.1
Beryllium	Be	4	2	2			2.2
Boron	B	5	2	3			2.3
Carbon	C	6	2	4			2.4
Nitrogen	N	7	2	5			2.5
Oxygen	O	8	2	6			2.6
Fluorine	F	9	2	7			2.7
Neon	Ne	10	2	8			2.8
Sodium	Na	11	2	8	1		2.8.1
Magnesium	Mg	12	2	8	2		2.8.2
Aluminium	Al	13	2	8	3		2.8.3
Silicon	Si	14	2	8	4		2.8.4
Phosphorus	P	15	2	8	5		2.8.5
Sulphur	S	16	2	8	6		2.8.6
Chlorine	Cl	17	2	8	7		2.8.7
Argon	Ar	18	2	8	8		2.8.8
Potassium	K	19	2	8	8	1	2.8.8.1
Calcium	Ca	20	2	8	8	2	2.8.8.2

SUMMARY

The atoms of all elements are made up of three kinds of particles. These are:

- protons, of mass 1 u and electric charge +e
- neutrons, of mass 1 u, uncharged
- electrons, of mass 0.0005 u and electric charge −e.

The protons and neutrons make up the nucleus at the centre of the atom. The electrons circle the nucleus in orbits. Groups of orbits with the same energy are called shells. The 1st shell can hold 2 electrons; the 2nd shell can hold 8 electrons; the 3rd shell can hold 18 electrons. The arrangement of electrons in an atom is called the electron configuration.

CHECKPOINT

❶ Silicon has the electron configuration (2.8.4). What does this tell you about the arrangement of electrons in the atom? Sketch the arrangement. (See Figures 7.4B and 7.4C for help.)

❷ Sketch the arrangement of electrons in the atoms of (a) He (b) C (c) F (d) Al (e) Mg. (See Table 7.2 for atomic numbers.)

❸ Copy this table, and fill in the missing numbers.

Particle	Mass number	Atomic number	Number of ...		
			protons	neutrons	electrons
Nitrogen atom	14	7	-	-	-
Sodium atom	23	-	-	-	11
Potassium atom	39	-	19	-	-
Uranium atom	235	92	-	-	-

FIRST THOUGHTS

7.5 The Periodic Table

The physical and chemical properties of elements and the arrangement of electrons in atoms of the elements; they all fall into place in the Periodic Table.

Let us see how the electron configurations of the elements tie in with their chemical reactions. Some interesting patterns emerge when the elements are taken in the order of their atomic numbers and then arranged in rows. A new row is started after each noble gas (see Table 7.3). The arrangement is called the Periodic Table. You will have seen copies of the Periodic Table on the walls of chemistry laboratories. It simplifies the job of learning about the chemical elements.

Table 7.3 ● A section of the Periodic Table

	Group 1	Group 2	Group 3	Group 4	Group 5	Group 6	Group 7	Group 0
Period 1	H (1)							He (2)
Period 2	Li (2.1)	Be (2.2)	B (2.3)	C (2.4)	N (2.5)	O (2.6)	F (2.7)	Ne (2.8)
Period 3	Na (2.8.1)	Mg (2.8.2)	Al (2.8.3)	Si (2.8.4)	P (2.8.5)	S (2.8.6)	Cl (2.8.7)	Ar (2.8.8)
Period 4	K (2.8.8.1)	Ca (2.8.8.2)						

Elements which have the same number of electrons in the outermost shell fall into vertical columns. The eight vertical columns of elements are called **groups**. The group number is the number of electrons in the outermost shell, except for Group 0, in which the elements have a full shell of 8 electrons. The horizontal rows of elements are called **periods**. The first period contains only hydrogen and helium. The second period contains the elements lithium to neon. The complete Periodic Table is shown in Figure 7.5A and in the *Key Science: Chemistry Teacher's Guide*.

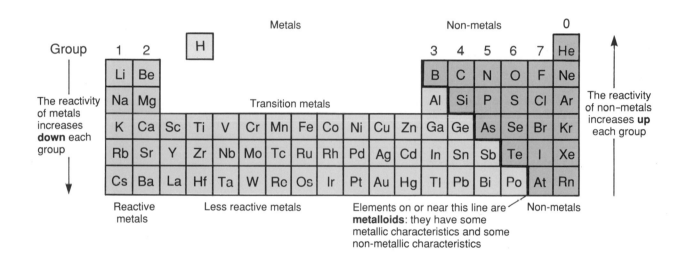

Figure 7.5A ● The Periodic Table

Patterns in the Periodic Table

1 Metallic elements appear at the left of the table.
2 Non-metallic elements appear at the right of the table.
3 There are borderline elements, such as silicon and germanium.
4 Group 1 is a set of very reactive metals called the **alkali metals**. The reactivity increases from top to bottom of the group.
5 Group 7 is a set of very reactive non-metals called the **halogens**. The reactivity decreases from top to bottom of the group.
6 Group 0 is a set of unreactive gases called the **noble gases**.
7 The metallic elements 21–30, 39–48 and 72–80 are known as **transition elements**.

● The noble gases

The elements in group 0 are helium, neon, argon, krypton, xenon and radon. These elements are called the noble gases. They are present in air (see Topic 14). The noble gases exist as single atoms, e.g. He, Ne. Their atoms do not combine in pairs to form molecules as do the atoms of most gaseous elements, e.g. O_2, H_2. For a long time, no-one was able to make the noble gases take part in any chemical reactions. In 1960, however, two of them, krypton and xenon, were made to combine with the very reactive element, fluorine. *Why are the noble gases so exceptionally unreactive?* Chemists came to the conclusion that it is the full outer shell of electrons that makes the noble gases unreactive.

● The alkali metals

The elements in group 1 are lithium, sodium, potassium, rubidium and caesium. They are a set of very similar metallic elements (see Table 7.4). They are known as the **alkali metals** on account of the strongly alkaline nature of their hydroxides. They all have one electron in the outer shell.

Table 7.4 ● The alkali metals (with iron for comparison)

| Element | Symbol | m.p. (°C) | b.p. (°C) | Density (g/cm³) | Hardness | Reaction with ... | | |
						air	water	non-metallic elements
Lithium	Li	180	1336	0.53	Softer than iron	All burn vigorously to form an oxide of formula M_2O (M = symbol for metal)	All are stored under oil. They react vigorously with cold water to give hydrogen and the hydroxide MOH. The hydroxides are all strong alkalis.	All combine with non-metals to form salts (and oxides). The salts are crystalline ionic solids. The alkali metals are the cations (positive ions) in the salts.
Sodium	Na	98	883	0.97				
Potassium	K	64	759	0.86				
Rubidium	Rb	39	700	1.53				
Caesium	Cs	29	690	1.9				
Iron	Fe	1530	3000	7.86		Burns to form an oxide	Reacts slowly with water to form rust. Reacts quickly with steam to give hydrogen and an oxide	Combines when heated with most non-metals. Form salts in which iron is the cation (positive ion).

Note: Reactivity increases down the group.

● The halogens

The elements in Group 7 are fluorine, chlorine, bromine and iodine. These elements are a set of very reactive non-metallic elements (see Table 7.5). They are called **the halogens** because they react with metals to form salts. (Greek: halogen = salt-former)

Table 7.5 ● The halogens

Element	Symbol	m.p. (°C)	b.p. (°C)	Colour	Reaction with metals to form salts	Reaction with hydrogen to form the compound HX	
Fluorine	F	−223	−188	Pale yellow	Dangerously reactive	Explodes	
Chlorine	Cl	−103	−35	Yellow-green	Readily combines to form chlorides.	Explodes in sunlight	Reactivity decreases down the group
Bromine	Br	−7	59	Red-brown	Combines when heated to form bromides	Reacts when heated	
Iodine	I	114	184	Purple-black	Combines when heated to form iodides. The halogens are the anions (negative ions) in their salts.	Reaction is not complete	

Who's behind the science

The history of the Periodic Table

The person who has the credit for drawing up the Periodic Table is a Russian scientist called Dimitri Mendeleev. He extended the work of a British chemist called John Newlands. In 1864, 63 elements were know to Newlands. He arranged them in order of relative atomic mass. When he started a new row with every eighth element, he saw that elements with similar properties fell into vertical groups. He spoke of 'the regular periodic repetition of elements with similar properties'. This gave rise to the name 'periodic table'. Sometimes, there were misfits in Newlands' table, for example, iron did not seem to belong where he put it with oxygen and sulphur. Newlands' ideas were not accepted.

Mendeleev had an idea about the troublesome misfits. He realised that many elements had yet to be discovered. Instead of slotting elements into positions where they did not fit he left gaps in the table. He expected that when further elements were discovered they would fit the gaps. He predicted that an element would be discovered to fit into the gap he had left in the table below silicon.

Table 7.6 ● The predictions which Mendeleev made for the undiscovered element which he called 'eka-silicon' (below silicon) compared with the properties of germanium

	Mendeleev's predicted properties for eka-silicon, Ek (1871)	The properties of germanium Ge (discovered in 1886)
Appearance	Grey metal	Grey-white metal
Density	~ 5.5 g/cm^3	5.47 g/cm^3
Relative atomic mass	73.4	72.6
Melting point	~ 800 °C	958 °C
Reaction with oxygen	Forms the oxide EkO$_2$. The oxide EkO may also exist.	Forms the oxide GeO$_2$ GeO also exists

By the time Mendeleev died in 1907, many of the gaps in the table had been filled by new elements. The noble gases were unknown to Mendeleev when he drew up his table. The first of them was discovered in 1894. As the noble gases were discovered, they all fell into place between the halogens and the alkali metals. They formed a new group, Group 0. This was a spectacular success for the Periodic Table.

SUMMARY

The chemical properties of elements depend on the electron configurations of their atoms. The unreactive noble gases all have a full outer shell of 8 electrons.

The reactive alkali metals have a single electron in the outer shell.

The halogens are reactive non-metallic elements which have 7 electrons in the outer shell: they are 1 electron short of a full shell.

In the Periodic Table, elements are arranged in order of increasing atomic (proton) number in 8 vertical groups. The horizontal rows are called the periods.

CHECKPOINT

❶ (a) Which of the alkali metals float on water?
  (b) Which of the alkali metals can be cut with a knife made of iron?
  (c) Which of the alkali metals melt at the temperature of boiling water?
  (d) Why would it be very dangerous to put an alkali metal into boiling water?
  (e) What is another name for sodium chloride?
  (f) What would you expect rubidium chloride to look like?
  (g) The alkali metals are kept under oil to protect them from the air. Which substances in the air would attack them?
  (h) What pattern can you see in (i) the melting points (ii) the boiling points of the alkali metals?

❷ (a) Which of the halogens is (i) a liquid and (ii) a solid at room temperature (20 °C)?
  (b) Why is fluorine not studied in school laboratories?
  (c) Write the formulas for (i) sodium iodide (ii) potassium fluoride.

❸ (a) When was germanium discovered?
(b) How do the properties of germanium compare with Mendeleev's prediction?
(c) How long did Mendeleev have to wait to see if he was right?
(d) In which group of the Periodic Table does germanium come?
(e) What important articles are manufactured from germanium and silicon?

7.6  Isotopes

FIRST THOUGHTS

There are two sorts of chlorine atom and three sorts of hydrogen atom. You can find out the difference between them in this section.

Atoms of the same element all contain the same number of protons, but the number of neutrons may be different. Forms of an element which differ in the number of neutrons in the atom are called **isotopes**. For example, the element chlorine, with relative atomic mass 35.5, consists of two kinds of atom with different mass numbers.

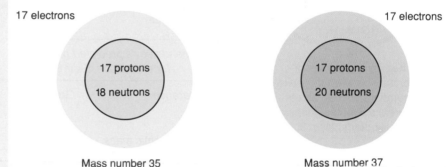

17 electrons — 17 protons, 18 neutrons — Mass number 35

17 electrons — 17 protons, 20 neutrons — Mass number 37

Figure 7.6A ● The isotopes of chlorine

Since the chemical reactions of an atom depend on its electrons, all chlorine atoms react in the same way. The number of neutrons in the nucleus does not affect chemical reactions. The different forms of chlorine are isotopes. Their chemical reactions are the same. In any sample of chlorine, there are three chlorine atoms with mass 35 u for each chlorine atom with mass 37 u so the average atomic mass is

$$\frac{(3 \times 35) + 37}{4} = 35.5 \text{ u}$$

RESOURCE – ACTIVITY – PACK

This is why the relative atomic mass of chlorine is 35.5.
Isotopes are shown as:

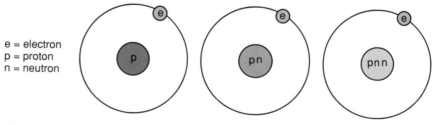

e = electron
p = proton
n = neutron

Figure 7.6B ● The isotopes of hydrogen

SUMMARY

Isotopes are atoms of the same element which differ in the number of neutrons. They contain the same number of protons (and therefore the same number of electrons).

The isotopes of chlorine are written as $^{35}_{17}\text{Cl}$ and $^{37}_{17}\text{Cl}$. The isotopes of hydrogen are $^{1}_{1}\text{H}$, $^{2}_{1}\text{H}$ and $^{3}_{1}\text{H}$. They are often referred to as hydrogen-1, hydrogen-2 and hydrogen-3. Hydrogen-2 is also called deuterium, and hydrogen-3 is also called tritium. Carbon has three isotopes, carbon-12, carbon-13 and carbon-14.

CHECKPOINT

❶ Hydrogen, deuterium and tritium are isotopes.
   (a) Copy and complete this sentence.
      Isotopes are _____ of an element which contain the same number of _____
      and _____ but different numbers of _____ .
   (b) Copy and complete the table.

	Hydrogen	Deuterium	Tritium
Atomic number			
Mass number			

   (c) Write the formula of the compound formed when deuterium reacts with oxygen.
   (d) Explain why isotopes have the same chemical reactions.

❷ Write the symbol with mass number and atomic number (as above) for each of the
   following isotopes.
   (a) oxygen with 8 protons and 8 neutrons
   (b) argon with 18 protons and 22 neutrons
   (c) bromine with 35 protons and 45 neutrons
   (d) chromium with 24 protons and 32 neutrons

❸ Two of the atoms described below have similar chemical properties. Which two are
   they?
   Atom X contains 9 protons and 10 neutrons.
   Atom Y contains 13 protons and 14 neutrons.
   Atom Z contains 17 protons and 18 neutrons.
   Explain your answer.

❹ Strontium-90 is a radioactive isotope formed in nuclear reactors. It can be
   accumulated in the human body because it follows the same chemical pathway
   through the body as another element X which is essential for health. After referring
   to the position of strontium in the Periodic Table, say which element you think is X.

❺ Sodium has the electron arrangement (2.8.1).
   (a) Draw and label a diagram to show how the protons, neutrons and electrons are
      arranged in a sodium atom.
   (b) Explain why sodium is electrically uncharged.
   (c) Rubidium is in the same group of the Periodic Table as sodium. Would you
      expect rubidium to be a metal or a non-metal?
   (d) How would you expect rubidium to be stored?
   (e) What products would you expect to be formed when rubidium reacts with cold
      water? How could you test each of these products?

TOPIC 8 — RADIOACTIVITY

8.1 What is radioactivity?

> Radioactivity: what is it?
> How was it discovered?
> What are α-, β-, γ-radiation?
> These are some of the things
> you will find out in this section

The changes which Marie Curie described as radioactivity are **nuclear changes**. A nuclear change is quite different from a chemical change. In a chemical change, bonds between atoms are broken, and new bonds are made, but the nuclei of the the atoms stay the same. In a nuclear change, new atoms of different elements are formed because the nuclei change. Marie Curie and her fellow scientists knew their work was revolutionary, but none of them could have imagined how it would change the world.

Isotopes which give off radioactivity are said to be **radioactive**. They are **radioisotopes**. The atoms of radioactive isotopes each have an unstable nucleus. The nucleus becomes stable by emitting small particles and energy. The particles and energy are called **radioactivity**; the breaking-up process is **radioactive decay**.

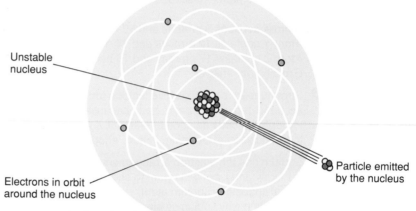

Unstable nucleus

Electrons in orbit around the nucleus

Particle emitted by the nucleus

Figure 8.1A ● Radioactive decay

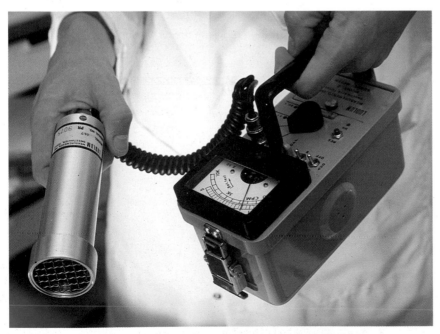

Figure 8.1B ● Using a Geiger–Müller tube.
(NOTE: experiments on radioactivity must only be done by a teacher.)

Investigating radioactivity

One way to detect radioactivity is to use a Geiger–Müller counter (named after its inventors). When radioactive particles enter it, a Geiger–Müller counter gives out a clicking sound. The number of clicks per minute is called the count rate. Even when there is no radioactive source nearby, the counter still clicks occasionally. This is due to **background radioactivity** from certain building materials and from cosmic radiation (from space). When you measure the count rate of a radioactive source, you have to subtract the background count.

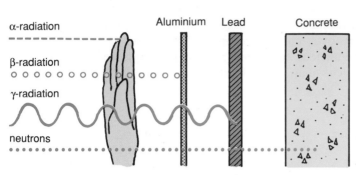

Figure 8.1C ● The penetrating power of radiation

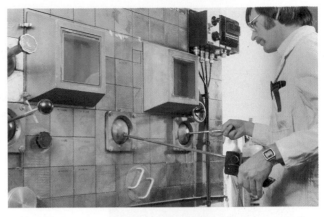

Figure 8.1D ● Lead bricks shield workers from radiation

You will find a computer simulation helpful to your understanding of radioactivity.

There are three types of radioactivity: **alpha-radiation (α-radiation)**, **beta-radiation (β-radiation)** and **gamma-radiation (γ-radiation)**. They differ in penetrating power (see Figure 8.1C). A thin metal foil with stop α-radiation, but a metal plate about 5 mm thick is needed to stop β-radiation, and γ-radiation will penetrate several centimetres of lead. γ-radiation passes through the skin and can penetrate bone. It can cause burns and cancer. People who work with sources of γ-radiation protect themselves by building a wall of lead bricks between themselves and the source (see Figure 8.1D).

The nature of α, β and γ-radiations

Beam of radioactivity enters a magnetic field

α-radiation is deflected by the magnetic field

γ-radiation and neutrons are undeflected by the magnetic field

A magnetic field at right angles to the plane of the paper

β-radiation is deflected in the opposite direction to α-radiation

Figure 8.1E ● Radiations in a magnetic field

What are these mysterious radiations? This was the question that Lord Rutherford and other scientists tackled early in this century. Within ten years of Henri Becquerel's discovery of radioactivity, scientists had unravelled the mystery. They discovered three new types of radiation by using a magnetic field to separate them (see Figure 8.1E). Scientists invented new instruments, like the Geiger–Müller counter and the cloud chamber (see Figure 8.1F) to use in their study.

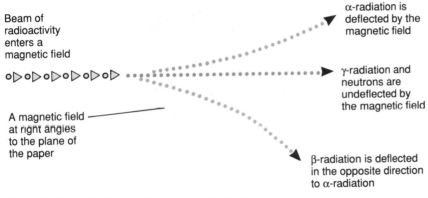

Perspex lid through which photographs can be taken

3 A radioactive source is placed in the chamber

1 The felt ring around the top of the chamber is soaked in ethanol. The air inside the chamber is therefore saturated with ethanol vapour

Light to illuminate the condensed droplets

Black metal base

Insulation in the base of the chamber

4 When a radioactive particle shoots through the chamber, it causes condensation. The visible trace of ethanol droplets which it leaves can then be photographed.

2 Dry ice (solid carbon dioxide) is placed under the metal base. It cools the air which becomes supersaturated

Figure 8.1F ● A cloud chamber

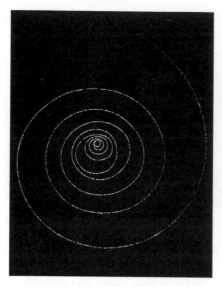

Figure 8.1G ● A β-particle track in a cloud chamber with a strong magnetic field

● α-radiation

α-radiation consists of α-particles. These are the nuclei of helium atoms; they consist of 2 protons and 2 neutrons and carry a positive charge. A nucleus of an α-emitting isotope changes from an unstable nucleus to a more stable nucleus by ejecting 2 protons and 2 neutrons as a single particle. The stable nucleus has 2 protons less: it has a different atomic number: it is a nucleus of an atom of a different element. An example of α-emission is

$$^{228}_{90}\text{Th} \quad \rightarrow \quad ^{4}_{2}\alpha \quad + \quad ^{224}_{88}\text{Ra}$$

An isotope of thorium (90 protons + 138 neutrons) → An α-particle (2 protons + 2 neutrons) + An isotope of radium (88 protons + 136 neutrons)

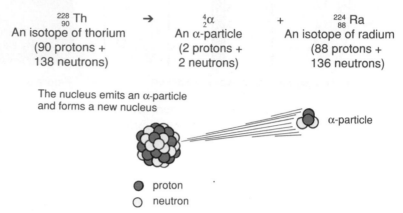

The nucleus emits an α-particle and forms a new nucleus

α-particle

● proton
○ neutron

Figure 8.1H ● α-emission

● β-radiation

β-radiation consists of electrons. A nucleus of a β-emitting isotope changes from an unstable nucleus to a stable nucleus by changing a neutron into a proton plus an electron and ejecting the electron from the nucleus. The new nucleus has one more proton than before and a different atomic number. It is a nucleus of an atom of a different element. An example of β-emission is

$$^{40}_{19}\text{K} \quad \rightarrow \quad ^{0}_{-1}\beta \quad + \quad ^{40}_{20}\text{Ar}$$

An isotope of potassium (19 protons + 21 neutrons) → A β-particle (an electron) + An isotope of argon (20 protons + 20 neutrons)

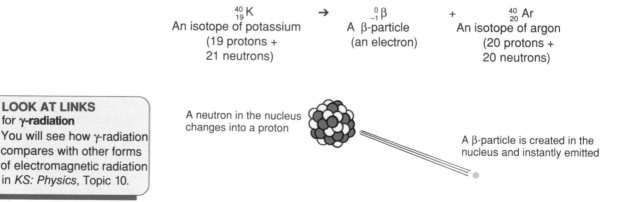

A neutron in the nucleus changes into a proton

A β-particle is created in the nucleus and instantly emitted

Figure 8.1I ● β-emission

LOOK AT LINKS
for γ-radiation
You will see how γ-radiation compares with other forms of electromagnetic radiation in *KS: Physics*, Topic 10.

SUMMARY

Radioactivity is caused by nuclear changes. Some isotopes have unstable nuclei. An unstable nucleus changes into a stable nucleus by emitting an α-particle or a β-particle or γ-radiation or a combination of these. A Geiger–Müller counter is used to detect and measure radioactivity.

● γ-radiation

γ-radiation is high-energy electromagnetic radiation. After an unstable nucleus has emitted an α- or a β-particle , it sometimes has surplus energy. It emits this energy as γ-radiation.

γ-radiation

The nucleus emits γ-radiation so that it can lose surplus energy

Figure 8.1J ● γ-emission

CHECKPOINT

❶ Henri Becquerel discovered radioactivity when he developed an unused photographic plate and found an image of a key on it. What type of radioactivity was emitted by the uranium salts?

❷ In an investigation to find out what type of radioactivity was emitted from a given source, the following measurements were made with a Geiger–Müller counter.

Source	Average count rate (counts per minute)
No source present	29.5
Source 20 mm from the tube and…	
1 no absorber present	385.2
2 a sheet of metal foil between the source and the tube	387.2
3 a 10 mm thick aluminium plate between the source and the tube	32.4

(a) What was the count rate due to the background activity?
(b) What was the count rate due to the source?
(c) What type of radioactivity did the source emit?

❸ Atomic nuclei are composed of protons and neutrons. Work out the number of protons and the number of neutrons in each of the following atoms.

(a) $^{238}_{92}$U   (b) $^{234}_{91}$Pa   (c) $^{227}_{89}$Ac

❹ (a) $^{238}_{92}$U is an α-emitter. When a uranium-238 nucleus emits an α-particle, it forms a new nucleus. Which of the following list is the new nucleus?

$^{234}_{92}$U $^{233}_{90}$Th $^{234}_{90}$Th $^{228}_{90}$Th $^{231}_{91}$Th

Write an equation for the nuclear change.

(b) $^{234}_{91}$Pa is a β-emitter. From the list in part (a), identify the nucleus formed when this change happens. Write the equation for the change.

8.2 Radioactive decay

- You will find a computer simulation of radioactive decay interesting.
- You can also use IT for data capture and display. A radioactive source, a VELA and an oscilloscope will show a graph of radioactivity against time.

Radioactivity is caused by unstable nuclei emitting particles and energy as they change into more stable nuclei. Each unstable nucleus is said to **decay** or **disintegrate** when it emits a radioactive particle.

The **activity** of a radioactive source is the number of its nuclei that disintegrate per second. The unit of activity is the **becquerel**, Bq.

> 1 Bq = 1 disintegration per second.

Suppose a radioactive isotope decays to form a non-radioactive product. The number of radioactive atoms decreases with time as the unstable nuclei of the radioactive isotope decay to form stable nuclei. *How does the activity of the radio-isotope change with time?* You can refer back to Figure 8.1B to see how a Geiger–Müller counter can be used to find out. The results of the investigation can be plotted as a graph of count rate against time.

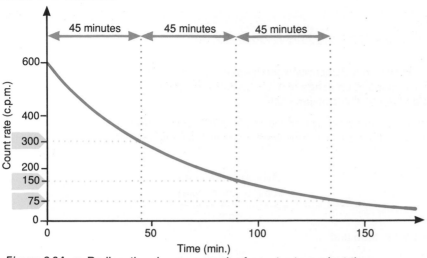

Figure 8.2A ● Radioactive decay: a graph of count rate against time

You can see from the graph that:
• the time taken for the count to fall from 600 to 300 c.p.m is 45 minutes,
• the time taken for the count to fall from 300 to 150 c.p.m is 45 minutes,
• the time taken for the count to fall from 150 to 75 c.p.m is 45 minutes.

The time for the activity to fall to half its value is the same, no matter what the original activity is. This time is called the **half-life**. The half-life of the isotope in this example is 45 minutes.

The disintegration of a radioactive nucleus is a **random** process. No-one can say when an individual nucleus will suddenly split up. However, for a large number of nuclei, you can predict how many nuclei will disintegrate in a certain time. This is a bit like throwing dice. You can't predict what number you will get with a single throw, but if you threw 1000 dice you would expect one-sixth of them to come up with the number 4 and so on.

Suppose you have a sample of material which contains 1000 radioactive nuclei and that 10 per cent disintegrate per hour. After one hour, 100 nuclei will have disintegrated, leaving 900. During the next hour, 90 (10% of 900) will disintegrate to leave 810. The table shows the number of radioactive nuclei remaining as each hour passes.

Investigating Radioactive Sources (program)

Use this program to check your knowledge of radioactivity and half-life. You will be asked to study three experiments with different radioactive sources.

Table 8.1 ● The number of radioactive nuclei that decay every hour

Time from the start (hours)	0	1	2	3	4	5	6	7
No. of radioactive nuclei present	1000	900	810	729	656	590	530	477
No. of nuclei that decay in the next hour	100	90	81	73	66	59	53	48

Figure 8.2B shows how the number of radioactive nuclei changes with time. The rate of decay is proportional to the number of radioactive nuclei left. The half-life can be read from the graph. It is the time taken for the number of nuclei to fall from, say, 1000 to 500.

SUMMARY

The activity of a radioisotope is the number of its nuclei that disintegrate per second. The half-life of a radioisotope is the time taken for its activity to fall to half its original value.

Figure 8.2B ● Half-life

CHECKPOINT

❶ Sodium-24 is a radioisotope used in medicine. Its half-life is 15 hours. A solution containing 8.0 mg of the isotope is prepared. What mass of isotope remains after (a) 15 hours (b) 30 hours (c) 5 days?

❷ Cobalt-60 is a radioisotope made by placing cobalt in a nuclear reactor. It has a half-life of 5 years. The activity of a piece of cobalt-60 is 32.0 kBq. How long would it take for its activity to fall to (a) 16.0 kBq (b) 1.0 kBq?

❸ In Figure 8.2B, how long would it take for the count rate to fall from 800 c.p.m to 100 c.p.m?

❹ The following measurements were made in an experiment using a Geiger–Müller tube near a radioactive source.

Time (hours)	0	0.5	1.0	1.5	2.0	2.5
Count (c.p.m.)	510	414	337	276	227	188

The background count was 30 c.p.m.
(a) The initial count rate from the source alone was 480 c.p.m (510–30). This is the corrected count rate. Work out the corrected count rates for the other readings.
(b) Plot a graph of the count rate (on the vertical axis) against time.
(c) Use your graph to find the half-life of the source.

8.3 Making use of radioactivity

FIRST THOUGHTS

When a new discovery is made, scientists are keen to find ways of using the discovery both in research and in everyday life. Radioactivity is a good example of a fascinating discovery for which many useful applications have been found.

Carbon-14 dating

Carbon is made of the isotopes carbon-12, carbon-13 and carbon-14. The isotope carbon-14 is radioactive. It has a half-life of 5700 years. Carbon-14 is present in the carbon dioxide which living trees use in photosynthesis. After a tree dies, it can take in no more carbon-14. The carbon-14 already present decays slowly; carbon-12 does not change. The ratio of the

LOOK AT LINKS
for **radiocarbon dating**
You will now understand better what was said about radioactive dating of rocks in Topic 2.5. You will meet the importance of dating fossils for the theory of evolution in *KS: Biology*, Topic 21.

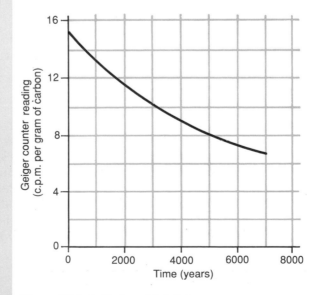

Figure 8.3A ● Carbon-14 dating

amount of carbon-14 left in the wood to the amount of carbon-14 in living trees can be used to tell the age of the wood. Animals take in carbon-14 in their food while they are alive. After their death, the proportion of carbon-14 in their bones tells how long it is since they died. A carbon-14 decay curve is shown in Figure 8.3A.

Other radio-isotopes

Radioactive isotopes differ enormously in their half-lives. Some are listed in Table 8.2.

Table 8.2 ● Some radioactive isotopes

Isotope	Radiation	Half-life	Isotope	Radiation	Half-life
Uranium-238	α	5000 million years	Iodine-131	β	8 days
Uranium-235	α	700 million years	Sodium-24	β	15 hours
Plutonium-239	α, γ	24 000 years	Bromine-82	β, γ	36 hours
Carbon-14	β	5700 years	Uranium-239	β	24 minutes
Strontium-90	β	59 years	Strontium-93	β, γ	8 minutes
Caesium-137	β, γ	30 years	Barium-143	β	12 seconds
Cobalt-60	γ	5 years	Polonium-213	α	4×10^{-6} s

Scientists have found many uses for radioactive isotopes. Figures 8.3 B–E show some of them.

Figure 8.3B ● Separating radioactive isotopes for medical use

● *Medical uses*

Radioactivity can be used to penetrate the body and kill cancerous tissue. Cobalt-60 and caesium-137 are often used for this purpose. They emit γ-rays. The dose of radiation must be carefully calculated to destroy cancerous tissue and leave healthy tissue alone. *Why is a γ-emitter used, rather than an α- or a β-emitter for this job?*

Radiation is used to destroy germs on medical instruments. It is more convenient than boiling, and it is also more efficient. Cobalt-60 is often used. *Why is a γ-emitter better than an emitter of α- or β-radiation for this job?*

The thyroid gland in the throat takes iodine from food and stores it. To find out whether the thyroid is working correctly, a patient is given food containing iodine-131, a β-emitter. The radioactive iodine can be detected as it passes through the body. The half-life of iodine-131 is 8 days. After a few weeks, there will be little left in the patient's body. *Can you explain why a β-emitter is better than a γ-emitter for this purpose?*

Figure 8.3C ● Radioactivity in agriculture

● Agricultural Research

This research worker is studying the uptake of fertilisers by plants. He has used a fertiliser containing phosphorus-32. This isotope is a β-emitter with a half-life of 14 days. By measuring the radioactivity of the leaves, the scientist can find out how much fertiliser has reached them.

● Industrial Uses

This research worker is measuring engine wear. The pistons of this engine have been in a nuclear reactor. Some of the metal atoms have become radioactive. As the engine runs, the pistons wear away and radioactive atoms enter the lubricating oil. The more the engine wears, the more radioactive the oil becomes. You can tell how well a lubricating oil reduces engine wear by timing the uptake of radioactivity.

Figure 8.3E shows the production of metal foil (e.g. aluminium baking foil). The detector measures the amount of radiation passing through the foil. If the foil is too thick, the detector reading drops. The detector sends a message to the rollers, which move closer together to make the foil thinner. *Why must the source be a β-emitter, not an α- or a γ-emitter? Why does it need to have a long half-life? Can you suggest a suitable isotope (see Table 8.2)?*

Figure 8.3D ● Measuring engine wear

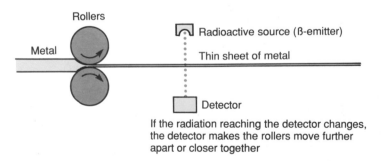

If the radiation reaching the detector changes, the detector makes the rollers move further apart or closer together

Figure 8.3E ● The manufacture of metal foil

Radioactivity and food

Food spoilage is a serious problem. About 20% of the world's food is lost through spoilage. The major cause is the bacteria, moulds and yeasts which grow on food. Some bacteria produce waste products which are toxic to people and cause the symptoms of food poisoning (sickness and diarrhoea). There are thousands of cases of food poisoning every year and some are fatal. Now there is an answer.

Irradiation of food with γ-rays kills 99% of disease-carrying organisms. These include *Salmonella*, which infects a lot of poultry, and *Clostridium*, the cause of botulism, which is often fatal. Spices, which are likely to contain micro-organisms as they are imported from tropical countries, can be irradiated with no loss of flavour. Irradiating potatoes is useful because it stops them sprouting without affecting the taste. The treatment is not suitable for all foods. Red meats turn brown and develop an

unpleasant taste, eggs develop a smell, shrimps turn black and tomatoes go soft.

Some people fear that irradiated food will be radioactive. In fact, foods contain a natural low level of radioactivity, and the treatment increases this level only slightly. The dose of radiation which the food receives is carefully calculated. By the time the food is eaten, the extra radioactivity has decayed. The best proof that irradiation is safe is that you can not detect it.

CHECKPOINT

❶ In 1985, a body was found in a peat bog in Norfolk. The reading which scientists obtained for the carbon-14 radioactivity of the body was 9 c.p.m. per gram of carbon.
  (a) Archaeologists called the body Pete Marsh. Where do you think they got the idea for the name?
  (b) Refer to Figure 8.3A, which shows a decay curve for carbon-14. How long ago did Pete Marsh die?

❷ Testing the filling of cans on a production line.
  Can A is full, but something has gone wrong on the production line, and Can B is only partly filled. Can you design a scheme for using a radioactive source and a detector to tell whether the cans coming off the production line are completely filled? Say whether you would use a source of α or β or γ radiation and say what kind of half-life would be suitable. Choose an isotope from Table 8.2 which could be used.

❸ Detecting leaks.
  Imagine that you are a scientist working for a water company. An underground pipe 2 km long is leaking. You want to find out where the leak is without digging up the pipe. You decide to add a radioactive isotope to the water at one end of the pipe and then drive slowly along the route which the pipe takes, testing the ground with a Geiger–Müller counter. If you detect radioactivity you will know where the underground leak is.

  Which radioisotope will you choose from Table 8.2? You will need:
  • an element which forms a water-soluble compound,
  • radiation which can penetrate the soil,
  • a half-life which will give you time to drive slowly along the route,
  • to avoid contaminating the water for long.

❹ A firm which manufactures plastic syringes gets an order from a hospital. The hospital wants the syringes to be sterile. The firm cannot sterilise the syringes by heating because this would soften the plastic. What can the firm do? (See Table 8.2 for details of radioactive isotopes.)

❺ An engineer is planning a method of checking the stability of an oil rig which is to be used in the North Sea. She decides to mix a radioactive isotope with the concrete which fixes the legs of the oil rig to the sea bed. Then she can install a detector to find out whether there is any movement in the concrete.

  Which isotope from Table 8.2 should she choose? Concrete contains calcium compounds so the isotope should be chemically similar to calcium. It should have a suitable half-life.

8.4 The dangers of radioactivity

As well as being useful, radioactivity can be dangerous. People who work with it must understand the science in what they are doing. This sign is the radioactivity hazard sign.

We make good use of radioactivity. However, large doses of radiation are dangerous. Exposure to a high dose of radiation burns the skin. Delayed effects are damage to the bones and the blood. People who are exposed to a lower level of radiation for a long time may develop leukaemia (a disease of the blood cells) and cancer. When radioactive elements get inside the body, they are very dangerous. They irradiate the body organs near them, and the risk of cancer is very great. Even an emitter of α-rays, the rays with least penetrating power, can do immense damage if it gets inside the body. Fairly low doses of radioactivity can damage human **genes**. This may result in the birth of deformed babies. People who work with radioactive sources take precautions to protect themselves.

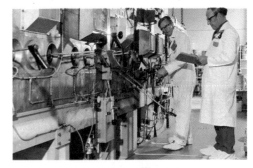

(b) These workers are using long handled tools to distance themselves from radioactive sources

(a) This worker is wearing protective clothing so that radioactive material does not get onto his hands or clothes

(c) A film badge measures the dose of radiation its wearer receives

Figure 8.4A ● Precautions

● They didn't know what the symbol meant

In the city of Goiania in Brazil in 1987, two junk collectors broke into a disused medical clinic and stole a heavy cylindrical object, which they took to be lead. They sold the metal cylinder to a junk yard. It emitted an eerie blue light from narrow slits. The manager of the junk yard hammered the head off the cylinder. Inside lay a capsule containing a powdery blue substance which stuck to the skin and glowed. It was caesium-137 a radioactive isotope which is used in cancer therapy. Massive doses of the γ-radiation from caesium-137 cause leukaemia, bleeding, sterility and cataracts. The dealer did not know this: he thought the bluish powder was so pretty and shiny that he gave away bits of it to his neighbours. Almost immediately, those who touched it became sick and feverish. One man stored some under his bed because he wanted to see it glow in the dark. Several children rubbed it on their bodies, and their skin became blistered and burnt.

SUMMARY

Radioactive isotopes are used in medicine, in research and in industry. Workers handling radioactive materials take precautions to shield themselves from radiation.

As a result of handling caesium-137, 20 people were taken to hospital, and 50 more people were put under medical observation, 25 homes were evacuated, and a large area of the city was cordonned off while the authorities cleared the area of radioactivity.

CHECKPOINT

❶ Why is it essential for people who handle radioactive material to wear gloves?

❷ Why do people build lead walls to protect them from γ-emitters but not from α-emitters?

❸ Why are α-emitters dangerous if they get on to your hands?

❹ The radioisotope iodine-131 is used for medical diagnosis. It has a half-life of 8 days. On 1 January, a doctor injects a patient with a solution containing 0.04 g of iodine-131. What is the maximum mass of iodine-131 that could be left in the patient's body on 2 January? Why is the actual mass less than this?

FIRST THOUGHTS

8.5 The nuclear bomb

Splitting the atom was a marvellous achievement. Who could have foreseen the horrifying use that would be made of this discovery?

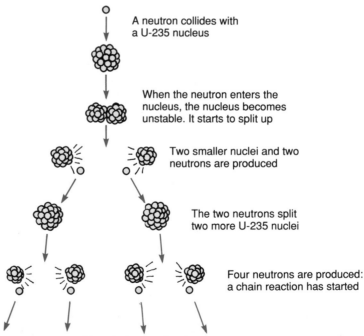

A neutron collides with a U-235 nucleus

When the neutron enters the nucleus, the nucleus becomes unstable. It starts to split up

Two smaller nuclei and two neutrons are produced

The two neutrons split two more U-235 nuclei

Four neutrons are produced: a chain reaction has started

Figure 8.5A ● The chain reaction in uranium fission

A German scientist called Otto Hahn made history in 1939. He split the atom! Hahn had been experimenting on firing neutrons at different types of nuclei. When neutrons struck uranium nuclei, some nuclei of uranium-235 split into two new nuclei and two neutrons. **Nuclear fission** (nucleus splitting) had occurred. The electrons of the original uranium-235 atom divided themselves between the two nuclei and two new atoms were formed. At the same time, an enormous amount of energy was released.

The energy given out was much greater than the energy given out in a chemical reaction. The energy which is released in nuclear fission is called **nuclear energy** (and also called **atomic energy**). The discovery that energy on a grand sale could be obtained from nuclear fission was made in 1939. Soon afterwards, the Second World War broke out. Scientists on both sides began trying to invent an 'atom bomb', a bomb which would release nuclear energy.

Figure 8.5A shows what happens in a block of uranium-235. The fission of one U-235 nucleus produces two neutrons. These two neutrons split two more uranium nuclei, producing four neutrons. A chain reaction is set off. In a large block of uranium-235, it results in an explosion. In a small block of uranium-235, an explosion does not occur because many neutrons escape from the surface before producing fission. A nuclear bomb consists of two blocks of uranium-235, each smaller than the critical mass. The bomb is detonated by firing one block into the other to make a single block which is larger than the critical mass. The detonation is followed by an atomic explosion.

Figure 8.5B ● Hiroshima after the bomb

On 6 August, 1945, a uranium bomb was dropped on the Japanese city of Hiroshima. It destroyed 60 percent of the city, and killed 140 000 people.

A huge mushroom-shaped cloud rose to a height of 10 km. A plutonium bomb was dropped on the city of Nagasaki three days later. Half the city was destroyed and thousands of people were killed in the blast. There had been air raids before which had destroyed large parts of cities and killed thousands of people. This time, the damage had been done in seconds by a single bomb. One nuclear bomb had done as much damage as 20 000 tonnes of TNT. There was a new horror as the unknown danger of nuclear radiation and fall-out revealed itself.

Where was Otto Hahn, who first split the uranium atom, when the bomb fell on Hiroshima? He was one of a group of German scientists who were interned in Britain. He was distressed by the thought of the great misery which the bomb had caused. Hahn told the officer in charge of the internment camp that when he first saw that fission might lead to an atomic bomb he had thought of suicide.

Figure 8.5C ● A CND rally

Thousands of people received a large dose of radiation, from which they never recovered. The damage was so terrible that no nation has used nuclear weapons since. Many nations have huge stockpiles of nuclear weapons. The Campaign for Nuclear Disarmament (CND) wants nations to destroy their stockpiles in case an accident brings about another nuclear explosion.

Some of the scientists who worked on nuclear fission foresaw a new age, the 'atomic age', in which the energy of the atomic nucleus could be used for great benefit to mankind. They did not want to usher in the atomic age with a bomb. One group of scientists proposed that the USA should demonstrate the new bomb on a desert or on a barren island and then tell Japan to surrender or face atomic bombs. The military decision to drop the bombs was the result of the huge casualties suffered in the war against Japan. It was estimated that a million more lives would be lost in an invasion of Japan. Even after the first bomb was dropped, Japan refused to surrender. After the second bomb, the emperor himself ended the war, inspite of an attempt by a group of army officers to depose him and continue the fight.

8.6 Nuclear reactors

FIRST THOUGHTS

This section shows how nuclear power stations utilise the energy of fission (splitting atomic nuclei) and explores the hope of harnessing the energy of fusion (joining atomic nuclei) in the future.

Nuclear reactors obtain energy from the same reaction as the nuclear bomb: the fission of uranium-235. In reactors, fission is carried out in a controlled way. Reactors use naturally occurring uranium, which is a mixture of uranium-235 and uranium-238. Uranium-235 only undergoes fission with slow neutrons. Figure 8.6A shows a nuclear reactor. Neutrons from the fuel rods go into the graphite core, where they collide with graphite atoms and lose kinetic energy. The graphite is called a 'moderator' because it slows down the neutrons. The neutrons then pass into the fuel rods and cause fission. The boron rods control the rate of fission by absorbing some of the neutrons. The heat generated by nuclear fission warms a coolant fluid which circulates through the moderator.

Sensors in the core of a nuclear reactor register the temperature. The information is fed into a computer which logs the data. If the temperature rises too high, the computer sounds an alarm and activates safety procedures. Only computers can calculate with the speed needed to monitor a reactor safely.

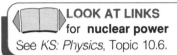

LOOK AT LINKS
for **nuclear power**
See *KS: Physics*, Topic 10.6.

The coolant may be water (as in pressurised water reactors) or a gas, e.g. carbon dioxide (as in gas cooled reactors). The heat is used to turn water into steam. The steam drives a turbine and generates electricity.

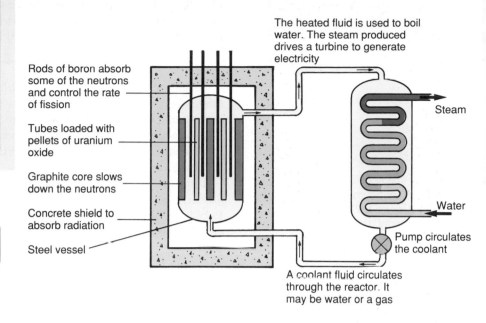

Rods of boron absorb some of the neutrons and control the rate of fission

Tubes loaded with pellets of uranium oxide

Graphite core slows down the neutrons

Concrete shield to absorb radiation

Steel vessel

The heated fluid is used to boil water. The steam produced drives a turbine to generate electricity

Steam

Water

Pump circulates the coolant

A coolant fluid circulates through the reactor. It may be water or a gas

Figure 8.6A ● A nuclear reactor

IT'S A FACT

The first nuclear reactor was built in an old squash court under a football stadium in Chicago in 1942. In charge was an Italian, Enrico Fermi, who had fled from Fascism in his native country to the USA in 1938. He assured the scientists present that the chain reaction would not get out of hand. It was held in check by a control rod. The control rod was pulled further and further out of the reactor. The neutron count rate rose and rose. Finally, Fermi ordered the reactor to be shut down. The scientists and engineers who saw the demonstration were impressed.

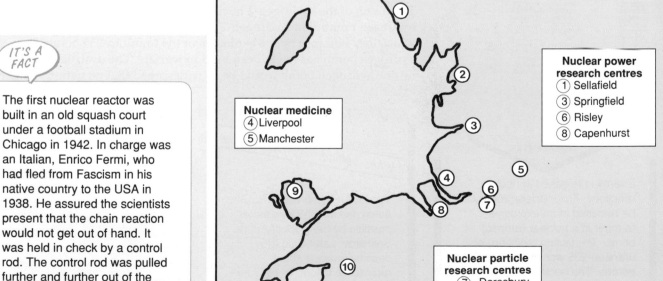

Nuclear power station
② Heysham ⑨ Wylfa ⑩ Trawsfynydd

Nuclear power research centres
① Sellafield
③ Springfield
⑥ Risley
⑧ Capenhurst

Nuclear medicine
④ Liverpool
⑤ Manchester

Nuclear particle research centres
⑦ Daresbury

Figure 8.6B ● The nuclear industry in North-west England and North Wales

8.7 Nuclear fusion

One glass of water could provide the same amount of energy as 1 tonne of petrol! The source of this energy is nuclear fusion. When two atoms of hydrogen-2 (deuterium) collide at high speed, the nuclei fuse together.

$$_1^2\text{H} + _1^2\text{H} \longrightarrow _2^3\text{He} + _0^1\text{n}$$

An isotope of helium and a neutron are produced. A large amount of energy is released when two light nuclei fuse together. It gives more hydrogen-2 atoms the energy they need to fuse with other atoms. Thus a chain reaction starts. There is no shortage of hydrogen-2 because it occurs in water as $^2\text{H}_2\text{O}$. Nuclear reactors of the future may use the fusion process as a source of nuclear energy.

There are enormous technical difficulties with fusion. The hydrogen-2 atoms must be heated to a very high temperature before they will fuse. Repulsion between the two positive charges sets in when the nuclei get close. If the atoms are moving fast enough, this repulsion can be overcome. The reactor must be made of materials which will withstand very high temperatures. Scientists are still working on the problems of obtaining energy from fusion.

The products of fusion are not radioactive. However, the metal structure in which fusion takes places does become radioactive by interaction with the neutrons produced in fusion. Fusion reactors of the future would, like fission reactors, involve the disposal of both high-level and low-level radioactive waste.

The Sun obtains its energy from the fusion of hydrogen-2 atoms. In the Sun, the temperature is about 10 million °C, and the hydrogen-2 atoms have enough energy to fuse.

The hydrogen bomb

The fusion of the hydrogen-2 nuclei is the source of energy in the **hydrogen bomb**. The hydrogen-2 atoms are raised to the temperature at which they will fuse by the explosion of the uranium-235 bomb. The hydrogen bomb has never been used in warfare. The destruction caused by one hydrogen bomb would be so widespread that no nation has dared to use it.

SUMMARY

Energy is released in nuclear reactions. This nuclear energy can be released in an explosive manner in a nuclear (atomic) bomb. The fission (splitting) of uranium-235 was used in nuclear bombs. The nuclear energy from the fission of uranium-235 is released in a safe, regulated manner in a nuclear power station. The fusion (joining together) of small nuclei also releases energy. The fusion of hydrogen-2 nuclei is ·the source of energy in the hydrogen bomb.

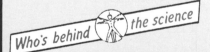

Who's behind the science

The fusion bomb which Russia detonated in 1953 was made possible by the work of Andrei Sakharov. Later Sakharov became anxious about radioactive fallout. In 1957 he began to campaign for a stop to tests on nuclear weapons. In 1968, he was sacked from his post in a research institute and later interned as a dissident. After his release in 1986, he continued to be active in politics until his death in 1989.

8.8 ⚛ Disposal of radioactive waste

Everyone who uses radioactive materials has to find a solution to the problem of disposing safely of radioactive waste.

Radioactive waste comes from uranium mines, nuclear power stations, hospitals and research laboratories. It must be disposed of in some place where it is not a health hazard. The method used for waste disposal depends on whether the radioactivity is low-level, intermediate-level or high-level.

Low-level waste

Power stations produce a lot of slightly radioactive cooling water. This passes through long pipes out to sea. It is discharged 1–2 km from the shore.

Laboratory equipment and protective clothing are placed in metal containers. These are buried (see Figure 8.8A).

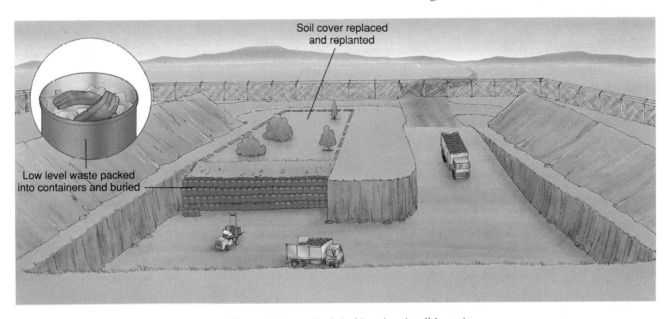

Figure 8.8A ● Burial of low-level solid waste

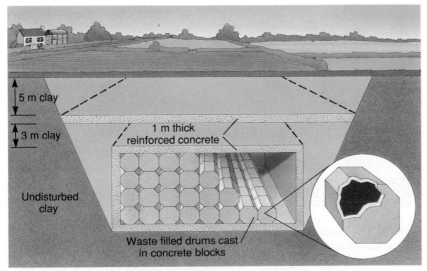

Figure 8.8B ● Burial of intermediate-level solid waste

Intermediate-level waste

Nuclear power stations produce large quantities of intermediate waste. Figure 8.8B shows one method of disposal. Drums containing waste are cast in concrete and buried under 8 m of clay. Another method is to bury drums of waste deep underground in disused mines.

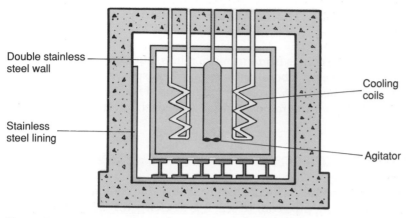

Figure 8.8C ● A storage tank for high-level liquid waste

Computers monitor the radioactivity and temperature of stores of high-level waste and keep a record of the measurements.

High-level waste

Nuclear reactors contain fuel rods (see Figure 8.6A). After a time, the fuel rods must be replaced. The spent fuel rods are highly radioactive. They are stored in cooling ponds of water until they are less radioactive. Then they are removed to a special treatment plant where they are dissolved in acid. The solution must be stored and cooled until it becomes less radioactive (see Figure 8.8C).

Liquid waste can be **vitrified** (turned into glass). France stores vitrified high-level waste, and Britain plans to start using this method. Figure 8.8D shows deep underground burial of steel canisters containing vitrified waste. There is a shortage of land sites for this purpose. British Nuclear Fuels Ltd is planning to store radioactive waste in tunnels under the sea (see Figure 8.8E).

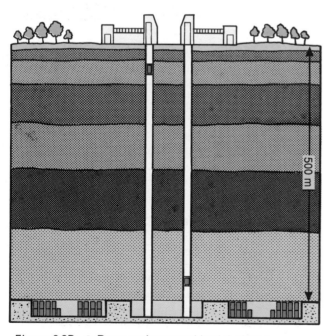

Figure 8.8D ● Deep underground burial of high level radioactive waste

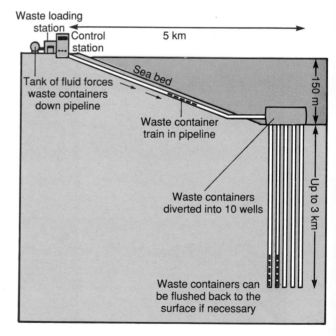

Figure 8.8E ● An undersea tunnel for the storage of radioactive waste

CHECKPOINT

❶ Why is a clay site chosen for the burial of intermediate-level waste (see Figure 8.8B) rather than a sandy soil?

❷ Some countries use disused salt mines for the storage of radioactive waste. Why do they feel sure that such sites will be dry? Why is it necessary to store the waste in dry conditions?

❸ When British Nuclear Fuels planned to build a site like that in Figure 8.8D in Humberside, people all over the county protested strongly. They said they did not want a 'nuclear dump in our back yard'. Posters read, 'Say no to a shallow grave'. The plan was dropped.

(a) Why do you think that people did not want a site of this kind in the county?

(b) Is it fair to describe the site as 'a shallow grave'?

(c) What will happen if every county refuses to store radioactive waste?

(d) Waste from nuclear power stations contains plutonium. How long does plutonium have to be stored before it has lost (i) half its radioactivity (ii) three quarters of its radioactivity? (See Table 8.3)

(e) Compose a letter to your local newspaper. Explain *either* why you think your county should accept a burial site for radioactive waste *or* why you think your county should resist such a plan.

FIRST THOUGHTS

8.9 ⚛ Safety

As you read this section, try to weigh up the advantages of nuclear power against the dangers.

Background radioactivity

Radioactive materials occur naturally in rocks, in soil and in the air. The low level of radioactivity which they give out is called **background radioactivity**. Figure 8.9A shows the radiation we receive from natural sources and from artificial sources.

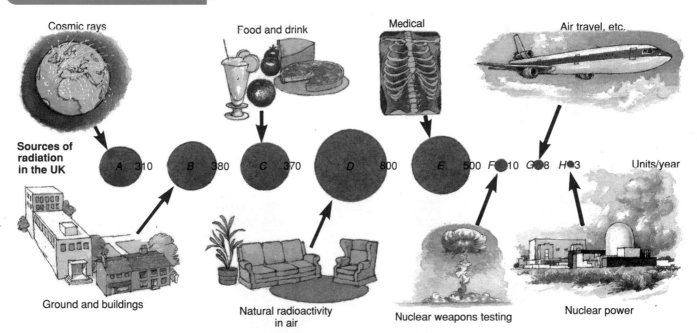

Cosmic rays Food and drink Medical Air travel, etc.

Sources of radiation in the UK

A 310 *B* 380 *C* 370 *D* 800 *E* 500 *F* 10 *G* 8 *H* 3 Units/year

Ground and buildings Natural radioactivity in air Nuclear weapons testing Nuclear power

Figure 8.9A ● Sources of radiation in the UK. The unit is the microsievert, a measure of the effect of radioactivity on cells

Accidents

● Windscale

They call it 'the day the reactor caught fire'. It happened at Windscale in Cumbria on 11 October 1957. The power station is now called Sellafield. The reactor overheated and graphite rods caught fire. They managed to put out the fire by flooding the reactor with water. Radioactive isotopes were blown from the reactor over the Lake District. In fact, no-one was injured in the accident.

LOOK AT LINKS
for **genetic damage**
The effect of radiation on genetic material is discussed further in *KS: Biology*, Topic 19.

RESOURCE
ACTIVITY
PACK

● *Three Mile Island*

A bigger accident happened in Three Mile Island in the USA in 1979. The reactor was a pressurised water reactor. The pumps which fed cold water into the reactor stopped and the temperature of the reactor shot up. After two hours, the operators rectified the fault. It took a week for the temperature to fall. The reactor was crippled. A cloud of radioactive substances fell over the island. Over three million litres of cooling water were radioactive. Luckily, no-one was injured in the accident. There may be long-term effects on health, however. In 1985, permission was given to reopen the plant. The public have staged such huge demonstrations against the reopening of the power station that work has not yet begun.

● *Chernobyl*

The worst nuclear accident happened in 1986 at the Chernobyl power station in the USSR. It was a pressurised water reactor with a graphite moderator, and a loss of cooling water caused the reactor to overheat. Steam reacted with graphite to produce hydrogen, which exploded. The explosion set the building on fire. Firemen battled heroically to put out the blaze. Many of those who fought the fire have since died of leukaemia. The reactor also caught fire. It burned for days while people tried to bring it under control. When the core of the reactor reached 5000 °C, they feared that it might melt the container and burn its way down into the earth. Helicopters managed to get near enough to drop sand and lead on the burning reactor to cool it.

Figure 8.9B ● The Chernobyl power station taken from a helicopter after the explosion

The roof of the reactor blew off in the explosion, and a cloud of radioactive material spread over the Ukraine in the USSR and drifted over other European countries. A week later, the cloud of radioactive material reached the UK.

From the area round Chernobyl, 135 000 people were evacuated. Thirty or more people died in the accident, and hundreds suffered from radiation sickness. People who received a smaller dose of radiation may develop leukaemia or cancer in the future. The final toll will not be known for many years.

Could Chernobyl happen here? The Chairman of the Central Electricity Generating Board answered this question in 1986. These are some of the points he made.

- The Chernobyl design is not used outside the USSR.
- The design is poor: the reactor is unstable when operated at low power.
- The operators at Chernobyl ignored some safety instructions.
- Chernobyl had no automatic fast-acting shutdown system to close the reactor if it became unsafe.
- The CEGB reactors all have computer-controlled safety systems which can shut down reactors if faults are detected.

CHECKPOINT

❶ Look at Figure 8.9A.
  (a) Background radiation is made up of sources A, B, C and D. What dose of radiation do these sources add up to?
  (b) Radiation from sources E, F, G and H is under our control. What dose of radiation do these sources add up to?
  (c) Which of the sources under our control adds the most to our total dose of radiation? What is the difficulty in reducing the dose from this source of radiation?
  (d) What is the total dose of radiation we receive in a year? What fraction of this comes from (i) nuclear power stations and (ii) nuclear weapons testing?

❷ Russia did not announce the accident at Chernobyl until after Sweden had detected an increase in atmospheric radioactivity. What do you think that a country which has a nuclear accident should do? If the neighbouring countries had known that a cloud of radioactive fallout was on the way, what precautions could they have taken?

❸ Strontium-90 is produced in nuclear reactors. It is radioactive. Why are babies especially at risk if an accident releases strontium-90 into the environment? (The clue is in the Periodic Table. Which elements resemble strontium? Which of these elements is present in milk?)

Further reading You will find more about nuclear power stations and nuclear bombs and missiles in *Extending Science 9: Nuclear Power* by R E Lee (Stanley Thornes (Publishers) Ltd).

TOPIC 9  IONS

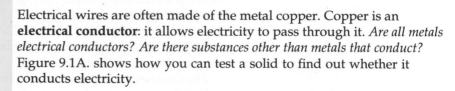

9.1 Which substances conduct electricity?

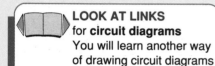

LOOK AT LINKS
for **circuit diagrams**
You will learn another way of drawing circuit diagrams in Topic 21 of *KS: Physics*.

SUMMARY

- **Solids.** Metallic elements, alloys and graphite (a form of carbon) conduct electricity.
- **Liquids.** Solutions of acids, alkalis and salts conduct electricity.

Electrical wires are often made of the metal copper. Copper is an **electrical conductor**: it allows electricity to pass through it. *Are all metals electrical conductors? Are there substances other than metals that conduct?* Figure 9.1A. shows how you can test a solid to find out whether it conducts electricity.

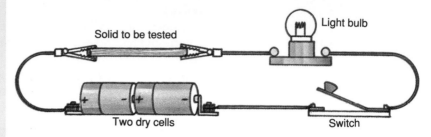

Figure 9.1A ● A testing circuit. If the solid conducts electricity, the bulb lights

Figure 9.1B shows a beaker of liquid. The two graphite rods in the liquid are **electrodes**: they can conduct a direct electric current into and out of the liquid. *Draw a **circuit diagram** like Figure 9.1A showing how you could test the liquid to see whether it conducts electricity.*

You should start your study of this topic by doing some experiments on conduction. *Are your results like those in Table 9.1?*

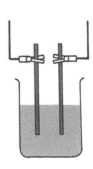

Figure 9.1B ● Test this liquid

Table 9.1 ● Electrical conductors

Solids	Liquids
Metallic elements	Mercury, the liquid metal
Alloys (mixtures of metals)	Solutions of acids, bases and salts
Graphite (a form of carbon)	Molten salts
(Solid compounds do not conduct.)	(Liquids such as ethanol and sugar solution do not conduct.)

9.2 Molten solids and electricity

LOOK AT LINKS
for **metals**
What enables metals to conduct? It is the metallic bond.
See Topic 19.2.

When a solid, e.g. a **metal**, conducts electricity, the current is carried by electrons. The battery forces the electrons through the conductor. The metal may change, for example it may become hot, but when the current stops flowing, the metal is just the same as before. When a metal conducts electricity, no chemical reaction occurs. When a molten salt conducts electricity, chemical changes occur, and new substances are formed.

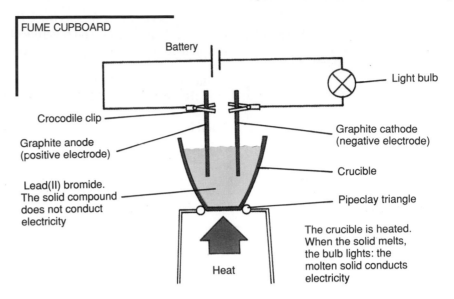

Figure 9.2A ● Passing a direct electric current through lead(II) bromide. This experiment should be done in a fume cupboard. (TAKE CARE: do not inhale the bromine vapour)

Figure 9.2A shows an experiment to find out what happens when a direct electric current passes through a molten **salt**. A salt is a compound of a metallic element with a non-metallic element or elements. Lead(II) bromide is a salt with a fairly low melting point. The container through which the current passes is called a **cell**. The rods which conduct electricity into and out of the cell are called **electrodes**. The electrode connected to the positive terminal of the battery is called the **anode**. The electrode connected to the negative terminal is called the **cathode**. The electrodes are usually made of elements such as platinum and graphite, which do not react with electrolytes.

When the salt melts, the bulb lights, showing that the molten salt conducts electricity. At the positive electrode (anode), bromine can be detected. It is a non-metallic element, a reddish-brown vapour with a very penetrating smell. (TAKE CARE: do not inhale bromine vapour) At the negative electrode (cathode), lead is formed. After cooling, a layer of lead can be seen on the cathode.

The experiment shows that lead(II) bromide has been split up by the electric current. It has been **electrolysed**. Compounds which conduct electricity are called **electrolytes**. Remember that all substances consist of particles (see Theme B, Topic 4). Since bromine goes only to the positive electrode, it follows that bromine particles have a negative charge. Since lead appears at the negative electrode only, it follows that lead particles have a positive charge. These charged particles are called **ions**. Positive ions are called **cations** because they travel towards the cathode. Negative ions are called **anions** because they travel towards the anode.

How do ions differ from atoms? A bromide ion, Br^-, differs from a bromine atom, Br, in having one more electron. The extra electron gives it a negative charge.

Bromine atom + Electron → Bromide ion
$$Br \quad + \quad e^- \quad → \quad Br^-$$

A lead(II) ion differs from a lead ion by having two fewer electrons. It therefore has a double positive charge, Pb^{2+}.

Lead atom → Lead ion + 2 Electrons
$$Pb \quad → \quad Pb^{2+} \quad + \quad 2e^-$$

Which electrode is which?
AnoDe
AD → ADD → + → positive
The anode is positive...
and the cathode is negative.

What happens when the ions reach the electrodes? They are **discharged**: they lose their charge. The positive electrode takes electrons from bromide ions so that they become bromine atoms.

Bromide ion → Bromine atom + Electron (taken by positive electrode)
$$Br^-(l) \rightarrow Br(g) + e^-$$

Then bromine atoms pair up to form bromine molecules:

$$2Br(g) \rightarrow Br_2(g)$$

The negative electrode gives electrons to the positively charged lead ions so that they become lead atoms.

Lead ion + 2 Electrons (taken from the negative electrode) → Lead atom
$$Pb^{2+}(l) + 2e^- \rightarrow Pb(l)$$

The electrons which are supplied to the anode by the discharge of bromine ions travel round the external circuit to the cathode. At the cathode, they combine with lead(II) ions (see Figure 9.2B).

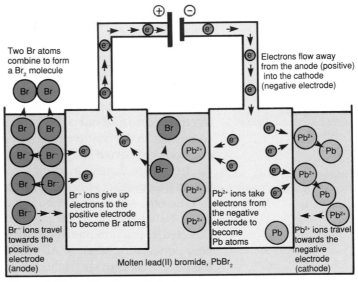

Figure 9.2B ● The flow of electrons

Lead(II) bromide is an uncharged substance because there are two Br^- ions for each Pb^{2+} ions. The formula is $PbBr_2$. *Why does solid lead(II) bromide not conduct electricity?*

In the solid salt, the ions cannot move. They are fixed in a rigid three-dimensional structure. In the molten solid, the ions can move and make their way to the electrodes.

You will have noticed that lead(II) ions have two units of charge, whereas bromide ions have a single unit of charge. By experiment, it is possible to find out what charge an ion carries. Table 9.2 shows the results of such experiments.

SUMMARY

Some compounds conduct electric current when they are molten. As they do they are electrolysed, that is, split up by the current. The explanation of electrolysis is that these compounds are composed of positive and negative ions.

Table 9.2 ● Some common ions

Positive ions			Negative ions	
+1	+2	+3	−1	−2
Hydrogen, H^+	Copper, Cu^{2+}	Aluminium, Al^{3+}	Bromide, Br^-	Oxide, O^{2-}
Sodium, Na^+	Iron(II), Fe^{2+}	Iron(III), Fe^{3+}	Chloride, Cl^-	Sulphide, S^{2-}
Potassium, K^+	Lead(II), Pb^{2+}		Iodide, I^-	Carbonate, CO_3^{2-}
	Magnesium, Mg^{2+}		Hydroxide, OH^-	Sulphate, SO_4^{2-}
	Zinc, Zn^{2+}		Nitrate, NO_3^-	

9.3 Solutions and conduction

The best way to begin this topic is by doing experiments to find out what types of solutions are electrolysed.

Who's behind the science

The scientist who did the first work on electrolysis was Michael Faraday. He was born in 1791, the son of a blacksmith. He received only a very basic education. When he was apprenticed to a London book binder, books on chemistry and physics came into his hands. He found them fascinating and led him to study science in his spare time. He attended a series of lectures given by Sir Humphry Davy, the director of the Royal Institution. Faraday wrote up the lectures and illustrated them with careful diagrams. He sent the notebook to Davy, asking for some kind of work in the laboratory. Davy agreed, and Faraday soon became a capable research assistant. On Davy's retirement, Faraday became director of the Royal Institution. Faraday is famous for his discovery of electromagnetic induction and for his work on the chemical effects of electric currents. In 1834, he put forward the theory that electrolysis could be explained by the existence of charged particles of matter. We now call them *ions*.

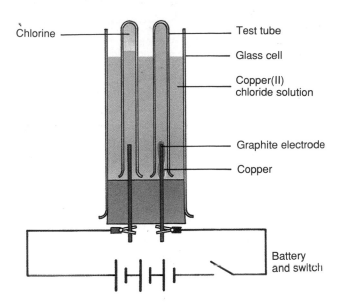

Figure 9.3A ● Electrolysis of copper(II) chloride

In Figure 9.3A a solution of the salt copper(II) chloride is being electrolysed. At the positive electrode (anode), bubbles of the gas chlorine can be seen. The negative electrode (cathode), becomes coated with a reddish brown film of copper.

Look at Figure 9.3A. At which electrode is copper deposited? Which kind of charge must copper ions carry? At which electrode is chlorine given off? Which kind of charge must chloride ions carry?

When they reach the electrodes, the ions are discharged.

At the negative electrode

Copper(II) ion + 2 Electrons (taken from the cathode) → Copper atom
$$Cu^{2+}(aq) + 2e^- \rightarrow Cu(s)$$

At the positive electrode

Chloride ion → Chlorine atom + Electron (given up at the anode)
$$Cl^-(aq) \rightarrow Cl(g) + e^-$$

Pairs of chlorine atoms then join to form molecules.

$$2Cl(g) \rightarrow Cl_2(g)$$

The electrons given to the anode by the discharge of chloride ions travel round the external circuit to the cathode. At the cathode, they combine with copper(II) ions.

To make sure you understand it, draw a diagram showing what happens to the ions and electrons in the electrolysis of copper(II) chloride solution. You can refer to Figure 9.2B for help.

Solid copper(II) chloride does not conduct electricity, but a solution of the salt in water does conduct. It is not the water in the solution that makes it conduct: experiment shows that water is a very poor electrical conductor. *What is the reason for the difference in behaviour between the solid and the solution?* In a solid, the ions are fixed in position, held together by strong attractive forces between positive and negative ions. In a solution of a salt, the ions are free to move (see Figure 9.3B).

The ions... which is which?
How to remember:
Current carries copper cations to cathode
Cations travel to the **cat**hode...
Anions travel to the **an**ode.

Water is a
non-conductor
because it is
not ionic

The crystalline solid
is a non-conductor
because the ions are
held in a rigid structure

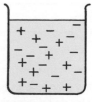

The solution is a
good conductor
because the ions
are now free to move

Figure 9.3B ● The ions must be free to move

SUMMARY

When a direct electric current is passed through solutions of some compounds, they are electrolysed, that is, split up by the current. Chemical changes occur in the electrolyte, and, as a result, new substances are formed.
Compounds which are electrolytes consist of positively and negatively charged particles called ions. In molten **ionic** solids and in solutions, the ions are free to move.

● *Safety matters*

Tap water conducts electricity. Never handle electrical equipment with wet hands. If the equipment is faulty, you run a much bigger risk of getting a lethal shock if you have wet hands.

Non-electrolytes and weak electrolytes

Some liquids and solutions do not conduct electricity. It follows that these **non-electrolytes** consist of molecules, not ions. Some substances conduct electricity to a very slight extent. They are **weak electrolytes**. These compounds consist mainly of molecules, which do not conduct. A small fraction of the molecules split up to form ions which do conduct. Such compounds are **partially ionised**.

CHECKPOINT

❶ A sodium atom can be written as $^{23}_{11}$Na.

The number of protons in a sodium atom is _____ and the number of electrons is _____ . The charge on a sodium atom is therefore _____ .

A sodium ion can be written as $^{23}_{11}$Na$^+$
The number of protons in a sodium ion is _____ . The charge on the ion is _____ and the number of electrons is therefore _____ .
The word equation for the formation of a sodium ion is:
sodium atom _____ electron _____ sodium atom
The symbol equation is: _____

❷ An atom of chlorine can be shown as $^{35}_{17}$Cl.
(a) How many (i) protons and (ii) electrons are there in a chlorine atom?
(b) What is the overall charge on a chlorine atom?

A chloride ion can be written as $^{35}_{17}$Cl$^-$
(c) What is the overall charge on a chloride ion?
(d) How many (i) protons and (ii) electrons does it contain?

Copy and complete the word equation for the formation of a chloride ion:
chlorine atom _____ electron _____ chloride ion
The symbol equation is: _____

❸ (a) Divide the following list into (i) electrical conductors, (ii) non-conductors.

copper, ethanol (alcohol), mercury, limewater, sugar solution, molten copper chloride, distilled water, sodium chloride crystals, molten sodium chloride, sodium chloride solution, dilute sulphuric acid

❹ Explain the words: electrolysis, electrolyte, electrode, anode, cathode, ion, anion, cation.

❺ Why are sodium chloride crystals not able to conduct electricity?

9.4 More examples of electrolysis

Sometimes the products formed when solutions are electrolysed are difficult to predict, but there are rules to help you.

Electrolysis
(program)

Use this program to see a computer simulation of electrolysis in action. You can study the movement of ions and the jobs done by the anode and cathode.

Sodium chloride solution

When molten sodium chloride (common salt) is electrolysed, the products are sodium and chlorine. When the aqueous solution of sodium chloride is electrolysed, the products are hydrogen (at the negative electrode) and chlorine (at the positive electrode). To explain how this happens, we have to think about the water present in the solution. Water consists of molecules, but a very small fraction of the molecules ionise into hydrogen ions and hydroxide ions.

Water → Hydrogen ions + Hydroxide ions
$$H_2O(l) \rightarrow H^+(aq) + OH^-(aq)$$

Hydrogen ions are attracted to the negative electrode as well as sodium ions. Sodium ions are more stable than hydrogen ions. It is easier for the negative electrode to give an electron to a hydrogen ion than it is for it to give an electron to a sodium ion. Sodium ions remain in solution while hydrogen ions are discharged to form hydrogen atoms. These atoms join in pairs to form hydrogen molecules.

$$H^+(aq) + e^- \rightarrow H(g)$$
$$2H(g) \rightarrow H_2(g)$$

Although the concentration of hydrogen ions in the solution is very low, it is kept topped up by the ionisation of more water molecules.

At the positive electrode, there are chloride ions and also hydroxide ions. The hydroxide ions have come, in very low concentration, from the ionisation of water molecules. Chloride ions are discharged while hydroxide ions remain in solution.

Copper(II) sulphate solution

When copper(II) sulphate is electrolysed, copper is deposited on the cathode, and oxygen is evolved at the anode. The oxygen comes from the water in the solution. At the positive electrode, there are hydroxide ions, OH^-, as well as sulphate ions, SO_4^{2-}. The hydroxide ions have come in low concentration from the ionisation of water molecules (see above). It is easier for the positive electrode to take electrons away from hydroxide ions than from sulphate ions, and hydroxide ions are discharged. The OH groups which are formed exist for only a fraction of a second before rearranging to give oxygen and water.

$$OH^-(aq) \rightarrow OH(aq) + e^-$$
$$4OH(aq) \rightarrow 2H_2O(l) + O_2(g)$$

Although only a tiny fraction of water molecules is ionised, once hydroxide ions have been discharged, more water molecules ionise to replace them with fresh hydroxide ions.

Dilute sulphuric acid

Figure 9.4A shows the electrolysis of dilute sulphuric acid.

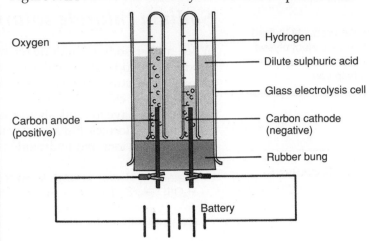

Figure 9.4A ● The electrolysis of dilute sulphuric acid

9.5 Which ions are discharged?

You will have seen that some ions are easier to discharge at an electrode than others.

LOOK AT LINKS
for **reactivity**
We will come back to the ease of discharge of metal ions in the reactivity series of metals in Topic 19.4.

Cations

Some metals are more **reactive** than others. The ions of very reactive metals are difficult to discharge. Sodium is a very reactive metal. When it reacts, sodium atoms form sodium ions. It is difficult to force a sodium ion to accept an electron and turn back into a sodium atom. Hydrogen ions are discharged in preference to sodium ions. The ions of less reactive metals, such as copper and lead, are easy to discharge.

Figure 9.5A ● Sodium ions do not want to accept electrons

Anions

Sulphate ions and nitrate ions are very difficult to discharge. When solutions of sulphates and nitrates are electrolysed, hydroxide ions are discharged instead, and oxygen is evolved.

SUMMARY

Water is ionised to a very small extent into $H^+(aq)$ ions and $OH^-(aq)$ ions. These ions are discharged when some aqueous solutions are electrolysed. The ions of very reactive metals, e.g. Na^+, are difficult to discharge. In aqueous solution, hydrogen ions are discharged instead. Hydrogen is evolved at the negative electrode. The anions SO_4^{2-} and NO_3^- are difficult to discharge. In aqueous solution, OH^- ions are discharged instead, and oxygen is evolved at the positive electrode.

CHECKPOINT

❶ Copy and complete this passage.

When molten sodium chloride is electrolysed, _____ is formed at the positive electrode and _____ is formed at the negative electrode.

When aqueous sodium chloride is electrolysed, _____ is formed at the positive electrode and _____ is formed at the negative electrode. The reaction that takes place at the positive electrode is

_____

and the reaction that takes place at the negative electrode is

_____

❷ A solution of potassium bromide is electrolysed. At one electrode a colourless gas is given off. At the other a brown vapour appears.
   (a) At which electrode does the brown vapour appear? Which ions have been discharged?
   (b) Which gas is produced at the other electrode? Which ions have been discharged?
   (c) Write equations for the discharge of the ions at each electrode.

FIRST THOUGHTS

9.6 Applications of electrolysis

Electrolysis is useful for:
- electroplating,
- extracting metals from their ores,
- manufacturing chemicals.

=== **SCIENCE AT WORK** ===

Do you know what happens when a group makes a recording? Sound waves activate a machine which cuts a groove in a soft plastic disc. This disc is coated with graphite so that it will conduct. Then it is electroplated with a thick layer of metal. The metal plating is prised off the plastic disc and used to stamp out copies of the original.

LOOK AT LINKS
for **rust**
Rusting, the problem and its prevention, is covered in Topic 19.11.

Electroplating

Some metals are prized for their strength and others for their beauty. Beautiful metals like silver and gold are costly. Often, objects made from less expensive metals are given a coating of silver or gold. The coating layer must stick well to the surface. It must be even and, to limit the cost, it should be thin. Depositing the metal by electrolysis is ideal. The technique is called **electroplating**.

Electroplating is used for protection as well as for decoration. You may have a bicycle with chromium-plated handlebars. The layer of chromium protects the steel underneath from rusting. Chromium does not stick well to steel. Steel is first electroplated with copper, which adheres well to steel, then it is nickel-plated and finally chromium-plated. The result is an attractive bright surface which does not corrode.

The rusting of iron is a serious problem. One solution to the problem is to coat iron with a metal which does not corrode. Food cans must not rust. They are made of iron coated with tin. Tin is an unreactive metal, and the juices in foods do not react with it.

Figure 9.6A ● Chromium plate

Electroplating is used to give a thin even layer of tin. A layer of zinc is applied to iron in the manufacture of 'galvanised' iron. Often, electroplating is employed.

The key is to be plated with nickel. It is made the negative electrode (cathode)

The electrolyte is a solution of a nickel salt, e.g. nickel sulphate

The positive electrode (anode) is made of the plating material, nickel. Nickel atoms ionise, replacing the nickel ions discharged from the solution

Figure 9.6B ● Electroplating

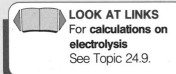

Extraction of metals from their ores

Reactive metals are difficult to extract from their ores. For some of them, electrolysis is the only method which works. Sodium is obtained by the electrolysis of molten dry sodium chloride, and aluminium is obtained by the electrolysis of molten aluminium oxide.

Electrolysis can also be used as a method of purifying metals. Pure copper is obtained by an electrolytic method.

Manufacture of sodium hydroxide

The three important chemicals, sodium hydroxide, chlorine and hydrogen are all obtained from the plentiful starting material, common salt. The method of manufacture is to electrolyse a solution of sodium chloride (common salt) in a diaphragm cell.

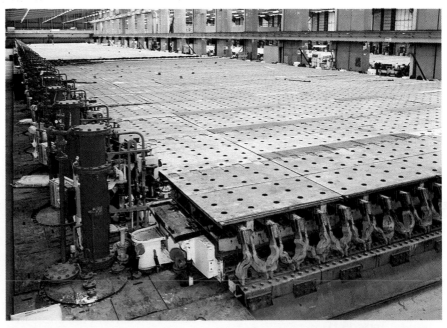

Figure 9.6C ● A room of diaphragm cells for the electrolysis of sodium chloride

LOOK AT LINKS
for **electrolysis**
In electrolysis (this topic), an electric current causes a chemical reaction to take place: electrical energy is converted into chemical energy.

In an electric cell, a chemical reaction inside the cell causes an electric current to flow through the external circuit, chemical energy is converted into electrical energy.
See Topic 20.

SUMMARY

The electrolysis of brine to give sodium hydroxide, chlorine and hydrogen is an important industrial process. It is carried out in the diaphragm cell.

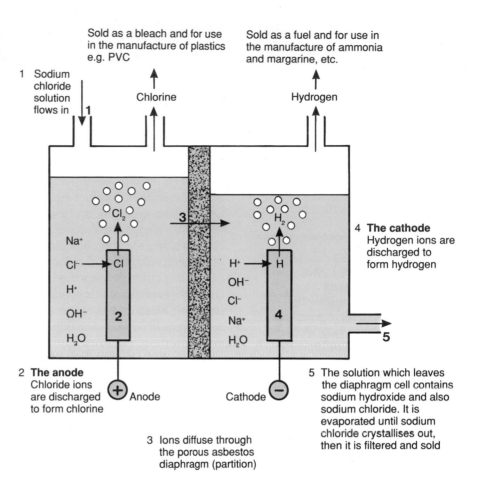

Sold as a bleach and for use in the manufacture of plastics e.g. PVC

Sold as a fuel and for use in the manufacture of ammonia and margarine, etc.

1 Sodium chloride solution flows in

Chlorine

Hydrogen

Na^+
Cl^-
H^+
OH^-
H_2O

Cl_2

3

H_2

H^+
OH^-
Cl^-
Na^+
H_2O

4 **The cathode**
Hydrogen ions are discharged to form hydrogen

5

2 **The anode**
Chloride ions are discharged to form chlorine   Anode

Cathode

5 The solution which leaves the diaphragm cell contains sodium hydroxide and also sodium chloride. It is evaporated until sodium chloride crystallises out, then it is filtered and sold

3 Ions diffuse through the porous asbestos diaphragm (partition)

Figure 9.6D ● An industrial diaphragm cell

CHECKPOINT

❶ (a) You are asked to electroplate a nickel spoon with silver. Draw the apparatus and the circuit you would use. Say what the electrodes are made of and what charge they carry. Say what electrolyte you could use.
   (b) Explain why silver plating is popular.

❷ Both paint and chromium plating are used to protect parts of a car body.
   (a) What advantages does paint have over chromium plating?
   (c) What advantages does chromium plating have over paint?
   (c) Why is chromium plating, rather than paint, used on the door handles?

❸ Many people like gold-plated watches and jewellery.
   (a) What two advantages do gold-plated articles have over cheaper metals?
   (b) Why do people choose gold plate rather than pure gold?
   (c) Why is the advantage of solid gold over gold plate?

TOPIC 10 THE CHEMICAL BOND

FIRST THOUGHTS

10.1 The ionic bond

What holds the atoms in a compound together? What holds the ions in a compound together? You can find out in this topic.

LOOK AT LINKS
for **electrons**
The arrangement of electrons, protons and neutrons in an atom is described in Topic 7.3.

LOOK AT LINKS
for **electrostatic attraction**
See *KS: Physics*, Topic 20.1.

Sodium chloride

Topic 9 dealt with electrolysis. You found that some compounds are electrolytes: they conduct electricity when molten or in aqueous solution. Other compounds are non-electrolytes. Electrolysis can be explained by the theory that electrolytes consist of small charged particles called **ions**. For example, sodium chloride consists of positively charged sodium ions and negatively charged chloride ions, $Na^+ Cl^-$.

Sodium chloride is formed when sodium (a metallic element) burns in chlorine (a non-metallic element). During the reaction, each sodium atom loses one **electron** to become a sodium ion. Each chlorine atom gains one electron to become a chloride ion.

$$\text{Sodium atom} \rightarrow \text{Sodium ion} + \text{Electron}$$
$$Na \rightarrow Na^+ + e^-$$

$$\text{Chlorine atom} + \text{Electron} \rightarrow \text{Chloride ion}$$
$$Cl + e^- \rightarrow Cl^-$$

When particles have opposite electric charges, a force of attraction exists between them. It is called **electrostatic attraction**. In sodium chloride, the sodium ions and chloride ions are held together by electrostatic attraction. The electrostatic attraction is the **chemical bond** in the compound, sodium chloride. It is called an **ionic bond** or **electrovalent bond**. Sodium chloride is an ionic or electrovalent compound. The compounds which conduct electricity when they are melted or dissolved are electrovalent compounds.

Figure 10.1A ● There is an attraction between sodium ions and chloride ions

A pair of ions, $Na^+ Cl^-$, does not exist by itself. It attracts other ions. The sodium ion attracts chloride ions, and the chloride ion attracts sodium ions. The result is a three-dimensional structure of alternate Na^+ and Cl^- ions. (Figure 10.1B) This is a **crystal** of sodium chloride. The crystal is uncharged because the number of sodium ions is equal to the number of chloride ions. The forces of attraction between the ions hold them in position in the structure. Since they cannot move out of their positions, the ions cannot conduct electricity. When the salt melts, the

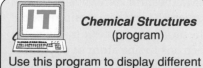
three-dimensional structure breaks down, the ions can move towards the electrodes and the molten solid can be electrolysed. There are millions of ions in even the tiniest crystal. The structures of ionic solids, such as sodium chloride, are described as **giant ionic structures**.

SUMMARY

Sodium and chlorine react to form positive sodium ions and negative chloride ions. The electrostatic attraction between these oppositely charged ions is an ionic bond or electrovalent bond.

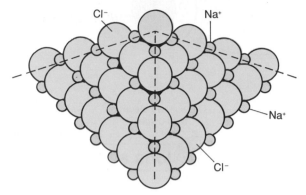

Figure 10.1B ● The structure of sodium chloride

CHECKPOINT

1 A sodium atom has 11 electrons, 11 protons and 12 neutrons.

(a) What is the charge on (i) an electron (ii) a proton and (iii) a neutron?
(b) What is the overall charge on a sodium atom?

A sodium ion has 10 electrons, 11 protons and 12 neutrons.
(c) What is the type of charge on a sodium ion?

2 A chlorine atom has 17 protons and 17 electrons.
(a) What is the overall charge on a chlorine atom?

A chlorine ion has 17 protons and 18 electrons.
(b) What is the type of charge on a chloride ion?

10.2 ⚛ Other Ionic compounds

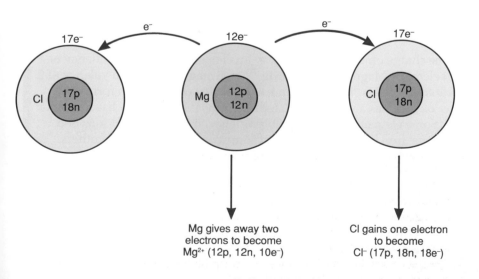

Figure 10.2A ● The formation of magnesium chloride

There are thousands of ionic compounds. In general, metallic elements give away electrons to form positive ions (cations). Non-metallic elements take up electrons to form negative ions (anions). Ionic compounds are formed when metallic elements give electrons to non-metallic elements.

Magnesium chloride is an ionic compound. It is formed when magnesium burns in chlorine. Each magnesium atom gives away two electrons. The cation formed therefore has two positive charges, Mg^{2+}. Since a chlorine atom wants only one electron, a magnesium atom has to combine with two chlorine atoms (see Figure 10.2A).

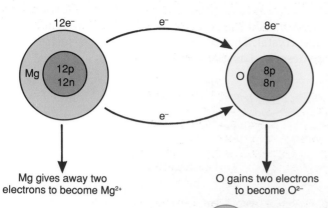

Mg gives away two
electrons to become Mg^{2+}

O gains two electrons
to become O^{2-}

Magnesium combines with oxygen to form magnesium oxide. One atom of magnesium gives two electrons to one atom of oxygen. The ions Mg^{2+} and O^{2-} are formed. The electrostatic attraction between them is an ionic bond (see Figure 10.2B).

Figure 10.2B ● The formation of an ionic bond between atoms of magnesium and oxygen

10.3 Ions and the Periodic Table

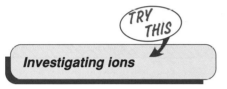

Investigating ions

You learned about the Periodic Table in Topic 7. Now you can find out how useful it is when it comes to remembering the charge on an ion.

Element	Symbol for ion	Group of Periodic Table
Sodium	Na^+	
Potassium	K^+	
Calcium	Ca^{2+}	
Magnesium	Mg^{2+}	
Aluminium	Al^{3+}	
Nitrogen	N^{3-}	
Oxygen	O^{2-}	
Sulphur	S^{2-}	
Chlorine	Cl^-	
Bromine	Br^-	

- Copy the table. Fill in the number of the Group in the Periodic Table to which each of the elements belongs.
- What is the connection between the charge on a cation and the number of the Group in the Periodic Table to which the element belongs?
- What is the connection between the charge on an anion and the number of the Group in the Periodic Table to which the element belongs?
- What would you expect to be the charge on (a) a barium ion (Group 2) and (b) a fluoride ion (Group 7)?

The charge on the ions formed by an element is called the valency of that element. Magnesium form Mg^{2+} ions: magnesium has a valency of 2. Sodium forms Na^+ ions: sodium has a valency of 1. Oxygen forms O^{2-} ions: oxygen has a valency of 2. Iodine forms I^- ions: iodine has a valency of 1. Some elements have a variable valency. Iron forms Fe^{2+} ions and Fe^{3+} ions: iron has a valency of 2 in some compounds and 3 in others.

Non-metallic elements often combine with oxygen to form anions, e.g. sulphate, SO_4^{2-} and nitrate, NO_3^-.

- **Metallic elements**
  Charge on cation = Number of the Group in the Periodic Table to which the element belongs = Valency of element
- **Non-metallic elements**
  Charge on anion = 8 – Number of Group in Periodic Table to which the element belongs = Valency of element

CHECKPOINT

❶ Copy and complete the following passage (lithium is an alkali metal in Group 1 and fluorine is a halogen in Group 7).

Lithium and fluorine combine to form _____ fluoride. When this happens, each lithium atom gives _____ to a fluorine atom. Lithium ions with a _____ charge and fluoride ions with a _____ charge are formed. The _____ ions and _____ ions are held together by a strong _____ attraction. This _____ attraction is a chemical bond. This type of chemical bond is called the _____ bond or _____ bond. A _____ dimensional structure of ions is built up.

❷ Sodium is a silvery-grey metal which reacts rapidly with water. Chlorine is a poisonous green gas. Sodium chloride is a white, crystalline solid, which is used as table salt. Explain how the sodium in sodium chloride differs from sodium metal. Explain how the chlorine in sodium chloride differs from chlorine gas.

❸ Say which of the following elements give (a) positive ions, (b) negative ions: calcium, potassium, sulphur, barium, zinc, oxygen, iron, iodine, lithium, fluorine, aluminium.
(The Periodic Table will help you.)

❹ There are two differences between
(a) a sodium ion, Na^+ (2.8), and a neon atom, Ne (2.8);
(b) a chloride ion, Cl^- (2.8.8), and an argon atom, Ar (2.8.8).
What are the differences?

FIRST THOUGHTS

10.4 Formulas of ionic compounds

How do you work out the formula of an ionic compound? You balance the charges on the ions, and suddenly formulas make sense!

Ionic and Molar Chemistry (program)

This program will help you to work out chemical formulas and to balance chemical equations. Also, use the program to study ionic and covalent bonding.

Ions are charged, but ionic compounds are uncharged. A compound has no overall charge because the sum of positive charges is equal to the sum of negative charges. In magnesium chloride, $MgCl_2$, every magnesium ion, Mg^{2+}, is balanced in charge by two chloride ions, $2Cl^-$.

Figure 10.4A ● The charges balance

This is how you can work out the formulas of electrovalent compounds.

Compound: Magnesium chloride

The ions present:	Mg^{2+} Cl^-
The charges must balance:	One Mg^{2+} ion needs two Cl^- ions.
Ions in the formula:	Mg^{2+} and $2Cl^-$ ions
The formula is:	$MgCl_2$

Compound: Sodium sulphate

The ions present:	Na^+ SO_4^{2-}
The charges must balance:	Two Na^+ are needed to balance one SO_4^{2-}
Ions in the formula:	$2Na^+$ and SO_4^{2-}
The formula is:	Na_2SO_4

Compound: Calcium hydroxide

The ions present:	Ca^{2+} OH^-
The charges must balance:	Two OH^- ions balance one Ca^{2+} ion
Ions in the formula:	Ca^{2+} and $2OH^-$
The formula is:	$Ca(OH)_2$

The brackets tell you that the 2 multiplies everything inside them. There are 2O atoms and 2H atoms, in addition to the 1Ca ion.

TRY THIS

Ionic compounds

Copy and complete these formulas:

Compound: Iron(II) sulphate

The ions present:	Fe^{2+} SO_4^{2-}
The charges must balance:	One Fe^{2+} ion balances __SO_4^{2-} ions
Ions in formula:	_____ and _____
The formula is:	_____

Compound: Iron(III) sulphate

The ions present:	Fe^{3+} SO_4^{2-}
The charges must balance:	__ Fe^{3+} balance __SO_4^{2-} ions
Ions in formula:	__ Fe^{3+} and __SO_4^{2-}
The formula is:	$Fe_2(SO_4)_3$

The formula contains __Fe, __S and __O.

You will notice that the sulphates of iron are named iron(II) sulphate and iron(III) sulphate. The roman numerals, II and III, show whether a compound contains iron(II) ions, Fe^{2+} or iron(III) ions, Fe^{3+}.

Table 10.1 shows the symbols and charges of some ions. From this table, you can work out the formula of any compound containing these ions.

RESOURCE ACTIVITY PACK

Table 10.1 ● The symbols of some common ions

Name	Symbol	Name	Symbol
Aluminium	Al^{3+}	Bromide	Br^-
Ammonium	NH_4^+	Carbonate	CO_3^{2-}
Barium	Ba^{2+}	Chloride	Cl^-
Calcium	Ca^{2+}	Hydrogencarbonate	HCO_3^-
Copper(I)	Cu^+	Hydroxide	OH^-
Copper(II)	Cu^{2+}	Iodide	I^-
Hydrogen	H^+	Nitrate	NO_3^-
Iron(II)	Fe^{2+}		
Iron(III)	Fe^{3+}	Oxide	O^{2-}
Lead(II)	Pb^{2+}	Phosphate	PO_4^{3-}
Magnesium	Mg^{2+}		
Mercury(II)	Hg^{2+}	Sulphate	SO_4^{2-}
Potassium	K^+		
Silver	Ag^+	Sulphide	S^{2-}
Sodium	Na^+		
Zinc	Zn^{2+}	Sulphite	SO_3^{2-}

SUMMARY

The formula of an ionic compound is worked out by balancing the charges on the ions.

❶ Write the formulas of the following ionic compounds.
(a) potassium chloride (b) potassium sulphate (c) ammonium chloride
(d) magnesium bromide (e) copper(II) hydroxide (f) zinc sulphate
(g) calcium carbonate (h) aluminium chloride (i) sodium hydrogencarbonate

❷ Write the formulas of the following ionic compounds.
(a) sodium hydroxide (b) calcium hydroxide (c) iron(II) hydroxide
(d) iron(III) hydroxide (e) aluminium oxide (f) iron (III) oxide

❸ Name the following.

(a) $AgBr$ (b) $AgNO_3$ (c) $Cu(NO_3)_2$ (d) $Al_2(SO_4)_3$ (e) $FeBr_2$ (f) $FeBr_3$ (g) PbO_2

(h) PbO (i) $Zn(OH)_2$ (j) NH_4Br (k) $Ca(HCO_3)_2$

FIRST THOUGHTS

10.5 The covalent bond

Some compounds are non-electrolytes because they consist of molecules, not ions. In this section, we think about the type of chemical bond in these molecular compounds.

Molecular compounds are formed by the combination of non-metallic elements. For example, carbon and oxygen combine to form carbon dioxide, CO_2. Hydrogen burns in chlorine to form hydrogen chloride, HCl. Hydrogen burns in oxygen to form water, H_2O. When non-metallic elements combine, both want to gain electrons; neither wants to form positive ions. Both elements can gain electrons by sharing electrons. When an atom of hydrogen and an atom of chlorine combine, the atoms come close enough for their outer electron shells to overlap. Figure 10.5A shows how overlapping of the electron shells holds the atoms together.

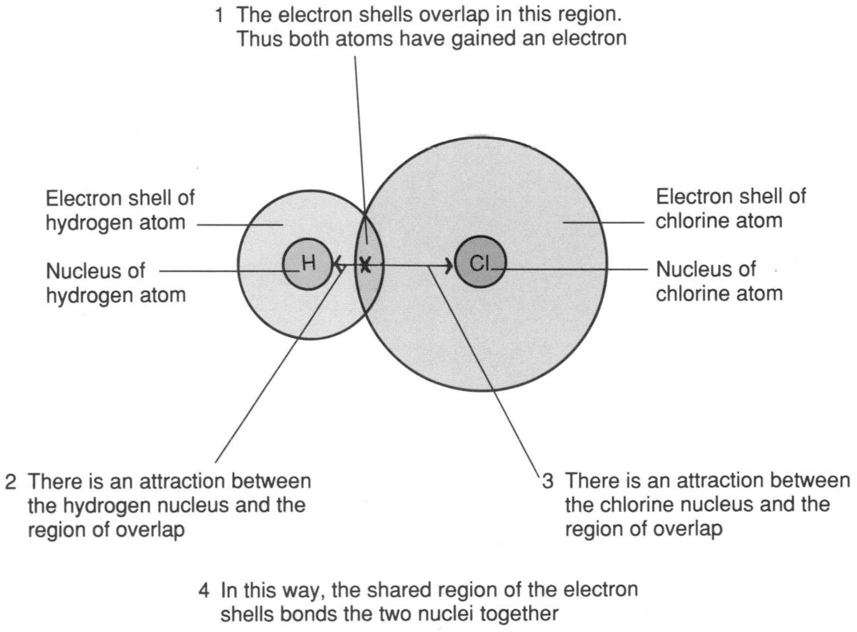

1 The electron shells overlap in this region. Thus both atoms have gained an electron

Electron shell of hydrogen atom

Nucleus of hydrogen atom

Electron shell of chlorine atom

Nucleus of chlorine atom

2 There is an attraction between the hydrogen nucleus and the region of overlap

3 There is an attraction between the chlorine nucleus and the region of overlap

4 In this way, the shared region of the electron shells bonds the two nuclei together

Figure 10.5A ● A molecule of hydrogen chloride

This type of chemical bond is called a **covalent bond**. Hydrogen chloride is a **covalent compound**. It is also described as a **molecular compound** because it consists of individual molecules.

Some atoms can form more than one covalent bond.

- A water molecule, H_2O, consists of an oxygen atom covalently bonded to two hydrogen atoms

$$H — O — H$$

- In a molecule of ammonia, NH_3, a nitrogen atom bonds to three hydrogen atoms

$$\begin{array}{c} H — N — H \\ | \\ H \end{array}$$

LOOK AT LINKS
for **methane**.
See Topic 25.1.

- In **methane** (North Sea gas), CH_4, a carbon atom forms covalent bonds with four hydrogen atoms

$$\begin{array}{c} H \\ | \\ H — C — H \\ | \\ H \end{array}$$

- In a molecule of oxygen, O_2, the atoms are joined by a **double bond**

$$O = O$$

- In carbon dioxide, CO_2, a carbon atom bonds to two oxygen atoms. Each of the bonds is a double bond

$$O = C = O$$

- In ethene, C_2H_4, there is a double bond between the carbon atoms

$$\begin{array}{cc} H & H \\ | & | \\ C & = C \\ | & | \\ H & H \end{array}$$

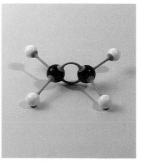

Figure 10.5B ● Covalent molecules
(a) water, H_2O (b) amonia, NH_3 (c) methane, CH_4 (d) oxygen, O_2
(e) carbon dioxide, CO_2 (f) ethene, C_2H_4

SUMMARY

Compounds which are non-electrolytes contain covalent bonds. In a covalent bond, the outer electron shells of two atoms overlap. The overlapping region is attracted to both atomic nuclei and therefore bonds the atoms.

Use a set of molecular models to build molecules of HCl, H_2O, NH_3, CH_4, $CH_2 = CH_2$ and CO_2 and others.

The shapes of covalent molecules

Is there an explanation for the different shapes of the molecules in Figure 10.5B? Why are the four bonds in methane, CH_4, arranged in space as you see them? Since the bonds are shared pairs of electrons (Figure 10.5C), they are negatively charged. As a result, they repel one another. So as to minimise the repulsion between them, the bonds take up positions in space which place the electron pairs as far apart as possible. Figure 10.5C shows how the bonds in CH_4 manage this. The bonds are directed towards the corners of a regular tetrahedron, with the carbon atom at the centre. (A tetrahedron has four faces. In a regular tetrahedron, all the sides and angles are equal.) The angle between each pair of bonds is 109.5°. If you make a model, you will see that making the angle between one pair of bonds greater than this decreases the angle between another pair of bonds. To achieve the maximum angle and the minimum repulsion between all the four bonds, an angle of 109.5° is the best angle.

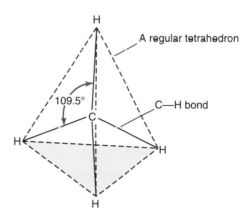

Figure 10.5C ● The arrangement of bonds in space in CH_4

In a molecule of ammonia, NH_3, there are three covalent bonds. An atom of nitrogen shares three electrons with three hydrogen atoms. There are three shared pairs of electrons. However, the nitrogen atom has five electrons in its outer shell. In total, therefore, there are three shared pairs of electrons and one unshared pair of electrons. The four electron pairs are directed towards the corners of a regular tetrahedron (see Figure 10.5D). The three bonds are as you see them in Figures 10.5B and D.

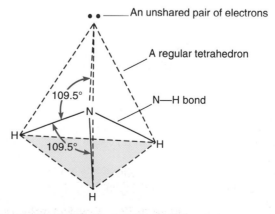

Figure 10.5D ● The arrangement of bonds in space in NH_3

In a molecule of water, H_2O, there are two covalent (electron pair) bonds. The oxygen atom, however, has 6 electrons in its outer shell. Two are used in bond formation, leaving 4 electrons unshared: 2 pairs of unshared electrons. The two bonds and the two unshared pairs of electrons occupy the corners of a regular tetrahedron (see Figure 10.5E).

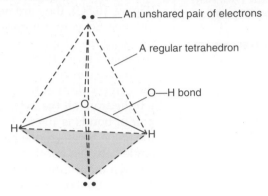

Figure 10.5E ● The arrangement of bonds in space in H_2O

In ethene, $CH_2{=}CH_2$, there is a double bond between the carbon atoms. The arrangement of four single bonds about a carbon atom is tetrahedral (see Figure 10.5C). You can use a model set with flexible bonds to assemble two $\overset{H}{\underset{H}{\diagdown}} C \diagdown$ groups with a tetrahedral arrangement of bonds. If you join two pairs of bonds between two carbon atoms, you will find that the four hydrogen atoms lie in the same plane (see Figure 10.5F).

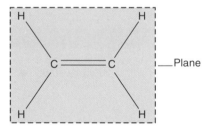

Figure 10.5F ● The arrangement of bonds in space in $CH_2{=}CH_2$

To make a model of a molecule of carbon dioxide, you need to fit four flexible bonds into a model carbon atom (with a tetrahedral arrangement of bonds). Then bend the bonds so that you can fit two of the bonds into each of two oxygen atoms. You will find that the molecule of CO_2 is linear (see Figure 10.5B).

CHECKPOINT

❶ Two atoms of chlorine combine to form a molecule, Cl_2. Neither chlorine atom wants to give an electron: both want to gain an electron. The figure shows a pair of chlorine atoms. They have come close enough together for their outer electron shells to overlap.
Explain how this satisfies the need for both chlorine atoms to gain electrons. Explain how the overlapping electron shells bond the atoms together.

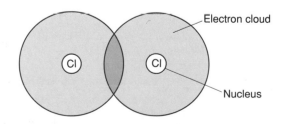

10.6 Forces between molecules

LOOK AT LINKS
for **allotropes**
The two allotropes of carbon, diamond and graphite, have different characteristics because they have different structures.
See Topic 5.8.

Covalent substances can be solids, liquids or gases. Gases consist of separate molecules. The molecules are far apart, and there are almost no forces of attraction between them.

In liquids there are forces of attraction between the molecules. In some covalent substances, the attractive forces between molecules are strong enough to make the substances solids. Ice is a solid. There are strong covalent bonds inside the covalent water molecules and weaker attractive forces between molecules (see Figure 10.6A). The forces of attraction between molecules hold them in a three-dimensional structure. It is described as a **molecular structure**.

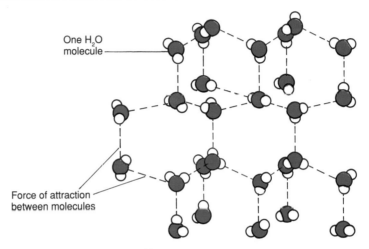

One H_2O molecule

Force of attraction between molecules

Figure 10.6A ● The structure of ice

SUMMARY

Some covalent substances are composed of individual molecules. Others are composed of larger units. These may be chains or layers or macromolecular structures.

Diamond is an **allotrope** of carbon. Diamond has a structure in which every carbon atom is joined by covalent bonds to four other carbon atoms (see Figure 5.8B). Millions of carbon atoms form a giant molecule. The structure is giant molecular or macromolecular.

Graphite, the other allotrope of carbon, has a layer structure (see Figure 5.8C). Strong covalent bonds join the carbon atoms within each layer. The layers are held together by weak forces of attraction.

CHECKPOINT

❶ (a) Why does diamond have a high melting point?
   (b) Why is it very difficult to scratch a diamond?
   (c) why does graphite rub off on to your hands?

❷ Solid iodine consists of shiny black crystals. Iodine vapour is purple. What is the difference in chemical bonding between solid and gaseous iodine?

10.7 Ionic and covalent compounds

What difference does the type of bond make to the properties of ionic and covalent compounds?
The physical and chemical characteristics of substances depend on the type of chemical bonds in the substances.

Table 10.2 ● Differences between ionic and covalent compounds

Ionic compounds	Covalent compounds
Ionic compounds consist of a giant structure of positive and negative ions.	Covalent compounds consist of separate uncharged molecules.
Ionic compounds are held together by strong ionic bonds – the attractive forces between positive ions and negative ions.	Strong covalent bonds join atoms inside the molecules. Weak bonds operate between molecules. Some covalent compounds form macromolecular structures, e.g. silica (sand) and ice.
Ionic compounds are electrolytes. They conduct electricity, when molten or in solution, and are split up in the process.	Covalent compounds are non-electrolytes. They do not conduct electricity
Ionic compounds have high melting points: they are solids at room temperature.	Covalent compounds are generally low melting point solids or liquids or gases. Substances with giant molecular structures have high melting and boiling points.

SUMMARY

The type of chemical bonds present, ionic or covalent, decides the properties of a compound, e.g. its physical state (s, l, g), boiling and melting points, electrolytic conductivity.

CHECKPOINT

❶ What kind of bonding would you expect between the following pairs of elements? (The Periodic Table on p 75 will help you.)

(a) K and Br   (c) Ca and S   (e) S and O
(b) Mg and I   (d) Ca and O   (f) F and O

❷

Solid	State	Melting point	Does it conduct electricity?
A	Solid	650 °C	Conducts when molten
B	Liquid	−20 °C	Does not conduct
C	Solid	700 °C	Does not conduct when molten
D	Solid	85 °C	Does not conduct when molten
E	Gas	−100 °C	Does not conduct

From the information in the table, say what you can about the chemical bonds in A, B, C, D and E.

❸ Below are five statements. Give a piece of evidence (e.g. a physical or chemical property of the substance) in support of each statement.
(a) An aqueous solution of sodium chloride contains ions.
(b) Copper exists as positive ions in copper(II) sulphate solution.
(c) Ethanol (alcohol) is a covalent compound.
(d) The forces between oxygen molecules are very weak.
(e) The forces between iodine molecules are stronger than the forces between oxygen molecules.

? THEME QUESTIONS

Element	Agron	Bromine	Calcium	Carbon	Chlorine	Gold	Hydrogen	Mercury	Phosphorus	Sulphur
Metal or non-metal	N-M	N-M	M	N-M	N-M	M	N-M	M	N-M	N-M
Melting point (°C)	−189	−7	850	3730 (sublimes)	−101	1060	−259	−39	44 (white) 590 (red)	113 (rh) 119 (mono)
Boiling point (°C)	−186	59	1490	4830	−35	2970	−252	357	280	445
Density (g/cm³)	0.0017	3.1	1.5	2.3 (gr) 3.5 (di)	0.003 017	19.3	0.000 083	13.6	1.8 (wh) 2.3 (red)	2.1 (rh) 2.0 (mono)

For questions 1–10 refer to the table of elements above.

1 Name the element with (a) the highest melting point (b) the lowest melting point.

2 Name the element with (a) the greatest density and (b) the lowest density.

3 Name the metal with (a) the lowest melting point and (b) the highest melting point.

4 Name the element which is liquid at room temperature and is (a) a metal (b) a non-metal.

5 Write down the names of all the elements which are (a) metals (b) non-metals.

6 Write down the names of all the elements which are (a) solids (b) liquids (c) gases.

7 Name the element which has the smallest temperature range over which it exists as a liquid.

8 Name the gaseous element which is (a) the most dense (b) the least dense.

9 Name the solid element which is (a) the most dense (b) the least dense.

10 Why are two sets of values given for (a) sulphur (b) phosphorus?

11 Copy out these equations into your book, and then balance them.
  (a) $H_2O_2(aq)$ → $H_2O(l) + O_2(g)$
  (b) $Fe(s) + O_2(g)$ → $Fe_3O_4(s)$
  (c) $Mg(s) + N_2(g)$ → $Mg_3N_2(s)$
  (d) $P(s) + Cl_2(g)$ → $PCl_3(s)$
  (e) $P(s) + Cl_2(g)$ → $PCl_5(s)$
  (f) $SO_2(g) + O_2(g)$ → $SO_3(g)$
  (g) $Na_2O(s) + H_2O(l)$ → $NaOH(aq)$
  (h) $KClO_3(s)$ → $KCl(s) + O_2(g)$
  (i) $NH_3(g) + O_2(g)$ → $N_2(g) + H_2O(g)$
  (j) $Fe(s) + H_2O(g)$ → $Fe_3O_4(s) + H_2(g)$

12 (a) Use the Periodic Table to answer the following questions:
  (i) Give the symbols for the elements carbon and sodium.
  (ii) Give the symbol for any inert gas.
  (iii) Give the symbol for any element in Group 7 (the Halogens)
  (iv) Give **one** reason why the symbol for helium is He and not H.
  (b) One molecule of carbon dioxide contains one atom of carbon and two atoms of oxygen. Its formula is written as CO_2.
  Write the formula of:
  (i) a molecule of sulphur dioxide, which has one atom of sulphur and two atoms of oxygen.
  (ii) a molecule of sulphuric acid, which contains two atoms of hydrogen, one atom of sulphur and four atoms of oxygen.

13 The element X has atomic number 11 and mass number 23. State how many protons and neutrons are present in the nucleus. Sketch the arrangement of electrons in an atom of X.

14 Atom A has atomic number 82 and mass number 204. Atom B has atomic number 80 and mass number 204. How many protons has atom A? How many neutrons has atom B? Are atoms A and B isotopes of the same element? Explain your answer.

15 Explain why Mendeleev fitted the elements into vertical groups in his Periodic Table. Why did he fit lithium, sodium, potassium, rubidium and caesium into the same group? Why did he leave some gaps in his table? The noble gases were discovered after Mendeleev had written his Periodic Table. What do they have in common (a) in their chemical reactions, (b) in their electronic configurations and (c) in their positions in the Periodic Table?

16 The structure of one molecule of an industrial solvent, dichloromethane, is shown below.

$$Cl - \overset{\displaystyle \overset{H}{|}}{\underset{\displaystyle \underset{H}{|}}{C}} - Cl$$

  (a) How many atoms are there in one molecule of dichloromethane?
  (b) How many bonds are there in one molecule of dichloromethane?

17 Phosphorus-32 is radioactive, with a half life of 14 days. A solution of sodium phosphate, containing phosphorus-32, gives a count rate of 6000 c.p.m in a Geiger–Müller counter. What will be the count rate after (a) 56 days (b) 140 days?

18 An engineer needs to put a radioactive isotope into an oil pipeline to investigate a leak. The isotope chosen must have a suitable half-life and emit a suitable type of radiation. Which of the following list would be suitable?

Isotope	Emission	Half-life
Copper-29	β	13 hours
Iodine-131	β, γ	8 days
Nitrogen-16	β, γ	7 seconds
Phosphorus-32	β	14 days
Sulphur-35	β	87 days
Sodium-24	β, γ	15 hours
Thallium-208	β, γ	3 minutes

Give reasons for your choice.

19 Why are people more worried about the accident risk posed by nuclear power stations than the accident risk in using cars?
This is not an easy question, and perhaps you would like to form a group to discuss it.

20 Argon-44 is a radioactive isotope. The amount of argon-44 in a sample was measured every ten minutes over a period of one hour. The results are shown in the table.

Radioactive argon remaining in the sample (%)	100	58	33	19	10	6	4
Time (min)	0	10	20	30	40	50	60

(a) Explain the meaning of 'radioactive' when used to describe an atom.
(b) On graph paper, draw a graph to show how the percentage of argon-44 in the sample changed during the hour.
(c) Use your graph to find the time when only 50% of the argon-44 was left in the sample.
(d) Some of the waste from nuclear power stations contains radioactive isotopes with very long half-lives. These isotopes give out large amounts of radiation.
(i) Explain why getting rid of this waste is a problem.
(ii) How could this waste be stored safely?
(LEAG)

21 Describe how you could use a battery and a torch bulb to test various materials to find out whether they are electrical conductors.
Divide the following list into conductors and non-conductors:
silver, steel, polythene, PVC, brass, candle wax, lubricating oil, dilute sulphuric acid, petrol, alcohol, sodium hydroxide solution, sugar solution, limewater

22 (a) Name three types of substance that conduct electricity.
(b) Why can molten salts conduct electricity while solid salts cannot?
(c) Why can solutions of salts conduct electricity while pure water cannot?
(d) Explain why metal articles can be electroplated but plastic articles cannot.

23 Name the substances A to H in the table. Write equations for the discharges of the ions which form these substances.

Electrolyte	Anode	Cathode
Copper(II) chloride solution	A	B
Sodium chloride solution	C	D
Dilute sulphuric acid	E	F
Sodium hydroxide solution	G	H

24 Your aunt runs a business making souvenirs. She asks your advice on how to electroplate a batch of small brass medallions with copper. Describe how this could be done. Draw a diagram of the apparatus and the circuit she could use.

25 In the electrolysis of potassium bromide solution, what element is formed (a) at the positive electrode (b) at the negative electrode? Why does the solution around the negative electrode become alkaline?

26 (a) Explain why the ionic compound $PbBr_2$ melts at a higher temperature than the covalent compound CH_2Br_2.
(b) You can smell the compound $CHCl_3$, chloroform. You cannot smell the compound KCl. What difference in chemical bonding is responsible for the difference?

27 The diagram shows apparatus that could be used in a laboratory to find out the effect of an electric current on an aqueous solution of sodium chloride.

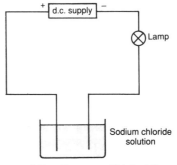

(a) On a copy of the diagram, (i) label the cathode, (ii) show the direction of flow of the electrons.
(b) What **two** observations would show that the solution conducts an electric current and that a chemical reaction is taking place?
(c) For the product formed at the positive electrode (i) give its name (ii) write the word equation for its formation.
(d) Give **one** reason why sodium is **not** a product when an electric current is passed through aqueous sodium chloride solution.
(LEAG)

THEME D
Making Materials

The Earth's crust and atmosphere supply us with many different materials. The study of the way materials behave is the science of chemistry. One of the jobs which chemists do is to find methods of separating useful substances from the Earth's crust. Industry uses these substances, e.g. metals, limestone and salt, as raw materials for the manufacture of the machines, buildings, fuels, medicines and possessions which we need. Chemists also work out ways for changing the substances which are found in nature into new substances. These changes are called chemical reactions.

TOPIC 11 METHODS OF SEPARATION

11.1 Raw materials from the Earth's crust

FIRST THOUGHTS

The Earth provides all the raw materials we use. The problem is to separate the substances we want from the mixture of substances which makes up the Earth's crust.

The Earth's crust and atmosphere provide us with all the raw materials that we use: metals, oil, salt, sand, limestone, coal and many other resources. We use these substances as the raw materials for the manufacture of the houses, clothing, tools, machines, means of transport, medicines and all the other goods which we need. Few useful raw materials are found in a pure form in the Earth's crust. It is usual to find raw materials that we want mixed up with other materials. Chemists have worked out methods of separating substances from mixtures. Table 11.1 summarises some of them.

Table 11.1 ● Methods of separating substances from mixtures

Mixture	Type	Method
Solid + Solid		Make use of a difference in properties, e.g. solubility or magnetic properties
Solid + Liquid	Mixture	Filter
Solid + Liquid	Solution	Crystallise to obtain the solid Distil to obtain the liquid
Liquid + Liquid	Miscible (form one layer)	Use fractional distillation
Liquid + Liquid	Immiscible (form two layers)	Use a separating funnel
Solid + Solid	In solution	Use chromatography

Make a Million
(program)

Try this game. You are invited to produce new materials using different chemical processes and the necessary raw materials. Your task is to make it a financial success.

11.2 Pure substances from a mixture of solids

Dissolving one of the substances

There are vast deposits of **rocksalt** in Cheshire. A salt mine is shown in Figure 11.2A. One method of mining salt is to insert charges of explosives in the rock face and then detonate them.

Rocksalt is crushed and used for spreading on the roads in winter to melt the ice. For many uses, pure salt (sodium chloride) is needed. It can be obtained by using water to dissolve the salt in rocksalt, leaving the rock and other impurities behind.

SUMMARY

A soluble substance can be separated from a mixture by dissolving it to leave insoluble substances behind.

Figure 11.2A ●
Winsford salt mine

11.3 Solute from solution by crystallisation

While the solvent is evaporating, dip a cold glass rod into the solution from time to time. When small crystals form on the rod, take the solution off the water bath and leave it to cool

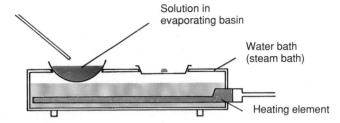

Solution in evaporating basin

Water bath (steam bath)

Heating element

Figure 11.3A ● Evaporating a solution to obtain crystals of solute

A laboratory method of evaporating a solution until it crystallises is shown in Figure 11.3A. The salt industry uses large scale evaporators which run non-stop. Figure 11.3B shows another method.

SUMMARY

A solute crystallises out of a saturated solution. If a solution is unsaturated, some of the solvent must be evaporated before crystals form.

Figure 11.3B ● Salt pans in the Canary Islands. Sea water flows into the 'pans' and much of the water evaporates in the hot sun. When the brine has became a saturated solution, salt crystallises. The sea water is pumped through sluice gates where the salt crystals are filtered out.

11.4 Filtration: separating a solid from a liquid

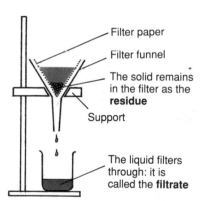

Filter paper

Filter funnel

The solid remains in the filter as the **residue**

Support

The liquid filters through: it is called the **filtrate**

Figure 11.4A ● Filtration

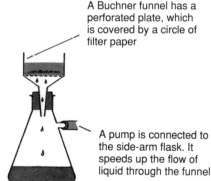

A Buchner funnel has a perforated plate, which is covered by a circle of filter paper

A pump is connected to the side-arm flask. It speeds up the flow of liquid through the funnel

Figure 11.4B ● Filtration under reduced pressure

Filtration can be used to separate a solid and a liquid. The filter must hold back solid particles but be fine enough to allow liquid to pass through. Figure 11.4A shows a laboratory apparatus for filtering through filter paper. Figure 11.4B shows a faster method: filtering under reduced pressure. In Topic 16.3, you will see how important filtration is in the purification of our drinking water.

11.5 Centrifuging

LOOK AT LINKS
for **protein in food**
See Topic 31.6.

Biochemists have developed methods of growing bacteria as a source of **high-protein food**. When bacteria are allowed to grow in a warm solution of nutrients, they multiply so fast that they can double in mass every 20 minutes. From time to time, the harvest of bacteria is separated from the nutrient solution. Bacteria are too small to be separated by filtration. They are so small that they are **suspended** (spread out) in the liquid. They do not settle to the bottom as heavier particles would do, and they do not dissolve to form a solution. They can be separated by the method of **centrifuging** (centrifugation). The suspension of bacteria and liquid is placed inside a centrifuge and spun at high speed. The motion causes bacteria to separate from the suspension and sink to the bottom of the centrifuge tubes. Figure 11.5A shows a small laboratory centrifuge.

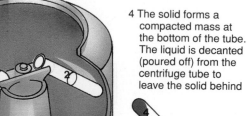

1 The suspension is poured into a glass tube inside the centrifuge

2 Another tube is used to balance the first

3 As the centrifuge spins, solid particles settle to the bottom of the tube

4 The solid forms a compacted mass at the bottom of the tube. The liquid is decanted (poured off) from the centrifuge tube to leave the solid behind

Figure 11.5A ● Centrifuging a suspension

> **SUMMARY**
>
> • Filtration will separate a solid from a liquid.
> • Centrifuging will separate a suspended solid from solution.

CHECKPOINT

1 Describe how you would obtain both the substances in the following mixtures of solids.
  (a) A and B: Both A and B are soluble in hot water, but only A is soluble in cold water.
  (b) C and D: Neither C nor D is soluble in water. C is soluble in ethanol, but D is not.

2 In Trinidad and some other countries, tar trickles out of the ground. It is a valuable resource. The tar is mixed with sand and gravel. Suggest how tar may be separated from these substances.

3 Blood consists of blood cells and a liquid called plasma. When a sample of blood is taken from a person, the blood cells slowly settle to the bottom. How can the separation of blood cells from plasma be speeded up?

4 Suggest how you could obtain the following:
  (a) iron filings from a mixture of iron filings and sand,
  (b) wax from a mixture of wax and sand,
  (c) sand and gravel from a mixture of both,
  (d) rice and salt from a mixture of both.

11.6 Distillation

> **LOOK AT LINKS**
> for **drinking water**
> Why is water more important than food?
> See Theme F, Topic 25.2.

Sometimes you need to separate a solvent from a solution. In some parts of the world, drinking water is obtained from sea water. The method of **distillation** is employed. Figure 11.6A shows a laboratory scale distillation apparatus. The processes that take place are:
• in the distillation flask, **vaporisation**: liquid ➜ vapour
• in the condenser, **condensation**: vapour ➜ liquid.
Vaporisation followed by condensation is called **distillation**.

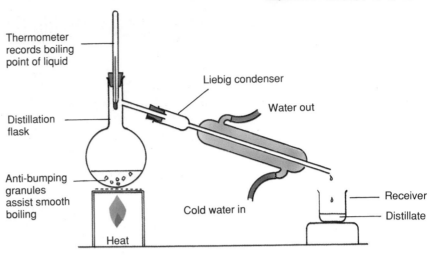

Thermometer records boiling point of liquid

Liebig condenser

Water out

Distillation flask

Anti-bumping granules assist smooth boiling

Cold water in

Heat

Receiver

Distillate

Figure 11.6A ● A laboratory distillation apparatus

SUMMARY

Distillation is used to separate a solvent from a solution.

CHECKPOINT

❶ Why do some countries in the Persian Gulf obtain drinking water by distilling sea water? You will need to consider (a) the other sources of water and (b) the cost of fuel for heating the still.

❷ Imagine that you are cast away on a desert island. You have two ways of obtaining drinking water. One is by separating pure water from sea water. The other is by collecting dew.

(a) Think of a way of obtaining pure water from sea water. You do not have a proper distillation apparatus, but you have some matches and an empty petrol tin which were in the lifeboat. You find wood, bamboo canes, palm trees and coconuts on the island. Make a sketch of your design. With other members of your class, make a display of your sketches.

(b) Every night on the island there is a heavy dew. How can you collect some of this dew? You have a sheet of plastic from the lifeboat. Again, sketch your idea. Then you can make another class display.

(c) Write a letter to a 10 year-old, telling how you survived the shipwreck and how you obtained drinking water until you were rescued.

11.7 Separating liquids

Using a separating funnel to separate immiscible liquids

When some liquids are added to each other, they do not mix: they form separate layers. They are said to be **immiscible**. They can be separated by using a separating funnel.

1 The mixture of immiscible liquids is poured in. It settles into two layers (or more) as the liquids do not mix

2 The tap is opened to let the bottom layer run into a receiver

3 The tap is closed and the receiver is changed. The tap is opened to let the top layer run out

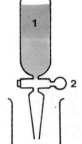

SUMMARY

Immiscible liquids can be separated by means of a separating funnel.

Figure 11.7A ● A separating funnel

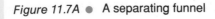

11.8 Fractional distillation

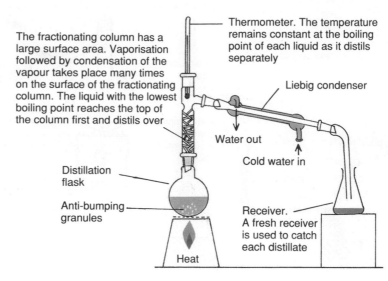

The fractionating column has a large surface area. Vaporisation followed by condensation of the vapour takes place many times on the surface of the fractionating column. The liquid with the lowest boiling point reaches the top of the column first and distils over

Thermometer. The temperature remains constant at the boiling point of each liquid as it distils separately

Liebig condenser

Water out

Cold water in

Distillation flask

Anti-bumping granules

Receiver. A fresh receiver is used to catch each distillate

Heat

Figure 11.8A ● Fractional distillation

An industrial fractional distillation plant uses a temperature sensor and a microcomputer to monitor the course of the distillation.

When liquids dissolve in one another to form a solution, instead of forming separate layers, they are described as **miscible**. Distillation can be used to separate a mixture of miscible liquids. Whisky manufacturers want to separate ethanol (alcohol) from water in a solution of these two liquids. They are able to do this because the two liquids have different boiling points: ethanol boils at 78 °C, and water boils at 100 °C. Figure 11.8A shows a laboratory apparatus which will separate a mixture of ethanol and water into its parts or **fractions**. The process is called **fractional distillation**. The temperature rises to 78 °C and then stays constant while all the ethanol distils over. When the temperature starts to rise again, the receiver is changed. At 100 °C, water starts to distil, and a fresh receiver is put into position to collect it.

Figure 11.8B ● An industrial distillation plant

Continuous fractional distillation

Crude petroleum oil is not a very useful substance. By fractional distillation, it can be separated into a number of very useful products (see Figure 11.8C). In the petroleum industry, fractional distillation is made to run continuously (non-stop). Crude oil is fed in continuously, and the separated fractions are run off from the still continuously. These fractions are not pure substances. Each is a mixture of substances with similar boiling points. The fractions with low boiling points are collected from the top of the fractionating column. Fractions with high boiling points are collected from the bottom of the column.

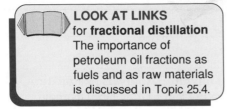

LOOK AT LINKS
for **fractional distillation**
The importance of
petroleum oil fractions as
fuels and as raw materials
is discussed in Topic 25.4.

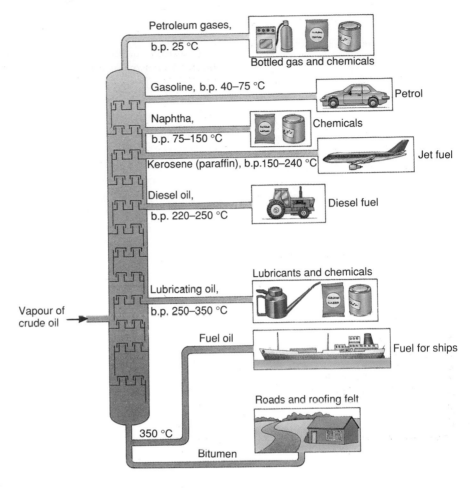

Figure 11.8C ● Continuous fractional distillation of crude oil

SUMMARY

Liquids are separated from a
solution by fractional distillation.
The process can be made to run
continuously, e.g., in the
fractionation of crude petroleum oil.

CHECKPOINT

❶ Your 12 year-old sister wants to know what you have been doing in your
science lessons. Write a technical report, in words which she can
understand, explaining *why* the petroleum industry distils crude oil and *how*
fractional distillation works.

❷ Your sister pours vinegar into the bottle of cooking oil by mistake. She asks
you to help her to put things right. How could you separate the two liquids?

❸ In Hammond Innes' novel, *Ice Station Zero*, a vehicle breaks down because
the villains have put sugar in the petrol tank. Explain how you could separate
the sugar from the petrol.

❹ After a collision at sea, thousands of litres of oil escape from a tanker. A
salvage ship sucks up a layer of oil mixed with sea water from the surface of
the sea.
(a) Describe how you could separate the oil from the sea water.
(b) Say why it is important to be able to do this.

11.9 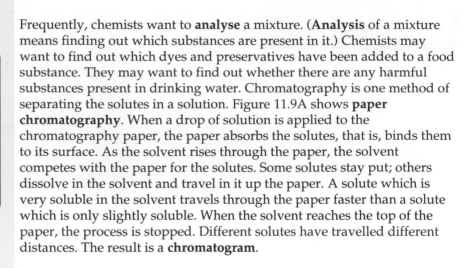 Chromatography

Frequently, chemists want to **analyse** a mixture. (**Analysis** of a mixture means finding out which substances are present in it.) Chemists may want to find out which dyes and preservatives have been added to a food substance. They may want to find out whether there are any harmful substances present in drinking water. Chromatography is one method of separating the solutes in a solution. Figure 11.9A shows **paper chromatography**. When a drop of solution is applied to the chromatography paper, the paper absorbs the solutes, that is, binds them to its surface. As the solvent rises through the paper, the solvent competes with the paper for the solutes. Some solutes stay put; others dissolve in the solvent and travel in it up the paper. A solute which is very soluble in the solvent travels through the paper faster than a solute which is only slightly soluble. When the solvent reaches the top of the paper, the process is stopped. Different solutes have travelled different distances. The result is a **chromatogram**.

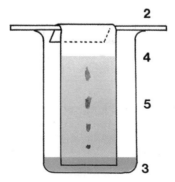

1 A drop of solution is touched on to the chromatography paper. The solvent evaporates. A spot of solute remains

2 The chromatography paper hangs from a glass rod. It must not touch the sides of the beaker

3 The spot of solute must be above the level of the solvent in the beaker

4 The solvent front. The solvent has travelled up the paper to this level

5 Spots of different substances present in the solute. As the substances travel through the paper at different speed, they become separated. The result is a chromatogram

Figure 11.9A ● Paper chromatography

Many solvents are used in chromatography. Ethanol (alcohol), ethanoic acid (vinegar) and propanone (a solvent often called acetone) are common. A chemist has to experiment to find out which solvent gives a good separation of the solutes. With a solvent other than water, a closed container should be used so that the chromatography paper is surrounded by the vapour of the solvent (see Figure 11.9B).

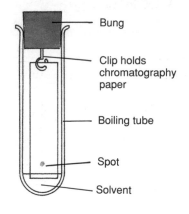

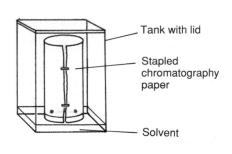

Figure 11.9B ● Chromatography with a solvent other than water

CHECKPOINT

❶ A chemist is asked to find out what substances are present in two mixtures, M_1 and M_2. He makes a chromatogram of the two mixtures and a chromatogram of some substances which he suspects may be present. The figure below shows his results.
Can you interpret his results? What substances are present in M_1 and M_2?

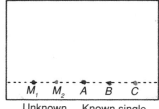

Unknown   Known single
mixtures   substances

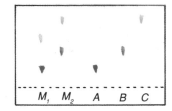

❷ Imagine that you are a detective investigating forged bank notes. Your investigations take you to a house where you find a printing press and some inks. Describe how you would find out whether these inks are the same as those used to make the bank notes.

❸ Dr Ecksplor discovers a strangely coloured orchid in Brazil. Professor Seeker finds an orchid of the same colour in Ecuador. The two scientists wonder whether the two flowers contain the same pigment. They feel sure that the pigment does not dissolve in water.
(a) Why do they feel sure that the pigment does not dissolve in water?
(b) Describe an experiment which they could do to find out whether the two orchids contain the same pigment.

❹ An analytical chemist has the task of finding out whether the red colouring in a new food product contains any dyes which are not permitted in foods. She runs a chromatogram on the food colouring and on some of the permitted red food dyes. The figure opposite shows her results. What conclusion can you draw from these chromatograms?

Note: Dye 1 = E122 Carmoisine, Dye 2 = E162 Beetroot red, Dye 3 = E128 Red 2G, Dye 4 = E120 Cochineal, Dye 5 = E124 Ponceau 4R, Dye 6 = E160 Capsorubin

Dyes which are allowed in foods are given **E numbers**.

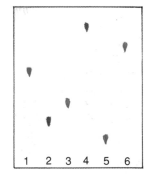

(a) Chromatogram of permitted dyes

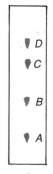

(b) Chromatogram of red food colouring

TOPIC 12  ACIDS AND BASES

12.1 Acids

FIRST THOUGHTS

o
o
o

Can you say what the elements in each of these sets of materials have in common?
List A: lime juice, vitamin C, stomach juice, vinegar
List B: toothpaste, household ammonia, Milk of Magnesia, lime

If you have remembered, well done! You will learn more about acids and bases in this topic.

LOOK AT LINKS
for **concentrated acids**
See Topic 29.11.

This symbol means **corrosive**. Concentrated acids always carry this symbol.

Industrial Chemistry Pack
(program)

Work in small groups. Using the first part of this program you can investigate the manufacture of sulphuric acid. What are the important factors involved?

Cells in the linings of our stomachs produce hydrochloric acid. This is a powerful acid. If a piece of zinc was dropped into hydrochloric acid of the concentration which the stomach contains, it would dissolve. In the stomach, hydrochloric acid works to kill bacteria which are present in foods and to soften foods. It also helps in **digestion** by providing the right conditions for the enzyme pepsin to begin the digestion of proteins.

Sometimes the stomach produces too much acid. Then the result is the pain of 'acid indigestion' and 'heartburn'. The many products which are sold as 'antacids' and indigestion remedies all contain compounds that can react with hydrochloric acid to neutralise (counteract) the excess of acid. In this chapter, we shall look at the properties (characteristics) of acids and the substances which react with them.

A sour taste usually shows that a substance contains an acid. The word 'acid' comes from the Latin word for 'sour'. Vinegar contains ethanoic acid, sour milk contains lactic acid, lemons contain citric acid and rancid butter contains butanoic acid. For centuries, chemists have been able to extract acids such as these from animal and plant material. They call these acids **organic acids**. As chemistry has advanced, chemists have found ways of making sulphuric acid, hydrochloric acid, nitric acid and other acids from minerals. They call these acids **mineral acids**. Mineral acids in general react much more rapidly than organic acids. We describe mineral acids as **strong acids** and organic acids as **weak acids**. Table 12.1 lists some common acids.

Solutions of acids (and other substances) can be **dilute solutions** or **concentrated solutions**. A dilute solution contains a small amount of acid per litre of solution. A concentrated solution contains a large amount of acid per litre of solution. Solutions of acids are always called, say, dilute hydrochloric acid or concentrated hydrochloric acid. Concentrated acids are very corrosive, and you need to know which type of solution you are dealing with.

Table 12.1 ● Some common acids

Acid	Strong or weak	Where you find it
Ascorbic acid	Weak	In fruits. Also called Vitamin C
Carbonic acid	Weak	In fizzy drinks: these contain the gas carbon dioxide which reacts with water to form carbonic acid.
Citric acid	Weak	In fruit juices, e.g. lemon juice
Ethanoic acid	Weak	In vinegar
Hydrochloric acid	Strong	In digestive juices in the stomach; also used for cleaning ('pickling') metals before they are coated
Lactic acid	Weak	In sour milk
Nitric acid	Strong	Used for making fertilisers and explosives
Phosphoric acid	Strong	In anti-rust paint; used for making fertilisers
Sulphuric acid	Strong	In car batteries; used for making fertilisers

What do acids do?

❶ **Acids have a sour taste**.
You will know the taste of lemon juice (which contains citric acid) and vinegar (which contains ethanoic acid). **Do not taste any of the strong acids**.

❷ **Acids change the colour of substances called indicators**.
For example, acids turn blue litmus red (see Table 12.4).

❸ **Acids react with many metals to produce hydrogen and a salt of the metal**.
Hydrogen is an element. It is a colourless, odourless (without smell) gas and is the least dense of all the elements. With air, hydrogen forms an explosive mixture, which is the basis of a test for hydrogen. If you put a lighted splint into hydrogen, you hear an explosive 'pop' (see Figure 12.1C).

Figure 12.1A ● Some common acids

Figure 12.1B ● The acid taste

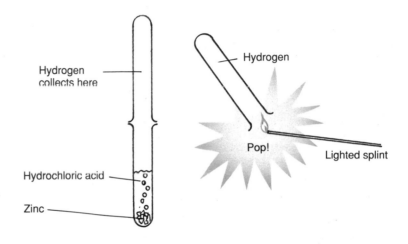

Figure 12.1C ● (a) Making hydrogen   (b) Testing for hydrogen

LOOK AT LINKS
for the **reactions of metals**
See Topic 19.3.

Some metals, e.g. copper, react very slowly with acids. Some metals, e.g. sodium, react dangerously fast. Much heat is given out, and the hydrogen which is formed may explode. Examples of metals which react at a safe speed are magnesium, zinc and iron.

Zinc   +   Sulphuric acid   →   Hydrogen   +   Zinc sulphate
$Zn(s)$ + $H_2SO_4(aq)$ → $H_2(g)$ + $ZnSO_4(aq)$

❹ **Acids react with carbonates to give carbon dioxide, a salt and water**.
Acid indigestion is caused by an excess of hydrochloric acid in the stomach. Some indigestion tablets contain magnesium carbonate. The reaction that takes place when magnesium carbonate reaches the stomach is

LOOK AT LINKS
for **bases**
See Topic 12.2.

Magnesium + Hydrochloric → Carbon + Magnesium + Water
carbonate   acid   dioxide   chloride
$MgCO_3(s)$ + $2HCl(aq)$ → $CO_2(g)$ + $MgCl_2(aq)$ + $H_2O(l)$

Other anti-acid tablets contain sodium hydrogencarbonate. In this case, the reaction that happens in the stomach is

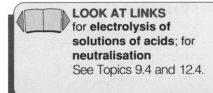
LOOK AT LINKS
for **electrolysis of solutions of acids**; for **neutralisation**
See Topics 9.4 and 12.4.

Sodium   + Hydrochloric → Carbon + Sodium + Water
hydrogencarbonate   acid   dioxide   chloride
$NaHCO_3(s)$ + $HCl(aq)$ → $CO_2(g)$ + $NaCl(aq)$ + $H_2O(l)$

Do anti-acid tablets contain carbonates and hydrogencarbonates? You can test some to find out. The test for carbon dioxide is shown in Figure 16.1D. When carbon dioxide is bubbled through limewater (calcium hydroxide solution), the limewater turns cloudy. Always wear safety glasses when working with acids and bases.

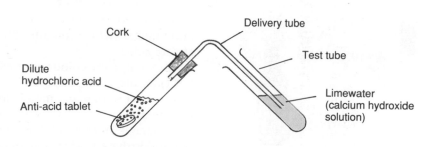

Cork

Delivery tube

Test tube

Dilute hydrochloric acid

Anti-acid tablet

Limewater (calcium hydroxide solution)

Figure 12.1D ● Testing for carbon dioxide

❺ **Acids neutralise bases**
(See Topic 12.4.)

What is an acid?

A Swedish chemist called Svante Arrhenius explained why different acids have so much in common. He put forward the theory that aqueous solutions of all acids contain hydrogen ions, $H^+(aq)$, in high concentration. By 'high' concentration, he meant a concentration much higher than that in water. This theory explains many reactions of acids, including **neutralisation**, **electrolysis of solutions** of acids and the reaction of metals with acids

Metal atoms + Hydrogen ions → Metal ions + Hydrogen molecules

$$M(s) + 2H^+(aq) \rightarrow M^{2+}(aq) + H_2(g)$$

Arrhenius gave this definition of an acid:

> **An acid is a substance that releases hydrogen ions when dissolved in water.**

A solution of a strong acid has a much higher concentration of hydrogen ions than a solution of a weak acid. A solution of a strong acid therefore reacts much more rapidly than a solution of a weak acid.

SUMMARY

Acids
• have a sour taste,
• change the colour of indicators,
• react with many metals to give hydrogen and a salt,
• react with carbonates and hydrogencarbonates to give carbon dioxide, a salt and water,
• react with bases (see Topic 12.2).

SUMMARY

Different acids have similar reactions. The reason is that solutions of all acids contain hydrogen ions in high concentration. The hydrogen ions are responsible for the typical reactions of acids.

CHECKPOINT

❶ Where in the kitchen could you find
  (a) a weak acid with a pleasant taste?
  (b) a weak acid with an unpleasant taste?
  (c) a weak acid with a very sour taste?

❷ What is the difference between a concentrated solution of a weak acid and a dilute solution of a strong acid?

❸ What is the difference between a concentrated solution of sulphuric acid and a dilute solution of sulphuric acid? Why do road tankers of sulphuric acid carry the sign shown opposite?

❹ *Toffee Recipe 1* Boil sugar and water with a little butter until the mixture thickens. Pour into a greased tray to set.
Toffee Recipe 2 Repeat Recipe 1. When the mixture thickens, add vinegar and 'bicarbonate of soda' (sodium hydrogencarbonate). Pour into a greased tray to set.
One of these recipes gives solid toffee. The other gives a honeycomb of toffee containing bubbles of gas. Which recipe gives the honeycomb? Why?

CORROSIVE

5 Rosie and Luke visited the underground caves in the limestone rock at Wookey Hole in Somerset. A guide told them that the chemical name for limestone is calcium carbonate. The children took a small piece of limestone rock to school and did an experiment with it. They put a piece of rock in a beaker and added dilute hydrochloric acid. Immediately, bubbles of gas were given off. (See the figures opposite)

(a) Why must you always wear safety glasses when working with acids?
(b) What is the name of the gas in the bubbles?
(c) What chemical test could they do to 'prove' what the gas was?

Rosie lit a candle, and then tilted the beaker so that gas poured on to it (see lower figure). The candle went out.

(d) What *two* things about the gas did this experiment tell the children?
(e) What could the gas be used for?

Their teacher prepared a gas jar of the gas. Luke added distilled water and shook the gas jar. Then he added litmus solution.

(f) What colour did the indicator turn? (See Table 12.1 and *What do acids do?* for help.)

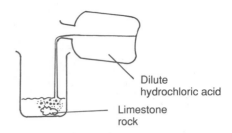

(a) Adding acid to limestone

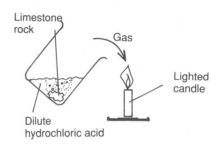

(b) Testing the gas on a lighted candle

12.2 Bases

Figure 12.2A ● Spreading lime

Bases are substances that neutralise (counteract) acids. Figure 12.2A shows a farmer spreading the base called 'lime' (calcium hydroxide) on a field. Some soils are too acidic to grow good crops. 'Liming' neutralises some of the acid and increases the **soil fertility**.

The product of the reaction between an acid and a base is a neutral substance, neither an acid nor a base, which is called a **salt**. Lime (calcium hydroxide) is a base. It reacts with nitric acid in the soil to form the salt calcium nitrate and water

Calcium hydroxide + Nitric acid ➔ Calcium nitrate + Water

A definition of a base is:

> **A base is a substance that reacts with an acid to form a salt and water only.**

Acid + Base ➔ Salt + Water

Soluble bases are called **alkalis**. Sodium hydroxide is an alkali; it reacts with an acid to form a salt and water.

Sodium hydroxide + Hydrochloric acid ➔ Sodium chloride + Water
$NaOH(aq)$ + $HCl(aq)$ ➔ $NaCl(aq)$ + $H_2O(l)$

LOOK AT LINKS
for **soil fertility**
See Topic 29.1.

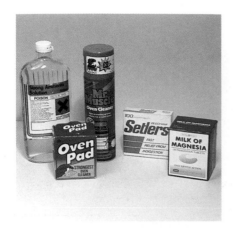

Figure 12.2B ● Some common bases

LOOK AT LINKS
for **soap**
You will see how soap and detergents work in Topics 16.9 and 16.10.

LOOK AT LINKS
for **weak acids** and **bases**,
See Topics 12.5 and 27.2.

Limewater (calcium hydroxide solution) is an alkali. A test for carbon dioxide is that it turns limewater cloudy. The reaction is an acid–base reaction to form a salt and water.

Carbon dioxide	+	Calcium hydroxide	→	Calcium carbonate	+	Water
$CO_2(g)$	+	$Ca(OH)_2(aq)$	→	$CaCO_3(s)$	+	$H_2O(l)$

The salt, calcium carbonate, appears as a cloud of insoluble white powder.

Table 12.2 ● Some common bases

Base	Strong or weak	Where you find it
Ammonia	Weak	In cleaning fluids for use as a degreasing agent; also used in the manufacture of fertilisers
Calcium hydroxide	Strong	Used to treat soil which is too acidic
Calcium oxide	Strong	Used in the manufacture of cement, mortar and concrete
Magnesium hydroxide	Strong	In anti-acid indigestion tablets and Milk of Magnesia
Sodium hydroxide	Strong	In oven cleaners as a degreasing agent; also used in soap manufacture

Table 12.2 lists some common bases. Different bases have a number of reactions in common.

❶ Bases neutralise acids (see previous page).
❷ Soluble bases can change the colour of **indicators**, e.g. turn red litmus blue (see Table 12.4).
❸ Soluble bases feel soapy to your skin. The reason is that soluble bases convert some of the oil in your skin into **soap**. Some household cleaning solutions, e.g. ammonia solution, use soluble bases as degreasing agents. They convert oil and grease into soluble soaps which are easily washed away.
❹ A solution of an alkali in water contains hydroxide ions, $OH^-(aq)$. This solution will react with a solution of a metal salt. Most metal hydroxides are insoluble. When a solution of an alkali is added to a solution of a metal salt, an insoluble metal hydroxide is **precipitated** from solution. (A precipitate is a solid which forms when two liquids are mixed.) For example

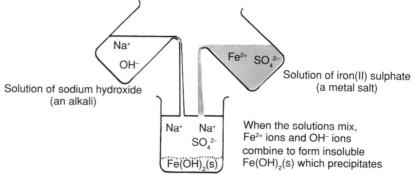

Figure 12.2C ● Precipitating an insoluble hydroxide

Iron(II) sulphate (solution)	+	Sodium hydroxide (solution)	→	Iron(II) hydroxide (precipitate)	+	Sodium sulphate (solution)
$FeSO_4(aq)$	+	$2NaOH(aq)$	→	$Fe(OH)_2(s)$	+	$Na_2SO_4(aq)$

A solution of a strong base contains a higher concentration of hydroxide ions than a solution of a weak base.

Table 12.3 ● Examples of bases

Metal oxides	Metal hydroxides	Alkalis (soluble bases)
Copper(II) oxide, CuO	Sodium hydroxide, NaOH	Sodium hydroxide, NaOH
Zinc oxide, ZnO	Magnesium hydroxide, $Mg(OH)_2$	Potassium hydroxide, KOH
		Calcium hydroxide, $Ca(OH)_2$
(Most metal oxides and hydroxides are insoluble)		Ammonia solution, $NH_3(aq)$

SUMMARY

Bases neutralise acids to form salts. Soluble bases are called alkalis. Metal oxides and hydroxides are bases. Alkalis change the colours of indicators. They are degreasing agents, and have a soapy 'feel'. Solutions of alkalis contain hydroxide ions, $OH^-(aq)$ in high concentration.

Ionic equations

There is another way of writing the equation of the reaction between iron(II) sulphate and sodium hydroxide. The reaction is the combination of iron(II) ions and hydroxide ions. The sodium ions and sulphate ions take no part in the reaction, and the equation can be written without them.

$$Fe^{2+}(aq) + 2OH^-(aq) \rightarrow Fe(OH)_2(s)$$

This is called an **ionic equation**.

CHECKPOINT

❶ (a) Write the names of the four common alkalis.
(b) Write the names of four insoluble bases.

❷ Four bottles of solution are standing on a shelf in the prep room. Their labels have come off and are lying on the floor. They read, Copper(II) sulphate, Iron(II) sulphate, Iron(III) sulphate and Zinc sulphate. The lab assistant knows that some insoluble metal hydroxides are coloured (see the table below).

Hydroxide	Formula	Colour
Copper(II) hydroxide	$Cu(OH)_2(s)$	Blue
Iron(II) hydroxide	$Fe(OH)_2(s)$	Green
Iron(III) hydroxide	$Fe(OH)_3(s)$	Rust
Magnesium hydroxide	$Mg(OH)_2(s)$	White
Zinc hydroxide	$Zn(OH)_2(s)$	White

She adds sodium hydroxide solution to a sample of each solution.
Her results are:
Bottle 1   White precipitate
Bottle 2   Rust-coloured precipitate
Bottle 3   Blue precipitate
Bottle 4   Green precipitate
Say which label should be stuck on each bottle.

❸ Write the formulas for the bases: calcium oxide, copper(II) oxide, zinc oxide, magnesium hydroxide, iron(II) hydroxide, iron(III) hydroxide.

12.3 Summarising the reactions of acids and bases

Figure 12.3A summarises the reactions of hydrochloric acid, a typical strong acid.

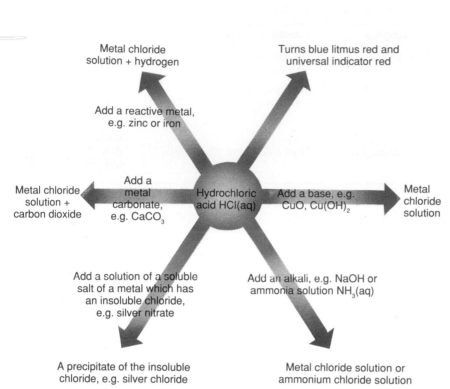

Figure 12.3A ● Reactions of hydrochloric acid

Figure 12.3B summarises the reactions of the strong base sodium hydroxide and the weak base ammonia.

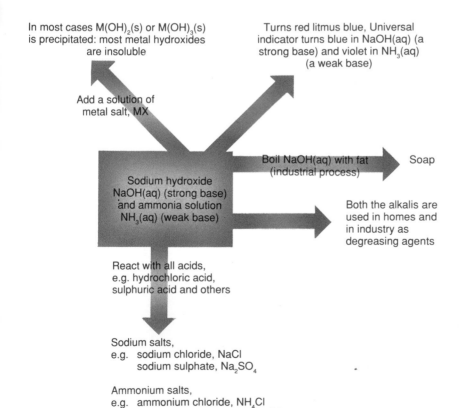

Figure 12.3B ● Some reactions of alkalis (soluble bases)

12.4 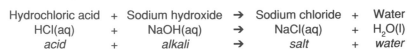 Neutralisation

What takes place when an acid neutralises a soluble base? An example is the reaction between hydrochloric acid and the alkali (soluble base) sodium hydroxide solution.

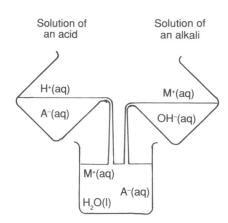

Solution of an acid

Solution of an alkali

H⁺(aq)

A⁻(aq)

M⁺(aq)

OH⁻(aq)

M⁺(aq)

A⁻(aq)

$H_2O(l)$

H⁺(aq) Ions and OH⁻(aq) ions have combined to form water molecules, $H_2O(l)$

The solution is a solution of the salt MA

Figure 12.4A ●
Acid + alkali → salt + water

Hydrochloric acid	+	Sodium hydroxide	→	Sodium chloride	+	Water
$HCl(aq)$	+	$NaOH(aq)$	→	$NaCl(aq)$	+	$H_2O(l)$
acid	+	*alkali*	→	*salt*	+	*water*

When the solutions of acid and alkali are mixed, hydrogen ions, $H^+(aq)$, and hydroxide ions, $OH^-(aq)$ combine to form water molecules.

$$H^+(aq) + OH^-(aq) \rightarrow H_2O(l)$$

Sodium ions, $Na^+(aq)$, and chloride ions, $Cl^-(aq)$ remain in the solution, which becomes a solution of sodium chloride. If you evaporate the solution, you obtain solid sodium chloride.

> **Neutralisation is the combination of hydrogen ions from an acid and hydroxide ions from a base to form water molecules. In the process, a salt is formed.**

What happens when an acid neutralises an insoluble base? An example is the reaction between sulphuric acid and copper(II) oxide

Sulphuric acid	+	Copper(II) oxide	→	Copper(II) sulphate	+	Water
$H_2SO_4(aq)$	+	$CuO(s)$	→	$CuSO_4(aq)$	+	$H_2O(l)$
acid	+	*base*	→	*salt*	+	*water*

Hydrogen ions and oxide ions, O^{2-}, combine to form water

$$2H^+(aq) + O^{2-}(s) \rightarrow H_2O(l)$$

The resulting solution contains copper(II) ions and sulphate ions: it is a solution of copper(II) sulphate. If you evaporate it, you will obtain copper(II) sulphate crystals.

SUMMARY

Neutralisation is the combination of hydrogen ions (from an acid) with hydroxide ions (from an alkali or an insoluble base) or oxide ions (from an insoluble base) to form water and a salt.

CHECKPOINT

❶ Bee stings hurt because bees inject acid into the skin. Wasp stings hurt because wasps inject alkali into the skin.

Your little brother is stung by a bee. You are in charge. What do you use to treat the sting: 'bicarbonate of soda' (sodium hydrogencarbonate), calamine lotion (zinc carbonate), vinegar (ethanoic acid) or Milk of Magnesia (magnesium carbonate)? Would you use the same treatment for a wasp sting? Give reasons for your answers.

❷ 'Acid drops' which you buy from a sweet shop contain citric acid.

• Put an acid drop in your mouth.
• Put some baking soda (sodium hydrogencarbonate) on your hand.
• Lick some of the baking soda into your mouth.

What happens to the taste of the acid drop? Why does this happen?

12.5 Indicators and pH

Table 12.4 ● The colours of some common indicators

Indicator	Acidic colour	Neutral colour	Alkaline colour
Litmus	Red	Purple	Blue
Phenolphthalein	Colourless	Colourless	Pink
Methyl orange	Red	Orange	Yellow

Universal indicator turns different colours in strongly acidic and weakly acidic solutions. It can also distinguish between strongly basic and weakly basic solutions (see Figure 12.5A). Each universal indicator colour is given a **pH number**. The pH number measures the acidity or alkalinity of the solution. For a neutral solution, pH = 7; for an acid solution, pH < 7; for an alkaline solution, pH > 7.

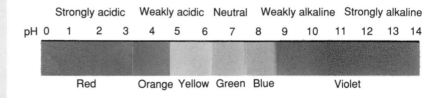

Figure 12.5A ● The colour of universal indicator in different pH solutions

SUMMARY

In general, mineral acids are stronger than organic acids. Some bases are stronger than others. Ammonia is a weak base. The pH of a solution is a measure of its acidity or alkalinity. Universal indicator turns different colours in solutions of different pH.

Figure 12.5B shows the pH values of solutions of some common acids and alkalis. Some salts do not have a pH value of 7. Carbonates and hydrogencarbonates are alkaline in solution.

IT'S A FACT

Why must you never use both ammonia as a degreaser and chlorine as a bleach on the same cleaning job?
Don't do it! You could get a dangerous reaction between them.

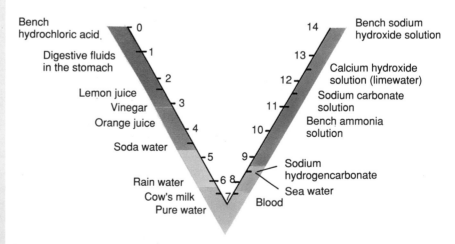

Figure 12.5B ● The pH values of some different solutions

SUMMARY

We use many acids and bases in everyday life. Tables 12.1 and 12.2 list some common acids and bases. The pH values of some solutions are given in Figure 12.5B.

CHECKPOINT

① Say whether these substances are strongly acidic, weakly acidic, neutral, weakly basic or strongly basic:
(a) cabbage juice, pH = 5.0 (b) pickled cabbage, pH = 3.0 (c) milk, pH = 6.5
(d) tonic water, pH = 8.2 (e) washing soda, pH = 11.5 (f) saliva, pH = 7.0
(g) blood, pH = 7.4

② You know what it feels like to be stung by a nettle. Is the substance which the nettle injects an acid or an alkali or neither? Think up a method of extracting some of the substance from a nettle and testing to see whether the extract is acidic or alkaline or neutral. If your teacher approves of your plan, try it out.

③ Figure 12.5A shows the colours of universal indicator. What colour would you expect universal indicator to be when added to each of the following?
(a) distilled water (b) lemon juice (c) household ammonia (d) battery acid
(e) oven cleaner

④

Liquid	pH value	Reaction with acid
A	1.0	None
B	8.5	Produces a salt, carbon dioxide and water
C	8.5	Produces a salt and water
D	13	Produces a salt and water

Uncle Harry is suffering from acid indigestion. Explain to him why it would not be a good idea to drink either Liquid A or Liquid D. Would you advise him to take Liquid B or Liquid C? Explain your choice.

⑤ The oven sprays which are sold for cleaning greasy ovens contain a concentrated solution of sodium hydroxide.
Why does sodium hydroxide clean the greasy oven?
Why does sodium hydroxide work better than ammonia?
What two safety precautions should you take to protect yourself when using an oven spray?
Why do domestic cleaning fluids contain ammonia, rather than sodium hydroxide?
Why do soap manufacturers use sodium hydroxide, rather than ammonia?

⑥

Crop	Wheat	Potatoes	Sugar beet
pH	6	9	7

The table shows the most suitable values of pH for growing some crops.
(a) Which crop grows best in (i) an acidic soil and (ii) an alkaline soil?
(b) A farmer wanted to grow sugar beet. On testing, he found that his soil had a pH of 5. Name a substance which could be added to the soil to make its pH more suitable for growing sugar beet.
(c) How does the substance you mention in (b) act on the soil to change its pH?

⑦ Describe a test which you could do in the laboratory to show that citric acid is a weaker acid than sulphuric acid.

⑧ Pair them up. Give the pH of each of the solutions listed.

Solution		pH
1 Ethanoic acid	A	7.0
2 Sodium chloride	B	1.0
3 Sulphuric acid	C	5.0
4 Ammonia	D	13.0
5 Sodium hydroxide	E	9.0

TOPIC 13 SALTS

13.1 Sodium chloride

> Sodium chloride, NaCl, is the salt which we call 'common salt' or simply 'salt'. The average human body contains about 250 g of sodium chloride.

Sodium chloride is essential for life: it enables muscles to contract, it enables nerves to conduct nerve impulses, it regulates osmosis and it is converted into hydrochloric acid, which helps digestion to take place in the stomach. Deprived of sodium chloride, the body goes into convulsions; then paralysis and death may follow. When we sweat, we lose both water and sodium chloride. We also **excrete** sodium chloride in urine. Our kidneys control the quantity of sodium chloride which we excrete. If we eat too much salt, our kidneys excrete sodium chloride; if we eat too little salt, our kidneys excrete water but no sodium chloride.

> IT'S A FACT
>
> A French army under Napoleon I invaded Russia in 1812. The harsh Russian winter forced the French army to retreat. Salt starvation was one of the hardships which Napoleon's soldiers endured on their retreat from Moscow. Shortage of salt lowered their resistance to disease and epidemics spread; it prevented their wounds from healing and led to infection and death.

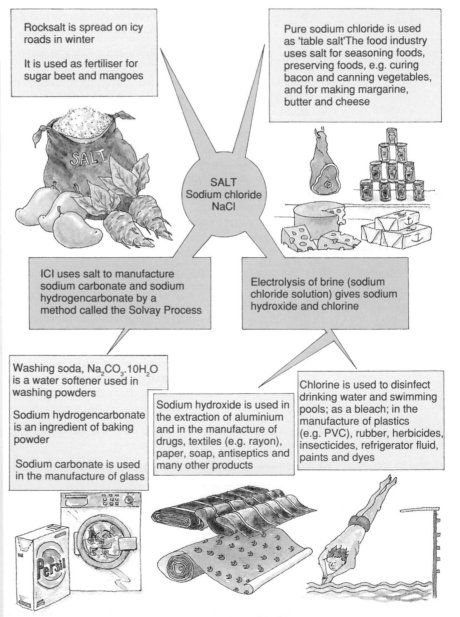

Rocksalt is spread on icy roads in winter

It is used as fertiliser for sugar beet and mangoes

Pure sodium chloride is used as 'table salt' The food industry uses salt for seasoning foods, preserving foods, e.g. curing bacon and canning vegetables, and for making margarine, butter and cheese

SALT
Sodium chloride
NaCl

ICI uses salt to manufacture sodium carbonate and sodium hydrogencarbonate by a method called the Solvay Process

Electrolysis of brine (sodium chloride solution) gives sodium hydroxide and chlorine

Washing soda, $Na_2CO_3.10H_2O$ is a water softener used in washing powders

Sodium hydrogencarbonate is an ingredient of baking powder

Sodium carbonate is used in the manufacture of glass

Sodium hydroxide is used in the extraction of aluminium and in the manufacture of drugs, textiles (e.g. rayon), paper, soap, antiseptics and many other products

Chlorine is used to disinfect drinking water and swimming pools; as a bleach; in the manufacture of plastics (e.g. PVC), rubber, herbicides, insecticides, refrigerator fluid, paints and dyes

Figure 13.1A ● Some of the uses of sodium chloride

13.2  Salts of some common acids

While some of the salts which we use occur naturally in the Earth's crust, others must be made by chemists.

Many salts can be made by neutralising acids. In a salt, the hydrogen ions in the acid have been replaced by metal ions or by the ammonium ion. Salts of hydrochloric acid are called **chlorides**. Salts of sulphuric acid are called **sulphates**. Salts of nitric acid are called **nitrates**. See Table 13.1.

Table 13.1 ● Salts of some common acids

Acids	Salts
Hydrochloric acid $HCl(aq)$	Sodium chloride, $NaCl$ Calcium chloride, $CaCl_2$ Iron(II) chloride, $FeCl_2$ Ammonium chloride, NH_4Cl
Sulphuric acid $H_2SO_4(aq)$	Sodium sulphate, Na_2SO_4 Zinc sulphate, $ZnSO_4$ Iron(III) sulphate, $Fe_2(SO_4)_3$
Nitric acid HNO_3	Sodium nitrate, $NaNO_3$ Copper(II) nitrate, $Cu(NO_3)_2$

13.3  Water of crystallisation

The crystals of some salts contain water. Such salts are called **hydrates**. Examples are:

Copper(II) sulphate-5-water, $CuSO_4.5H_2O$, *blue*
Cobalt(II) chloride-6-water, $CoCl_2.6H_2O$, *pink*
Iron(II) sulphate-7-water, $FeSO_4.7H_2O$, *green*

The water present in the crystals of hydrated salts gives them their shape and their colour. It is called **water of crystallisation**. The formula of the hydrate shows the proportion of water in the crystal, e.g. five molecules of water to each pair of copper ions and sulphate ions: $CuSO_4.5H_2O$.

When blue crystals of copper(II) sulphate-5-water are heated gently, the water of crystallisation is driven off. The white powdery solid that remains is anhydrous (without water) copper(II) sulphate.

Copper(II) sulphate-5-water	→	Water	+	Copper(II) sulphate
(blue crystals)	→	(vapour)	+	(white powder, *anhydrous*)
$CuSO_4.5H_2O(s)$	→	$5H_2O(g)$	+	$CuSO_4(s)$

If copper(II) sulphate crystals are left in the air, they slowly lose some or all of their water of crystallisation.

SUMMARY

- Some salts occur naturally, but most of them must be made by the chemical industry.
- Some salts crystallise with water of crystallisation. This gives the crystals their shape and, in some cases, their colour.

CHECKPOINT

❶ Name the salts with the following formulas: NaI, NH_4NO_3, KBr, $BaCO_3$, Na_2SO_4, $ZnCl_2$, $CrCl_3$, $NiSO_4$, $CaCl_2$, $MgCl_2$.

❷ Write formulas for the salts: potassium nitrate, potassium bromide, iron(II) chloride, iron(II) sulphate, iron(III) bromide, ammonium chloride.

❸ State the number of oxygen atoms in each of these formulas:
(a) Pb_3O_4 (b) $3Al_2O_3$ (c) $Fe(OH)_3$ (d) $MgSO_4.7H_2O$ (e) $Co(NO_3)_3.9H_2O$

④ The table shows how samples of four substances change in mass when they are left in the air.

Substance	Mass of sample fresh from bottle (g)	Mass of sample after 1 week in the air (g)
A	13.10	13.21
B	15.25	15.25
C	11.95	5.01

Which of the three substances in the table (a) loses water of crystallisation on standing (b) absorbs water from the air and (c) is unchanged by exposure to air?

⑤ Barbara: *Bother! The holes in the salt cellar are blocked again. I wonder why that happens.*
Razwan: It happens because sodium chloride absorbs water from the air.
Gwynneth: My mum puts a few grains of rice in the salt cellar. That stops it getting clogged.

(a) Explain why the absorption of water by sodium chloride will block the salt cellar.
(b) Explain how rice grains stop the salt cellar clogging.
(c) Describe an experiment which you could do to show that what you say in (b) is correct.

⑥ When a tin of biscuits is left open, do the biscuits absorb water vapour from the air or give out water vapour? Describe an experiment which you could do to find out.

FIRST THOUGHTS

13.4 Useful salts

Salts – who needs them? Farmers, doctors, dentists, industrial manufacturers, photographers: they all use salts in their work.

LOOK AT LINKS
for **NPK fertilisers**
See Topic 29.6.

● Agricultural uses

NPK fertilisers are mixtures of salts. They contain ammonium nitrate and ammonium sulphate to supply nitrogen to the soil. They contain phosphates to supply phosphorus and potassium chloride as a source of potassium.

Potato blight used to be a serious problem. It destroyed the potato crop in Ireland in 1846. Thousands of people starved and many more were forced to leave the country in search of food. Now, the fungus which causes potato blight can be killed by spraying with a solution of a simple salt, copper(II) sulphate. This fungicide is also used on vines to protect the grape harvest.

Figure 13.4A ● NPK fertiliser

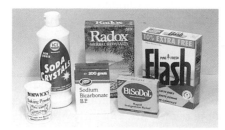

Figure 13.4B ● Washing soda, bath salts, baking powder

LOOK AT LINKS
for **water softeners**
See Topics 16.12 and 16.13.

SCIENCE AT WORK

Liquid crystals are a form of matter which has a regular structure like a solid and which flows like a liquid. Many liquid crystals change their colour with the temperature. A recent invention uses the behaviour of liquid crystals to diagnose appendicitis. A thin plastic film coated with a suitable liquid crystal is placed on the patient's abdomen. Since an inflamed appendix produces heat, it shows up immediately as a change in the colour of the film. Appendicitis can be difficult to diagnose by other methods because the pain is often felt some distance from the appendix. Can you think of some further uses of liquid crystal temperature strips?

LOOK AT LINKS
Temperature strips as thermometers are discussed in
KS: Physics, Topic 6.1.

● **Domestic uses**

Sodium carbonate-10-water is known as 'washing soda'. It is used as a **water softener**. It is an ingredient of washing powders and is also sold as bath salts.

Sodium hydrogencarbonate is known as 'baking soda'. It is added to self-raising flour. When heated at a moderate oven temperature, it decomposes.

Sodium hydrogencarbonate	→	Sodium carbonate	+	Carbon dioxide	+	Steam
$2NaHCO_3(s)$	→	$Na_2CO_3(s)$	+	$CO_2(g)$	+	$H_2O(g)$

The carbon dioxide and steam which are formed make bread and cakes 'rise'.

● **Medical uses**

Figure 13.4C shows someones leg being set in a plaster cast. Plaster of Paris is the salt calcium sulphate-$\frac{1}{2}$-water, $CaSO_4.\frac{1}{2}H_2O$. It is made by heating calcium sulphate-2-water, $CaSO_4.2H_2O$, which is mined. When plaster of Paris is mixed with water, it combines and sets to form a strong 'plaster cast'. It is also used for plastering walls.

People who are always tired and lacking in energy may be suffering from anaemia. This illness is caused by a shortage of haemoglobin (the red pigment which contains iron) in the blood. It can be cured by taking iron compounds in the diet. The 'iron tablets' which anaemic people may be given often contain iron(II) sulphate-7-water.

Patients who are suspected of having a stomach ulcer may be given a 'barium meal'. It contains the salt barium sulphate. After a barium meal, an X-ray photograph of the body shows the path taken by the salt. Being large, barium ions show up well in X-ray photographs.

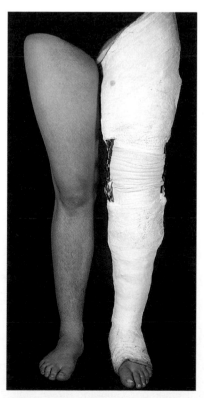

Figure 13.4C ● A patient having his leg set in a plaster cast

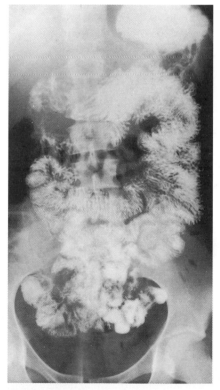

Figure 13.4D ● X-ray photograph of a patient after consuming a barium meal

LOOK AT LINKS
for **teeth**
You will find more about teeth and dental health in Topic 11.5 of *KS: Biology*.

TRY THIS

Fluoridation

Who's behind *the science*

The fluoride story started in 1901 when Fred McKay began work as a dentist in Colorado, USA. Many of his patients had mottled teeth or dark brown stains on their teeth. He could find no information about this condition and called it 'Colorado Brown Stain'. He became convinced that the stain was caused by something in the water. After Oakley in Idaho changed its source of drinking water, the children there no longer developed mottled or stained teeth. The drinking water in a number of towns was analysed and showed that teeth were affected when there were over 2 p.p.m (parts per million) of fluoride in the drinking water. It affected children's teeth more than adults' teeth. McKay also noticed that people with 'Colorado Brown Stain' had less dental decay than other patients!

You can use a microcomputer and a histogram program to display these results as a bar graph.

● *Fluorides for healthy teeth*

Why do dentists recommend that people use toothpastes containing the salt calcium fluoride? Tooth enamel reacts with calcium fluoride to form a harder enamel which is better at resisting attack by mouth acids. To make sure that everyone gets protection from tooth decay, many water companies add a small amount of calcium fluoride to drinking water. The concentration of fluoride ions must not rise above 1 p.p.m. Spending 7p per person each year on fluoridation saves the National Health Service £6 per person each year in dentistry. Some people are violently opposed to the fluoridation of water supplies. This is because they are worried about the effects of drinking too much fluoride. Worldwide, 230 million people drink fluoridated water.

In 1945, a British dentist called Weaver inspected the teeth of children from North Shields and South Shields. These two towns are on opposite banks of the River Tyne. Tables 13.2 and 13.3 show the results (rounded off) of Weaver's inspection of 1000 children aged 5 and 1000 children aged 12. The figures give the number of DMF (decayed, missing or filled) teeth per 1000 teeth. The numbers of teeth in each position in both upper and lower jaws and on both left and right sides have been added together. At the time, the water supplies contained:

North Shields 0.25 p.p.m fluoride, South Shields 1.40 p.p.m fluoride.

Table 13.2 ● Survey of 1000 children aged 5 years

| Position of tooth | Number of DMF teeth per 1000 teeth | |
	North Shields	South Shields
1	265	155
2	200	100
3	135	55
4	675	440
5	725	485

Table 13.3 ● Survey of 1000 children aged 12

| Position of tooth | Number of DMF teeth per 1000 teeth | |
	North Shields	South Shields
1	45	20
2	60	20
3	15	10
4	70	25
5	75	30
6	725	490
7	160	75

(a) On graph paper, draw a bar graph of the number of DMF teeth in each position for 5 year-olds in (i) North Shields and (ii) South Shields. Shade the bars for the two towns differently.
(b) Draw similar bar graphs for 12 year-olds in (i) North Shields and (ii) South Shields.
(c) What do you observe from your bar graphs about the teeth of children in North Shields compared with South Shields (i) at 5 years and (ii) at 12 years?
(d) What is the fundamental difference between the teeth of a 5 year-old and those of a 12 year-old?
(e) How did the fluoride content of the water in North Shields compare with that in South Shields?
(f) What can you deduce from these figures about the effect of fluoride in drinking water on the health of children's teeth?
(g) Remember 'Colorado Brown Stain'? What is the maximum safe fluoride level?

● Industrial uses

Sodium chloride has many uses in industry; see Figure 13.1A.

● Photography

Silver bromide, AgBr, is the salt which is used in black and white photography. Light affects silver bromide. A photographic film is a piece of celluloid covered with a thin layer of gelatin containing silver bromide. When you take a photograph, light falls on to some areas of the film. In exposed areas of film, light converts some silver ions, Ag^+, into silver atoms, which are black. These black atoms form a *latent image* of the object.

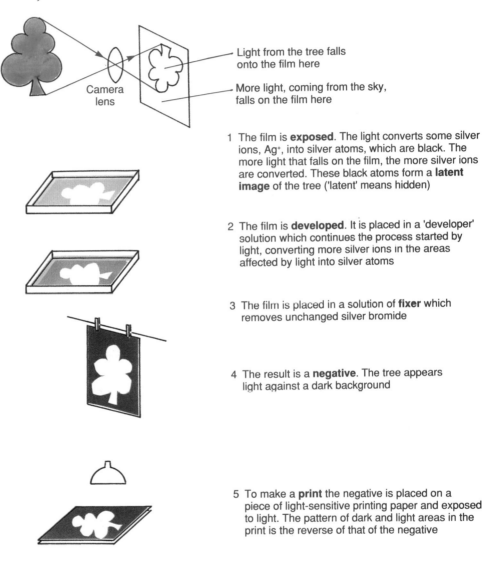

Light from the tree falls onto the film here

More light, coming from the sky, falls on the film here

Camera lens

1 The film is **exposed**. The light converts some silver ions, Ag^+, into silver atoms, which are black. The more light that falls on the film, the more silver ions are converted. These black atoms form a **latent image** of the tree ('latent' means hidden)

2 The film is **developed**. It is placed in a 'developer' solution which continues the process started by light, converting more silver ions in the areas affected by light into silver atoms

3 The film is placed in a solution of **fixer** which removes unchanged silver bromide

4 The result is a **negative**. The tree appears light against a dark background

5 To make a **print** the negative is placed on a piece of light-sensitive printing paper and exposed to light. The pattern of dark and light areas in the print is the reverse of that of the negative

Figure 13.4E ● Black and white photography

SUMMARY

Salts are used in agriculture, in the home, in medicine and in industry. Methods of making salts are therefore important.

Some of these useful salts are mined, e.g. sodium chloride, magnesium sulphate and calcium sulphate. Others must be made by the chemical industry, e.g. silver bromide and copper(II) sulphate.

13.5 Methods of making salts

Since salts are so useful, chemists have found methods of making them. This topic describes the methods.

The method which you choose for making a salt depends on whether it is soluble or insoluble. Soluble salts are made by neutralising an acid. Insoluble salts are made by adding two solutions. Table 13.4 summarises the facts about the solubility of salts.

Table 13.4 ● Soluble and insoluble salts

Salts	Soluble	Insoluble
Chlorides	Most are soluble	Silver chloride Lead(II) chloride
Sulphates	Most are soluble	Barium sulphate Calcium sulphate Lead(II) sulphate
Nitrates	All are soluble	None
Carbonates	Sodium and potassium carbonates	Most are insoluble
Ethanoates	All are soluble	None
Sodium salts	All are soluble	None
Potassium salts	All are soluble	None
Ammonium salts	All are soluble	None

Methods for making soluble salts

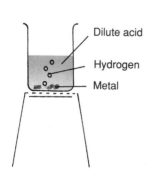

An acid is neutralised by adding a metal, a solid base or a solid metal carbonate or a solution of an alkali.

Method 1: Acid + Metal ➔ Salt + Hydrogen
Method 2: Acid + Metal oxide ➔ Salt + Water
Method 3: Acid + Metal carbonate ➔ Salt + Water + Carbon dioxide
Method 4: Acid + Alkali ➔ Salt + Water

The practical details of Methods 1–3 are as follows.

Step one   Add an excess (more than enough) of the solid to the acid (see Figure 13.5A).

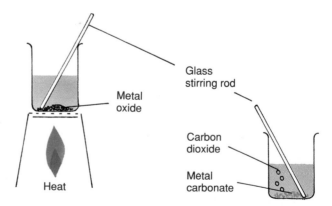

Method 1
Warm the acid.
Switch off the Bunsen.
Add an excess of the metal to the acid.
Wait until no more hydrogen is evolved.
The reaction is then complete.

Method 2
Add an excess of the metal oxide to the acid. Wait until the solution no longer turns blue litmus red. The reaction is then over

Method 3
Add an excess of the metal carbonate to the acid. Wait until no more carbon dioxide is evolved. The reaction is then over

Figure 13.5A ● Adding an excess of the solid reactant to the acid

Step two   Filter to remove the excess of solid (Figure 13.5B).

Step three   Gently evaporate the filtrate (Figure 13.5C).

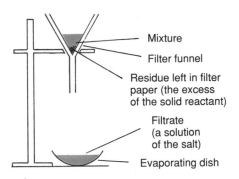

Figure 13.5B ● Filtering to remove the excess of solid reactant

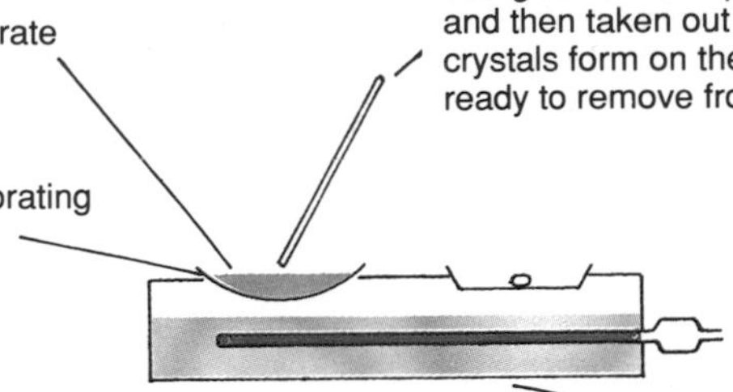

Figure 13.5C ● Evaporating the filtrate

Step four   As the solution cools, crystals of the salt form. Separate the crystals from the solution by filtration. Use a little distilled water to wash the crystals in the filter funnel. Then leave the crystals to dry.

● *Acid + alkali*

Method 4   Reaction between an acid and an alkali

Ammonia is an alkali. The preparation of ammonium salts is important because they are used as fertilisers. A laboratory method for making ammonium sulphate is shown in Figure 13.5D.

SUMMARY

Soluble salts are made by the reactions:
- Acid + Reactive → Salt + Hydrogen
  metal
- Acid + Metal → Salt + Water
  oxide
- Acid + Metal → Salt + Water + Carbon
  carbonate   dioxide

In these three methods, an excess of the solid reactant is used, and unreacted solid is removed by filtration.

1 Add the ammonia solution to the dilute sulphuric acid, stirring constantly

2 From time to time, remove a drop of acid on the end of the glass rod. Spot it on to a strip of indicator paper on a white tile. When the test shows that the solution has become alkaline, stop adding ammonia. You have now added an excess of ammonia

3 Evaporate the solution until it begins to crystallise. Leave it to stand. Filter to obtain crystals of ammonium sulphate

SUMMARY

Soluble salts can be made by the reaction:
- Acid + Alkali → Salt + Water

This method is used for ammonium salts. An excess of ammonia is added. The excess is removed when the solution is evaporated.

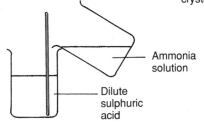

Figure 13.5D ● Making ammonium sulphate

Method for insoluble salts

Insoluble salts are made by mixing two solutions. For example, you can make the insoluble salt barium sulphate by mixing a solution of the soluble salt barium chloride with a solution of the soluble salt sodium sulphate. When the two solutions are mixed, the insoluble salt barium sulphate is precipitated (thrown out of solution). What do you think remains in solution? The method is called **precipitation**.

● *Precipitation*

The equation for the reaction is

Barium chloride	+	Sodium sulphate	→	Barium sulphate	+	Sodium chloride

$$BaCl_2(aq) + Na_2SO_4(aq) \rightarrow BaSO_4(s) + 2NaCl(aq)$$

An ionic equation can be written for the reaction

Barium ions	+	Sulphate ions	→	Barium sulphate

$$Ba^{2+}(aq) + SO_4^{2-}(aq) \rightarrow BaSO_4(s)$$

The ionic equation shows only the ions which take part in the precipitation reaction: the barium ions and sulphate ions.

SUMMARY

Insoluble salts are made by precipitation. A solution of a soluble salt of the metal is added to a solution of a soluble salt of the acid.

Remember

- All nitrates are soluble.
- All sodium, potassium and ammonium salts are soluble.

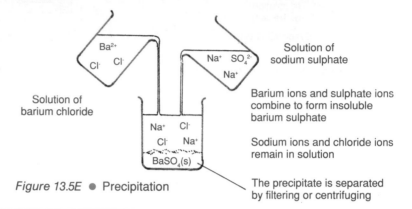

Figure 13.5E ● Precipitation

CHECKPOINT

❶ Refer to Figure 13.5A.
- (a) How do you know that all the acid has been used up
  - (i) in the reaction with a metal?
  - (ii) in the reaction with a metal oxide?
  - (iii) in the reaction with a metal carbonate?
- (b) Why is it important to make sure that all the acid is used up?
- (c) If some acid were left unneutralised in Step 1, what would happen to it in Step 3? How would it affect the crystals of salt formed in Step 4?
- (d) Why is it easier to remove an excess of base than an excess of acid?

❷ Complete the following word equations

- (a) magnesium + sulphuric acid → _____ sulphate + _____
- (b) zinc oxide + hydrochloric acid → _____ chloride + _____
- (c) calcium carbonate + hydrochloric acid → _____ + _____ + _____
- (d) nickel oxide + hydrochloric acid → _____ + _____
- (e) chromium oxide + sulphuric acid → _____ + _____
- (f) magnesium oxide + nitric acid → _____ nitrate + _____

❸ Write chemical equations for the reactions (a), (b) and (c) in Question 2.

❹ Refer to Table 13.4 for solubility
- (a) Strontium sulphate, $SrSO_4$, is insoluble. Which soluble strontium salt and which soluble sulphate could you use to make strontium sulphate?
- (b) What would you do to obtain a dry specimen of strontium sulphate?
- (c) Write a word equation for the reaction.
- (d) Write an ionic equation for the reaction (the strontium ion is Sr^{2+}).

❺ Barium carbonate, $BaCO_3$, is insoluble.
- (a) Name two solutions which you could mix to give a precipitate of barium carbonate.
- (b) Say what you would do to obtain barium carbonate from the mixture.
- (c) Write a word equation for the reaction.
- (d) Write an ionic equation (the barium ion is Ba^{2+}).

? THEME QUESTIONS

1 (a) What sort of mixture can be separated by
        (i) filtration, (ii) chromatography, (iii) centrifuging?
   (b) A mixture contains two liquids, A and B. A boils at
        75 °C, and B boils at 95 °C. Draw an apparatus you
        could use to separate A and B. Label the drawing.

2 The police are investigating a case of poison pen letters.
   The police chemist makes chromatograms from the ink
   in the letters and the ink in the pens of three suspects.
   The diagram below shows her results. What conclusion
   can you reach?

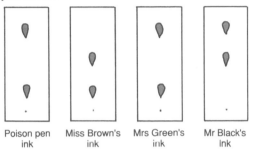

| Poison pen ink | Miss Brown's ink | Mrs Green's ink | Mr Black's ink |

3 The diagram shows the
   structure of a fractionating
   column in an oil refinery.
   (a) The technique used to
        separate the fractions
        is called:
        A chromatography
        B distillation
        C evaporation
        D filtration

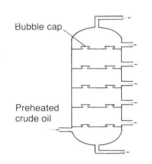

Bubble cap

Preheated
crude oil

The table gives information about the various fractions.

Fraction	Boiling point range (°C)	Number of carbon atoms in molecule
Bitumen	Above 350	More than 25
Fuel gases	Below 40	1–4
Gasoline	40–100	4–8
Heavy gas oil	300–350	20–25
Kerosene	160–250	10–16
Light gas oil	250–300	16–20
Naphtha	80–180	5–12

   (b) (i) Which fraction will collect at the top of the
        column and why?
        (ii) Which fraction will collect at the bottom of the
        column?
   (c) Petrol consists mainly of compounds containing
        eight carbon atoms. About 10% of petrol comes
        directly from the fractions containing these
        compounds. The remainder of the petrol is obtained
        by cracking.
        (i) Which **two** fractions in the table could be used
        directly for petrol?
        (ii) What is cracking and how is it done?   (LEAG)

4 The indicator phenolphthalein is colourless in neutral
   and acidic solutions and turns pink in alkaline solutions.
   Of the solutions A, B and C, one is acidic, one is
   alkaline, and one is neutral. Explain how, using only
   phenolphthalein, you can find out which solution is
   which.

5 Seven steel bars were placed in solutions of different pH
   for the same time. The table shows the percentage
   corrosion of the steel bars.

pH of solution	1	2	3	4	5	6	7
Percentage corrosion of steel bar	65	60	55	50	20	15	10

   (a) On graph paper, plot the percentage corrosion
        against the pH of the solution.
   (b) Read from your graph the percentage corrosion at
        pH 4.5.

6 Look at the information from the label of a bottle of
   concentrated orange squash.

   Concentrated orange squash
   Ingredients: Sugar, Water, Citric acid,
   Flavourings, Preservative E250, Artificial sweetener,
   Yellow colourings E102 and E103

   (a) Explain what is meant by 'concentrated'.
   (b) A sample of concentrated orange squash is mixed
        with water. A piece of universal indicator paper is
        dipped in. What colour will the paper turn?
   (c) Why does using universal indicator paper give a
        better result than adding universal indicator solution
        to the orange squash?
   (d) Describe a simple experiment you could do to
        prove that **only two** yellow substances are present
        in the concentrated orange squash.

7 Zinc sulphate crystals, $ZnSO_4.7H_2O$, can be made from
   zinc and dilute sulphuric acid by the following method.
   Step 1 Add an excess of zinc to dilute sulphuric acid.
            Warm.
   Step 2 Filter.
   Step 3 Partly evaporate the solution from Step 2.
            Leave it to stand.
   (a) Explain why an excess of zinc is used in Step 1.
   (b) How can you tell when Step 1 is complete?
   (c) Name the residue and the filtrate in Step 2.
   (d) Explain why the solution is partly evaporated in
        Step 3.
   (e) Would you dry the crystals by strong heating or by
        gentle heating? Explain your answer.
   (f) Write a word equation for the reaction. Write a
        symbol equation.

8 What method would you use to make the insoluble salt lead(II) carbonate? Explain why you have chosen this method. Say what starting materials you would need, and say what you would do to obtain solid lead(II) carbonate. (For solubility information, see Theme D, Table 13.7.)

9 Explain the following:
(a) Tea changes colour when lemon juice is added.
(b) Sodium sulphate solution is used as an antidote to poisoning by barium compounds.
(c) Washing soda takes some of the pain out of bee-stings.
(d) Toothpastes containing aluminium hydroxide fight tooth decay.
(e) Calcium fluoride is added to some toothpastes.
(f) Scouring powders often contain sodium hydroxide and powdered stone.

10 A student receives the following instructions: 'Make a solution of copper sulphate by neutralising dilute sulphuric acid, using the base copper oxide.'
(a) Explain the meanings of the terms:
solution, neutralising, dilute, base.
(b) Draw an apparatus which the student could use for carrying out the instructions.
(c) Describe exactly how the student should carry out the instructions.

11 Use the information in the table. Say how you could separate:
(a) aluminium and cobalt
(b) chromium and polyurethane.

Substance (in powder form)	Solubility in water	Solubility in ethanol	Magnetic properties
Aluminium	Insoluble	Insoluble	None
Cobalt	Insoluble	Insoluble	Magnetic
Chromium	Insoluble	Insoluble	None
Polyurethane	Insoluble	Soluble	None

12 Select the correct endings to the sentences.
(i) A solid can be purified by crystallisation because
  a it dissolves in cold water but not in hot water
  b it is insoluble in hot and cold water
  c it is very soluble in hot and cold water

d it is more soluble in hot water than in cold water
e it is more soluble in cold water than in hot water.
(ii) Two liquids can be separated by distillation if
  a they have different densities
  b they form two layers
  c they have different boiling points
  d their boiling points are the same
  e they have different freezing points.
(iii) A solvent can be separated from a solution because
  a the particles of the solvent (liquid) are smaller than those of the solute (dissolved substance)
  b the solvent is less dense than the solute
  c the solvent has a higher freezing point than the solute
  d the solvent evaporates more easily than the solute
  e the solvent condenses more easily than the solute.

13 The table shows the colours of some indicators.

Indicator	pH 1 2 3 4 5 6 7 8 9 10 11 12 13 14
Methyl orange	←Red→ ←——— Yellow ———→
Bromocresol green	←Yellow→ ←——— Blue ———→
Phenol red	←—Yellow—→ ←——Red——→
Phenolphthalein	←—Colourless—→ ←—— Red ——→

(a) What colour is a solution of pH 10 when a few drops of bromocresol green are added?
(b) What colour is a solution of pH 3.5 when a few drops of methyl orange are added?
(c) What colour is a solution of pH 6.5 when a few drops of phenol red are added?
(d) A solution turns yellow when either methyl orange or phenol red is added. What is the approximate pH of the solution?
(e) A solution is colourless when phenolphthalein is added and red when phenol red is added. What is the pH of the solution?
(f) A mixture of bromocresol green, phenol red and phenolphthalein is added to a solution of pH 10. What is the colour?
(g) A mixture of all four indicators is added to a strong acid. What is the colour?

— THEME E —

Air and Water

A modern industrial society makes plenty of goods to keep us free from squalor and starvation. Plenty has led to pollution. The air is polluted by objectionable gases, dust and smoke from our cars and factories. Rivers and lakes are polluted by waste from our factories, fields and our own bodies. The land is polluted by our rubbish tips. Urgent measures are needed to combat pollution. If we take the necessary steps, we can avoid disaster. We can reduce the pollutants in our atmosphere; we can make our rivers and lakes clean again; by making use of much of our rubbish, we can conserve the Earth's resources.

TOPIC 14 AIR

14.1 Oxygen the life-saver

FIRST THOUGHTS

As you study this topic, think about the vital importance of air as a source of:

- oxygen – which supports the life of plants and animals
- carbon dioxide – which supports plant life
- nitrogen – the basis of natural and synthetic fertilisers
- the noble gases – filling our light bulbs

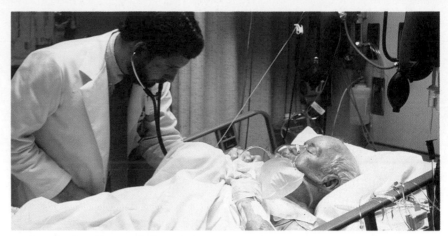

Figure 14.1A ● Oxygen in hospitals

Hospitals need oxygen for patients who have difficulty in breathing. Tiny premature babies often need oxygen. They may be put into an oxygen high pressure chamber. This chamber contains oxygen at 3–4 atmospheres pressure, which makes 15–20 times the normal concentration of oxygen dissolve in the blood. People who are recovering from heart attacks and strokes also benefit from being given oxygen. Patients who are having operations are given an anaesthetic mixed with oxygen.

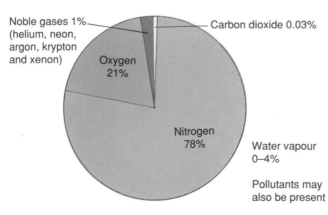

Noble gases 1%
(helium, neon, argon, krypton and xenon)

Carbon dioxide 0.03%

Oxygen 21%

Nitrogen 78%

Water vapour 0–4%

Pollutants may also be present

In normal circumstances, we can obtain all the oxygen we need from the air. Figure 14.1B shows the percentages of oxygen, nitrogen and other gases in pure, dry air. Water vapour and pollutants may also be present in air.

Figure 14.1B ● Composition of clean, dry air in percentage by volume

14.2 How oxygen and nitrogen are obtained from air

The method used by industry to obtain oxygen and nitrogen from air is **fractional distillation** of liquid air. Air must be cooled to −200 °C before it liquefies. It is very difficult to get down to this temperature. However, one way of cooling a gas is to compress it and then allow it to expand suddenly. Figures 14.2A and 14.2B show how air is first liquefied and then distilled.

LOOK AT LINKS
for **fractional distillation**
See Topic 11.8.

LOOK AT LINKS
How the industry uses the products of fractional distillation of air is discussed in Topics 14.3, 14.4 and 14.10.

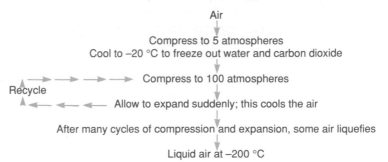

Figure 14.2A ● Liquefaction of air

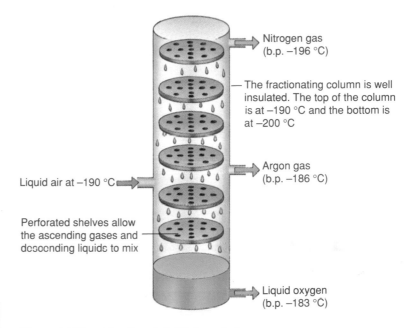

Figure 14.2B ● Fractional distillation of air

SUMMARY

Air is a mixture of nitrogen, oxygen, noble gases, carbon dioxide, water vapour and pollutants. The fractional distillation of liquid air yields oxygen, nitrogen and argon.

Oxygen, nitrogen and argon are redistilled. They are stored under pressure in strong steel cylinders. Industry finds many important uses for them.

CHECKPOINT

❶ The top of the fractionating column in Figure 14.2B is at –190 °C. Explain why nitrogen is a gas at the top of the column but oxygen is a liquid.

❷ The table lists nitrogen, oxygen and the noble gases.

Gas	Boiling point (°C)
Argon	–186
Helium	–269
Krypton	–153
Neon	–246
Nitrogen	–196
Oxygen	–183
Xenon	–108

(a) Which gas has (i) the lowest boiling point and (ii) the highest boiling point?
(b) List the gases which would still be gaseous at –200 °C.

14.3 Uses of oxygen

Oxygen is a colourless, odourless gas which is slightly soluble in water. All living things need oxygen for respiration. Breathing becomes difficult at a height of 5 km above sea level, where the air pressure is only half that at sea level. Climbers who are tackling high mountains take oxygen with them. Aeroplanes which fly at high altitude carry oxygen. Astronauts must carry oxygen, and even unmanned space flights need oxygen. Deep-sea divers carry cylinders which contain a mixture of oxygen and helium.

Industry has many uses for oxygen. Figure 14.3B shows an oxy-acetylene torch being used to weld metal at a temperature of about 4000 °C. The hot flame is produced by burning the gas ethyne (formerly called acetylene) in oxygen. Substances burn faster in pure oxygen than in air.

The steel industry converts brittle cast **iron** into strong **steel**. Cast iron is brittle because it contains impurities such as carbon, sulphur and phosphorus. These impurities will burn off in a stream of oxygen. Steel plants use one tonne of oxygen for every tonne of cast iron turned into steel. Many steel plants make their own oxygen on site.

The proper treatment of sewage is important for public health. Air is used to help in the decomposition of sewage. Without this treatment, sewage would pollute many rivers and lakes. One method of treating polluted lakes and rivers to make them fit for plants and animals to live in is to pump in oxygen.

IT'S A FACT

The introduction of the oxy-acetylene flame brought about a revolution in metal working. Industry moved out of the blacksmith era into the twentieth century with gas-welding and flame-cutting techniques.

LOOK AT LINKS
for **metals**
See Topic 19.

SUMMARY

Uses for pure oxygen are:
*   treating patients who have breathing difficulties
*   supporting high-altitude pilots, mountaineers, deep-sea divers and space flights
*   in steel making and other industries
*   treating polluted water

Figure 14.3A ● Astronaut using oxygen

Figure 14.3B ● Oxyacetylene welding being used on an automated production line

CHECKPOINT

❶ Explain why oxygen is used in (a) steelworks (b) metal working (c) sewage treatment (d) space flights and (e) hospitals. Say what advantage pure oxygen has over air for each purpose.

14.4 Nitrogen

So far, oxygen seems to be the important part of the air, but nitrogen has its uses too, as you will find out in this section.

LOOK AT LINKS
Sparks produced by static electricity can ignite powders.
See Topic 20.1 of *KS: Physics*.

Figure 14.4A ● Oil tanker delivering oil to a refinery. Once it is empty, the tanks will be purged with nitrogen

SUMMARY

Nitrogen is a rather unreactive gas. It is used to provide an inert (chemically unreactive) atmosphere.

Nitrogen is a colourless, odourless gas which is slightly soluble in water. It does not readily take part in chemical reactions. Many uses of nitrogen depend on its unreactive nature. Liquid nitrogen (below −196 °C) is used when an inert (chemically unreactive) refrigerant is needed. The food industry uses it for the fast freezing of foods.

Vets use the technique of artificial insemination to enable a prize bull to fertilise a large number of dairy cows. They carry the semen of the bull in a type of vacuum flask filled with liquid nitrogen.

Many foods are packed in an atmosphere of nitrogen. This prevents the oils and fats in the foods from reacting with oxygen to form rancid products. As a precaution against fire, nitrogen is used to purge oil tankers and road tankers. The silos where grain is stored are flushed out with nitrogen because dry grain is easily ignited.

Figure 14.4B ● Grain silos

14.5 The nitrogen cycle

LOOK AT LINKS
The importance of proteins as food is described in Topic 11.3 of *KS: Biology*.

Nitrogen is an essential element in **proteins**. Some plants have nodules on their roots which contain nitrogen-fixing bacteria. These bacteria **fix** gaseous nitrogen, that is, convert it into nitrogen compounds. From these nitrogen compounds, the plants can synthesise proteins. Members of the legume family, such as peas, beans and clover, have nitrogen-fixing bacteria.

Plants other than legumes synthesise proteins from nitrates. *How do nitrates get into the soil?* Nitrogen and oxygen combine in the atmosphere during lightning storms and in the engines of motor vehicles during combustion. They form nitrogen oxides (compounds of nitrogen and oxygen). These gases react with water to form nitric acid. Rain showers bring nitric acid out of the atmosphere and wash it into the ground, where it reacts with minerals to form nitrates. Plants take in these nitrates

Figure 14.5A ● Nodules containing nitrogen fixing bacteria on the roots of a legume

through their roots. They use them to synthesise proteins. Animals obtain the proteins they need by eating plants or by eating the flesh of other animals.

In the excreta of animals and the decay products of animals and plants, **ammonium salts** are present. Nitrifying bacteria in the soil convert ammonium salts into nitrates. Both nitrates and ammonium salts can be removed from the soil by **denitrifying bacteria**, which convert the compounds into nitrogen. To make the soil more fertile, farmers add both nitrates and ammonium salts as fertilisers. The balance of processes which put nitrogen into the air and processes which remove nitrogen from the air is called the **nitrogen cycle** (see Figure 14.5B).

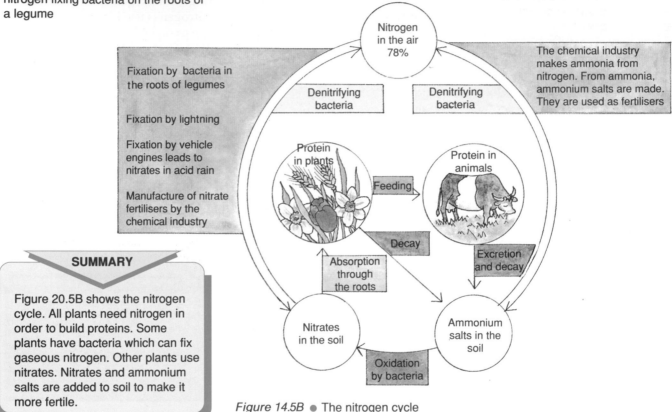

Figure 14.5B ● The nitrogen cycle

SUMMARY

Figure 20.5B shows the nitrogen cycle. All plants need nitrogen in order to build proteins. Some plants have bacteria which can fix gaseous nitrogen. Other plants use nitrates. Nitrates and ammonium salts are added to soil to make it more fertile.

CHECKPOINT

❶ *Process A*　Nitrogen-fixing bacteria convert nitrogen gas into nitrates.

Process B　Nitrifying bacteria use oxygen to convert ammonium compounds (from decaying plant and animal matter) into nitrates.

Process C　Denitrifying bacteria turn nitrates into nitrogen.

(a) Say where *Process A* takes place.
(b) Say what effect the presence of air in the soil will have on *Process B*.
(c) Say what effect waterlogged soil which lacks air will have on *Process C*.
(d) Explain why plants grow well in well-drained, aerated soil.
(e) A farmer wants to grow a good crop of wheat without using a fertiliser. What could he plant in the field the previous year to ensure a good crop?
(f) Explain why garden manure and compost fertilise the soil.

❷ Explain why nitrogen is used in (a) food packaging (b) oil tankers (c) hospitals and (d) food storage.

14.6 Carbon dioxide and the carbon cycle

The percentage by volume of carbon dioxide in clean, dry air is only 0.03%. Perhaps you think this makes carbon dioxide sound rather unimportant. Through studying this section, you may change your mind about the importance of carbon dioxide!

LOOK AT LINKS
For a fuller account of photosynthesis and respiration.
See Topic 11.1 of *KS: Biology.*

LOOK AT LINKS
A reaction like photosynthesis, in which energy is taken in, is called an endothermic reaction. A reaction like respiration, in which energy is given out, is called an exothermic reaction.
See Topics 26.1 and 26.2.

Plants need carbon dioxide, and animals, including ourselves, need plants. Plants take in carbon dioxide through their leaves and water through their roots. They use these compounds to **synthesise** (build) sugars. The reaction is called **photosynthesis** (photo means light). It takes place in green leaves in the presence of sunlight. Oxygen is formed in photosynthesis.

Photosynthesis (in plants)

catalysed by chlorophyll in green leaves

Sunlight + Carbon dioxide + Water → Glucose + Oxygen
(a sugar)

The energy of sunlight is converted into the energy of the chemical bonds in glucose.

Animals eat foods which contain starches and sugars. They inhale (breathe in) air. Inhaled air dissolves in the blood supply to the lungs. In the cells, some of the oxygen in the dissolved air oxidises sugars to carbon dioxide and water and energy is released. This process is called **respiration**. Plants also respire to obtain energy.

Respiration (in plants and animals)

Glucose + Oxygen → Carbon dioxide + Water + Energy

The processes which take carbon dioxide from the air and those which put carbon dioxide into the air are balanced so that the percentage of carbon dioxide in the air stays at 0.03%. This balance is called the **carbon cycle** (see Figure 14.6A).

SUMMARY

Figure 14.6A shows the carbon cycle. Photosynthesis is the process in which green plants use sunlight, carbon dioxide and water to make sugars and oxygen. Respiration is the process in which animals and plants oxidise carbohydrates to carbon dioxide and water with the release of energy.

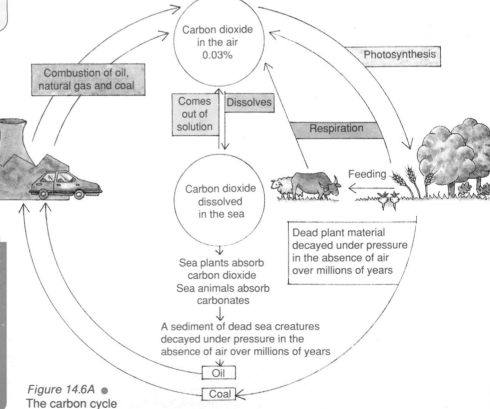

Figure 14.6A ●
The carbon cycle

❶ Name two processes which add carbon dioxide to the atmosphere.

❷ Suggest a place where you would expect the percentage of carbon dioxide in the atmosphere to be lower than average.

❸ Suggest two places where you would expect the percentage of carbon dioxide in the atmosphere to be higher than average.

14.7 The greenhouse effect

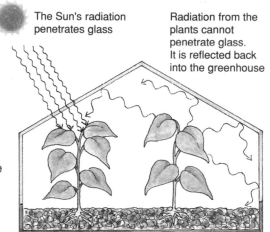

The Sun's radiation penetrates glass

Radiation from the plants cannot penetrate glass. It is reflected back into the greenhouse

Figure 14.7A ● A greenhouse

The Sun is so hot that it emits high-energy radiation. The Sun's rays can pass easily through the glass of a greenhouse. The plants in the greenhouse are at a much lower temperature. They send out infra-red radiation which cannot pass through the glass. The greenhouse therefore warms up (Figure 14.7A).

LOOK AT LINKS
Infra-red radiation and light are electromagnetic waves. See Topic 6.7 of *KS: Physics*.

LOOK AT LINKS
Other gases which contribute to the greenhouse effect are methane (see Topic 17.8), oxides of nitrogen (from vehicle exhausts; see Topic 17.7) and CFCs (from aerosols and refrigerators; see Topic 17.11).

Radiant energy from the Sun falls on the Earth and warms it. The Earth radiates heat energy back into space as infra-red radiation. Unlike sunlight, infra-red radiation cannot travel freely through the air surrounding the Earth. Both water vapour and carbon dioxide absorb some of the infra-red radiation. Since carbon dioxide and water vapour act like the glass in a greenhouse, their warming effect is called the greenhouse effect. Without carbon dioxide and water vapour, the surface of the Earth would be at −40 °C. Most of the greenhouse effect is due to

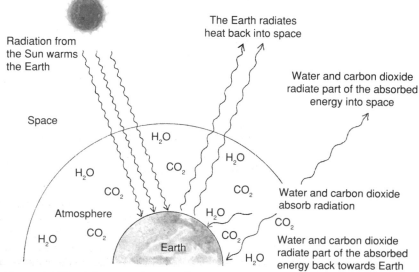

The Earth radiates heat back into space

Radiation from the Sun warms the Earth

Water and carbon dioxide radiate part of the absorbed energy into space

Space

Water and carbon dioxide absorb radiation

Atmosphere

Water and carbon dioxide radiate part of the absorbed energy back towards Earth

Earth

Figure 14.7B ● The greenhouse effect

Worldwide, 16 thousand million tonnes of carbon dioxide are formed each year by the combustion of fossil fuels!

Every month, an area of tropical forest equal to the size of Belgium is felled and burned in South America.

SUMMARY

Carbon dioxide and water vapour reduce the amount of heat radiated from the Earth's surface into space and keep the Earth warm. Their action is called the greenhouse effect. The percentage of carbon dioxide in the atmosphere is increasing, and the temperature of the Earth is rising. If it continues to rise, the polar ice caps could melt.

water vapour. The Earth does, however, radiate some wavelengths which water vapour cannot absorb. Carbon dioxide is able to absorb some of the radiation which water vapour lets through.

The surface of the Earth has warmed up by 0.75 °C during the last century. The rate of warming up is increasing. Unless something is done to stop the temperature rising, there is a danger that the temperature of the Arctic and Antarctic regions might rise above 0 °C. Then, over the course of a century or two, polar ice would melt and flow into the oceans. If the level of the sea rose, low-lying areas of land would disappear under the sea.

One reason for the increase in the Earth's temperature is that we are putting too much carbon dioxide into the air. The combustion of coal and oil in our power stations and factories sends carbon dioxide into the air. The second reason is that we are felling too many trees. In South America, huge areas of tropical forest have been cut down to make timber and to provide land for farming. In many Asian countries, forests have been cut down for firewood. The result is that worldwide there are fewer trees to take carbon dioxide from the air by photosynthesis. The percentage of carbon dioxide is increasing, and some scientists calculate that it will double by the year 2000 (see Figure 14.7C). This would raise the Earth's temperature by 2 °C.

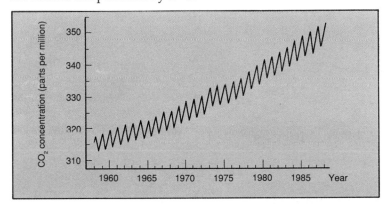

Figure 14.7C ● Atmospheric concentration of carbon dioxide (These measurements were made at the Mauna Loa Observatory in Hawaii)

CHECKPOINT

❶ The amount of carbon dioxide in the atmosphere is slowly increasing.
(a) Suggest two reasons why this is happening.
(b) Explain why people call the effect which carbon dioxide has on the atmosphere the 'greenhouse effect'.
(c) Why are some people worried about the greenhouse effect?
(d) Suggest two things which could be done to stop the increase in the percentage of carbon dioxide in the atmosphere.

❷ *Selima* Did you hear what Miss Sande said about the greenhouse effect making the temperature of the Earth go up?
Joshe I don't know what she's worried about. We wouldn't be here at all if it weren't for the greenhouse effect.
(a) What does Joshe mean by what he says? What would the Earth be like without the greenhouse effect?
(b) Is Joshe right in thinking there is no cause for worry?

❸ The burning of fossil fuels produces 16 000 million tonnes of carbon dioxide per year. Carbon dioxide is thought to increase the average temperature of the air. It is predicted that the effect of this increase in temperature will be to melt some of the ice at the North and South Poles. Describe the effects which this could have on life for people in other parts of the world.

14.8 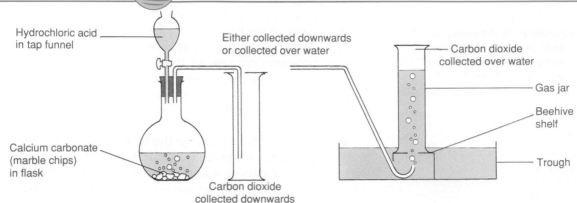 Carbon dioxide

Hydrochloric acid in tap funnel

Either collected downwards or collected over water

Carbon dioxide collected over water

Gas jar

Beehive shelf

Calcium carbonate (marble chips) in flask

Trough

Carbon dioxide collected downwards

Figure 14.8A ● A laboratory preparation of carbon dioxide

LOOK AT LINKS
Why is carbon dioxide used as a fire extinguisher?
See Topic 15.8.

Carbon dioxide can be made in the laboratory. Figure 14.8A shows one method. You will remember from Topic 12 that a carbonate reacts with an acid to form carbon dioxide, a salt and water. In this case

Calcium carbonate (marble chips)	+	Hydrochloric acid	→	Carbon dioxide	+	Calcium chloride	+	Water
$CaCO_3(s)$	+	$2HCl(aq)$	→	$CO_2(g)$	+	$CaCl_2(aq)$	+	$H_2O(l)$

Carbon dioxide can be collected over water. (*What does this tell you about the solubility of carbon dioxide?*) It can also be collected downwards. (*What does this tell you about the density of carbon dioxide?*)

In water, carbon dioxide dissolves slightly to form the weak acid carbonic acid, H_2CO_3. Under pressure, the solubility increases. Many soft drinks are made by dissolving carbon dioxide under pressure and adding sugar and flavourings. When you open a bottle of fizzy drink, the pressure is released and bubbles of carbon dioxide come out of solution. People like the bubbles and the slightly acidic taste.

When carbon dioxide is cooled, it turns into a solid. This is known as **dry ice** and as **Dricold**. It is used as a refrigerant for icecream and meat. When it warms up, dry ice sublimes (turns into a vapour without melting first).

SCIENCE AT WORK

Pop stars sometimes use Dricold on stage. As it sublimes, it cools the air on stage. Water vapour condenses from the cold air in swirling clouds.

SUMMARY

Carbon dioxide:
• is a colourless, odourless gas
• is denser than air
• dissolves slightly in water to form carbonic acid
• does not burn
• allows few materials to burn in it

Figure 14.8B ● Subliming carbon dioxide being poured from a gas jar

CHECKPOINT

❶ Why do icecream sellers prefer dry ice to ordinary ice?

❷ Why is carbon dioxide used by the soft drinks industry?

14.9 Testing for carbon dioxide and water vapour

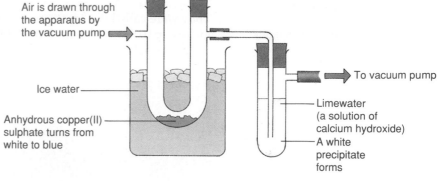

Figure 14.9A ● Testing for carbon dioxide and water vapour in air

Figure 14.9A shows how you can test for the presence of carbon dioxide and water vapour in air.

LOOK AT LINKS
for **copper sulphate**
See Topic 13.3.

● *Test for carbon dioxide*
A white precipitate is formed when carbon dioxide reacts with a solution of calcium hydroxide, **limewater**.

Carbon dioxide	+	Calcium hydroxide	→	Calcium carbonate	+	Water
		(limewater)		(white precipitate)		
$CO_2(g)$	+	$Ca(OH)_2(aq)$	→	$CaCO_3(s)$	+	$H_2O(l)$

SUMMARY

The carbon dioxide in the air turns limewater cloudy. The water vapour in the air turns anhydrous copper(II) sulphate from white to blue.

● *Test for water vapour*
Water turns anhydrous copper(II) sulphate from white to blue.

Copper(II) sulphate	+	Water	→	Copper(II) sulphate-5-water
(anhydrous, a white solid)				(blue crystals)
$CuSO_4(s)$	+	$5H_2O(l)$	→	$CuSO_4.5H_2O(s)$

CHECKPOINT

1 Malachite is a green mineral found in many rocks. How could you prove that malachite is a carbonate?

2 (a) How could you prove that whisky contains water?
(b) How could you show that whisky is not pure water? (Don't say 'Taste it': tasting chemicals is often dangerous.)

14.10 The noble gases

Deep-sea divers have to take their oxygen with them (see Figure 14.10A). If they take air in their cylinders, nitrogen dissolves in their blood. Although the solubility of nitrogen is normally very low, at the high pressures which divers experience the solubility increases. When the divers surface, their blood can dissolve less nitrogen than it can at high pressure. Dissolved nitrogen leaves the blood to form tiny bubbles. These cause severe pains, which divers call 'the bends'. The solution to this problem is to breathe a mixture of oxygen and **helium**. Since helium is much less soluble than nitrogen, there is much less danger of the bends.

Helium is a safe gas to use because it takes part in no chemical reactions. It is one of the **noble gases** (see Figure 14.1B).

For a long time, it seemed that the noble gases (helium, neon, argon, krypton and xenon) were unable to take part in any chemical reactions. They were called the 'inert gases'. Argon, the most abundant of them, makes up 0.09% of the air. Figure 14.2B shows how argon is obtained by the fractional distillation of liquid air.

Figure 14.10A ● A diver carries a mixture of oxygen and helium

Neon, argon, krypton and xenon are used in display lighting (see Figure 14.10B). The discharge tubes which are used as strip lights contain these gases at low pressure. When the gases conduct electricity, they glow brightly. Most electric light bulbs are filled with argon. The filament of a light bulb is so hot that it would react with other gases. Krypton and xenon are also used to fill light bulbs.

The low density of helium makes it useful. It is used to fill balloons and airships.

Figure 14.10B ● Neon lights

SUMMARY

The noble gases take part in few chemical reactions. Argon is used to fill light bulbs. Neon and other noble gases are used in illuminated signs. Helium is used to fill airships.

CHECKPOINT

❶ The first airships contained hydrogen. Modern airships use helium. The table gives some information about the two gases and air.

	Hydrogen	Helium	Air
Density (g/cm³)	8.3×10^{-5}	1.66×10^{-4}	1.2×10^{-3}
Chemical reactions with air	Forms an explosive mixture with air	No known chemical reactions	

(a) What advantage does helium have over hydrogen for filling airships?
(b) What disadvantage does hydrogen have compared with helium?
(c) Why is air not used for 'airships'?

TOPIC 15 *OXYGEN*

15.1 Blast-off

FIRST THOUGHTS

In this topic, you can find out about some of the elements and compounds that react with oxygen. Oxygen is a very reactive element, and many substances burn in it.

LOOK AT LINKS
for **rockets**
Why do rockets need to be so powerful? Topic 18.8 of *KS: Physics* will tell you.

On 16 July 1969, ten thousand people gathered at the Kennedy Space Centre in Florida, USA. They had come to watch the spacecraft *Apollo 11* lift off on its journey to the moon. While the *Saturn* rocket stood on the launch pad, its roaring jet engines burned 450 tonnes of kerosene in 1800 tonnes of pure oxygen. The thrust from the jets shot the rocket through the lower atmosphere, trailing a jet of flame. At a height of 65 km the first stage of the rocket separated. For six minutes, the second stage burned hydrogen in pure oxygen, taking the spacecraft to a height of 185 km. Then the second stage separated. The third stage, burning hydrogen in oxygen, put *Apollo 11* into an orbit round the Earth at a speed of 28 000 km/hour. From this orbit, *Apollo 11* headed for the moon. The energy needed to lift *Apollo 11* into space came from burning fuels (kerosene and hydrogen) in pure oxygen. Fuels burns faster in oxygen than they do in air. They therefore deliver more power. We shall return to this important reaction of burning in Topic 15.7.

Figure 15.1A ● Rocket launch

15.2 Test for oxygen

Who's behind the science

A new gas was discovered by the British chemist, Joseph Priestley, on 1 August 1786. He had heated mercury in air and found that it combined with part of the air to form a solid which he called 'red calx of mercury'. When Priestley heated 'red calx of mercury', he obtained mercury and a new gas. He tried breathing the new gas and found that it produced a 'light and easy sensation in his chest'. The gas came to be called oxygen. What do you think 'red calx of mercury' is called now?

Oxygen is a colourless, odourless gas, which is only slightly soluble in water. It is neutral. Oxygen allows substances to burn in it: it is a good **supporter of combustion**. One test for oxygen is to lower a glowing wooden splint into the gas. If the splint starts to burn brightly, the gas is oxygen.

Oxygen relights a glowing splint (see Figure 15.2A).

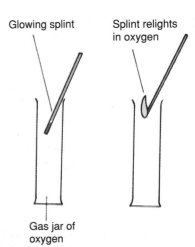

Glowing splint   Splint relights in oxygen

Gas jar of oxygen

Figure 15.2A ● Testing for oxygen

15.3 The percentage by volume of oxygen in air

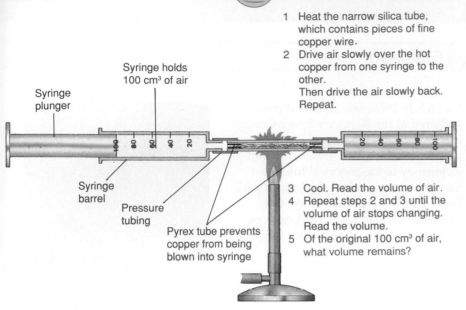

1 Heat the narrow silica tube, which contains pieces of fine copper wire.
2 Drive air slowly over the hot copper from one syringe to the other.
   Then drive the air slowly back. Repeat.

Syringe holds 100 cm³ of air

Syringe plunger

Syringe barrel

Pressure tubing

Pyrex tube prevents copper from being blown into syringe

3 Cool. Read the volume of air.
4 Repeat steps 2 and 3 until the volume of air stops changing. Read the volume.
5 Of the original 100 cm³ of air, what volume remains?

A method of finding out the percentage by volume of oxygen in air is shown in Figure 15.3A. Oxygen combines with copper to form copper(II) oxide. The other gases in the air do not react. The idea is to pass a measured volume of air over hot copper until the copper has taken all the oxygen out of the air. Then you can measure the new volume and find out how much air has been used up.

Figure 15.3A ● Finding the percentage by volume of oxygen in air

15.4 A method of preparing oxygen in the laboratory

LOOK AT LINKS
for **catalysts**
See Topic 23.7.

IT'S A FACT

The mass of oxygen in the atmosphere is 12 hundred million million tonnes!

RESOURCE -ACTIVITY- PACK

Hydrogen peroxide is a colourless liquid which decomposes to give oxygen and water.

Hydrogen peroxide	→	Oxygen	+	Water
$2H_2O_2(aq)$	→	$O_2(g)$	+	$2H_2O(l)$

Solutions of hydrogen peroxide are kept in stoppered brown bottles to slow down the rate of decomposition. If you want to speed up the formation of oxygen, you can add a **catalyst**. A substance which speeds up a reaction without being used in the reaction is called a catalyst. The catalyst manganese(IV) oxide, MnO_2, is often used.

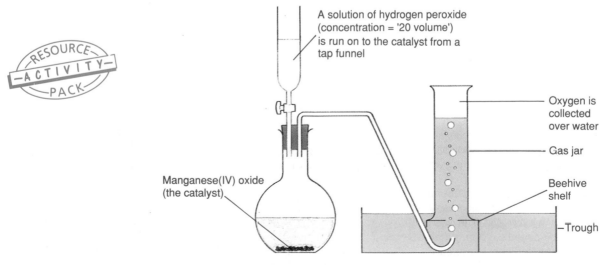

A solution of hydrogen peroxide (concentration = '20 volume') is run on to the catalyst from a tap funnel

Oxygen is collected over water

Gas jar

Beehive shelf

Trough

Manganese(IV) oxide (the catalyst)

Figure 15.4A ● A laboratory preparation of oxygen

15.5 The reactions of oxygen with some elements

Most elements combine with oxygen; many elements burn in oxygen.

Table 15.1 shows how some elements react with oxygen. The products of the reactions are **oxides**. An oxide is a compound of oxygen with one other element.

Table 15.1 ● How some elements react with oxygen

Element	Observation	Product	Action of product on water
Calcium (metal)	Burns with a red flame	Calcium oxide, CaO (a white solid)	Dissolves to give a strongly alkaline solution
Copper (metal)	Does not burn; turns black	Copper(II) oxide, CuO (a black solid)	Insoluble
Iron (metal)	Burns with yellow sparks	Iron oxide, Fe_3O_4 (a blue-black solid)	Insoluble
Magnesium (metal)	Burns with a bright white flame	Magnesium oxide, MgO (a white solid)	Dissolves slightly to give an alkaline solution, pH = 9
Sodium (metal)	Burns with a yellow flame	Sodium oxide, Na_2O (a yellow-white solid)	Dissolves readily to form a strongly alkaline solution, pH = 10
Carbon (non-metal)	Glows red	Carbon dioxide, CO_2, (an invisible gas)	Dissolves slightly to give a weakly acidic solution, pH = 4
Phosphorus (non-metal)	Burns with a yellow flame	Phosphorus(V) oxide, P_2O_5, (a white solid)	Dissolves to give a strongly acidic solution, pH = 2
Sulphur (non-metal)	Burns with a blue flame	Sulphur dioxide, SO_2, (a fuming gas with a choking smell)	Dissolves readily to form a strongly acidic solution, pH = 2

15.6 Oxides

A pattern can be seen in the characteristics of oxides. The oxides of metallic elements are **bases**. The bases which dissolve in water are called **alkalis**. Most of the oxides of non-metallic elements are acids, but some are neutral. Acids react with bases to form **salts**.

LOOK AT LINKS
for **salts**
See Topic 13.

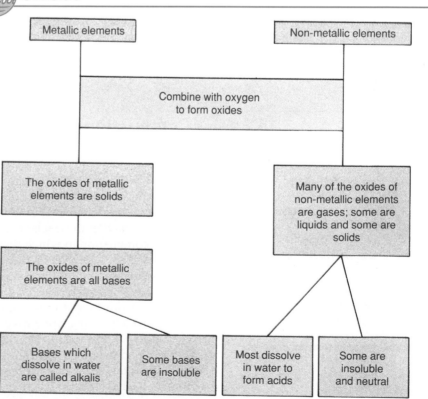

Figure 15.6A ● Oxides

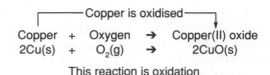

When an element combines with oxygen, we say it has been **oxidised**. Oxygen **oxidises** copper to copper(II) oxide. This reaction is an **oxidation**.

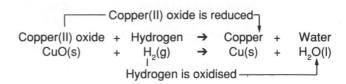

┌─────── Copper is oxidised ───────┐

Copper + Oxygen → Copper(II) oxide

$2Cu(s)$ + $O_2(g)$ → $2CuO(s)$

This reaction is oxidation

The opposite of oxidation is **reduction**. When a substance loses oxygen, it is **reduced**. Copper(II) oxide can be reduced by heating it and passing hydrogen over it.

┌──── Copper(II) oxide is reduced ┐

Copper(II) oxide + Hydrogen → Copper + Water

$CuO(s)$ + $H_2(g)$ → $Cu(s)$ + $H_2O(l)$

└──── Hydrogen is oxidised ────┘

You can see from the equation that oxidation and reduction occur together. Copper(II) oxide is reduced to copper while hydrogen is oxidised to water. Copper(II) oxide is called an **oxidising agent** because it gives oxygen to hydrogen. Hydrogen is called a **reducing agent** because it takes oxygen from copper(II) oxide.

SUMMARY

Most elements combine with oxygen to form oxides. Many elements burn in oxygen. The oxides of metals are basic. The oxides of non-metallic elements are acidic or neutral. An oxidising agent gives oxygen to another substance. A reducing agent takes oxygen from another substance.

CHECKPOINT

❶ (a) Describe two differences in the physical characteristics of metallic and non-metallic elements (see Topic 5.7 also).

    (b) State two differences between the chemical reactions of metallic and non-metallic elements (see Topic 12.1 also).

❷ Write word equations and balanced chemical equations for the combustion in oxygen of (a) sulphur (b) carbon (c) magnesium (d) sodium.

15.7  Combustion

LOOK AT LINKS

for the **combustion of food**
See Topic 14.6.

LOOK AT LINKS
for **petroleum oil**
See Topic 11.8.

In many oxidation reactions, energy is given out. The fireworks called 'sparklers' are coated with iron filings. When the iron is oxidised to iron oxide, you can see that energy is given out in the form of heat and light. An oxidation reaction in which energy is given out is called a combustion reaction. A combustion in which there is a flame is described as burning. Substances which undergo combustion are called fuels.

Daily, we make use of the combustion of fuels. In respiration, the **combustion of foods** provides us with energy. We use fuels to heat our homes, to cook our food, to run our cars and to generate electricity. Many of the fuels which we use are derived from petroleum oil. Petrol (used in motor vehicles), kerosene (used in aircraft and as domestic paraffin), diesel fuel (used in lorries and trains) and natural gas (used in gas cookers) are obtained from **petroleum oil**. These fuels are mixtures of **hydrocarbons**. Hydrocarbons are compounds of carbon and hydrogen

LOOK AT LINKS
for **carbon monoxide**
see Topic 17.6.
A burning cigarette
produces some carbon
monoxide.
See Topic 17.12.

only. It is important to know what products are formed when these fuels burn. Figure 15.7A shows how you can test the products of combustion of kerosene which is burned in paraffin heaters. **Do not use petrol in this apparatus**. You can burn a candle instead of kerosene. Candle wax is another hydrocarbon fuel obtained from crude oil.

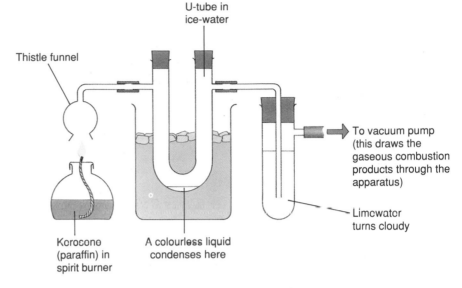

Figure 15.7A ● How to test the combustion products of a hydrocarbon fuel, e.g. kerosene or candle wax (*NOTE: do not use petrol*)

SUMMARY

The combustion of fuels is an oxidation reaction. The combustion of hydrocarbon fuels is a vital source of energy in our economy. These fuels burn to form carbon dioxide and water. If the supply of air is insufficient, the combustion products include carbon monoxide (a poisonous gas) and carbon (soot).

The combustion products are carbon dioxide and water. Other hydrocarbon fuels give the same products.

Hydrogen + Oxygen → Carbon dioxide + Water vapour

SUMMARY

Oxidation is the addition of oxygen to a substance. Combustion is oxidation with the release of energy. Burning is combustion accompanied by a flame. Respiration is combustion which takes place in living tissues. In respiration, food materials are oxidised in the cells with the release of energy.

If you burn a candle in this apparatus, you will see a deposit of carbon (soot) in the thistle funnel. This happens when the air supply is insufficient to oxidise all the carbon in the hydrocarbon fuel to carbon dioxide, CO_2. Another product of incomplete combustion is the poisonous gas carbon monoxide, CO. Because you cannot see or smell carbon monoxide, it is doubly dangerous. Many times, people have been poisoned by carbon monoxide while running a car engine in a closed garage. The engine could not get enough oxygen for complete combustion to occur. The exhaust gases from petrol engines always contain some carbon monoxide, some unburnt hydrocarbons and some soot, in addition to the harmless products: carbon dioxide and water.

CHECKPOINT

❶ (a) What type of compound is present in petrol?
  (b) What products are formed in combustion (i) if there is plenty of air and (ii) if there is a limited supply of air?

❷ In February 1988 newspapers carried a report of a woman who fell asleep in front of a fire and never woke up. Later, workmen removed three buckets full of birds' nesting materials from the chimney. What do you think had caused the woman's death?

❸ Why should you make sure the window is open if you use a gas heater in the bathroom?

Fire-extinguishers

> Sometimes fires get out of control and methods of extinguishing them are important.

The fire triangle in Figure 15.8A illustrates the three things which a fire needs: fuel, oxygen and heat. If one of these three is removed, the fire goes out.

Firefighters deal with four types of fire:

- **Class A** Materials such as wood, paper, cloth and plastics
- **Class B** Flammable liquids and gases such as petrol, cooking oil and natural gas
- **Class C** Electrical equipment
- **Class D** Metals

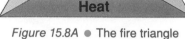

Figure 15.8A ● The fire triangle

● **Class A fires**

These fires are usually tackled by removing the heat side of the fire triangle. For every fuel, there is an ignition temperature. Below this temperature the fuel will not burn. Directing water on to the burning material cools it until it is below its ignition temperature.

Figure 15.8B shows a water fire-extinguisher. It contains a small cartridge of carbon dioxide under pressure. When someone operates the lever, the cartridge is punctured, and carbon dioxide is released. The pressure of the gas forces a powerful jet of water out of the nozzle.

In the older soda–acid extinguishers, carbon dioxide is generated in the cylinder from the reaction of an acid and a carbonate. Foam extinguishers contain a foam-stabiliser as well as water. Out of the nozzle comes a foam of water and bubbles of carbon dioxide. The foam-stabiliser stops the bubbles dispersing.

When a person's clothing is on fire, the best thing to do is to wrap them in a fire blanket. A fire blanket, which is usually made from glass fibre, keeps out air. If there is no fire blanket handy, you should lie the person on the floor and wrap a carpet or rug round them.

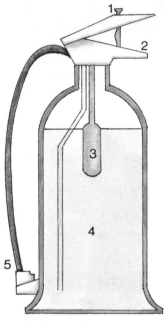

1 Remove the safety pin
2 Squeeze the lever
3 This action pierces the cartridge of pressurised carbon dioxide
4 The steel cylinder contains water
5 The pressure of carbon dioxide forces a jet of water out of the nozzle

Figure 15.8B ● A water fire-extinguisher

● **Class B fires**

Water is not used on burning gases and liquids. Burning petrol or oil floats on water, and the water makes things worse by spreading the fire. This makes fires at sea especially dangerous. A chip pan fire is easier to deal with (see Figure 15.8C).

No! Don't use water!
Switch off the gas or electricity
Wet a towel. Wring it out
Put the damp towel over the pan
Wait. Don't remove until the fire has been out for some time

Figure 15.8C ● How to put out a chip pan fire

Figure 15.8D ● A fireman demonstrates how using the wrong kind of extinguisher can make the problem worse – in this case he is using a water extinguisher on an oil fire

To extinguish burning oil or petrol, the oxygen side of the fire triangle must be removed. One way is to use a carbon dioxide extinguisher. Carbon dioxide does not support combustion. It is denser than air and a layer of carbon dioxide will form a 'blanket' over the fire and keep out oxygen. Starved of oxygen, the fire goes out.

Powder extinguishers can be used on burning liquids. They contain a powder, such as sodium hydrogencarbonate, which is decomposed by heat to release carbon dioxide. If this powder is thrown on to a fire, it generates carbon dioxide at the base of the fire, and is very effective. Workers on drilling rigs light gas flares. When they want to extinguish a flare, they tip a large quantity of dry powder extinguisher on to it.

Figure 15.8E ● Don't use water

● Class C fires

Electrical fires pose a problem. They cannot be extinguished with water extinguishers. If you direct a jet of water on to a smouldering piece of electrical equipment, a current of electricity can pass through the jet of water and give you an electric shock. You should switch off at the mains and use one of the carbon dioxide extinguishers.

● Class D fires

Burning metals are difficult fires to fight. Some metals, such as sodium and magnesium, react with cold water. When they are hot enough, iron and steel react with water. The reaction between metals and water produces the flammable gas hydrogen, which burns in air with an explosion. This is why firefighters never use water on burning metals.

On some metal fires, powder extinguishers can be used, but some metals react with carbon dioxide. There are gases other than carbon dioxide which do not support combustion and are denser than air. BCD extinguishers contain bromochlorodifluoromethane, $CBrClF_2$. On hot surfaces this liquid forms a dense vapour which excludes air. It is chemically unreactive, even at high temperatures.

SUMMARY

A fire needs fuel, oxygen and heat. Many fire-extinguishers work by cooling the fire. Others use carbon dioxide to exclude air. Metal fires are often extinguished with an inert liquid, e.g. BCD.

CHECKPOINT

❶ How would you put out these fires?
(a) A waste paper basket fire.
(b) A smouldering television set.
(c) A chip pan fire.
(d) A person whose lab coat was on fire.
(e) A child whose nightdress was on fire.

❷ What special difficulties do metal fires present to firefighters? How are metal fires extinguished?

❸ What special difficulties do fires in oil rigs present? How are they extinguished?

❹ A small fire in a school laboratory can often be put out with dry sand. Explain how this method works.

15.9  Rusting

Rusting is a costly nuisance. Methods of slowing down the process can save a lot of money.

Test whether iron nails are protected from rusting by (a) a coat of paint (b) a layer of grease.

Obtain some galvanised iron nails. Do an experiment to test whether they rust more slowly in brine than ordinary nails do.

LOOK AT LINKS
The important aspect of this work on rusting is to devise methods of preventing or at least slowing down the process. Methods of delaying rusting are described in Topic 19.11.

RESOURCE
–ACTIVITY–
PACK

SUMMARY

Rusting is the oxidation of iron and steel to iron(III) oxide.

TRY THIS

● **Iron and rust**

Many metals become corroded by exposure to the air. The corrosion of iron and steel is called rusting. Rust is the reddish brown solid, hydrated iron(III) oxide, $Fe_2O_3.nH_2O$ (The number of water molecules, n, varies.)

Rusting is an oxidation reaction:

$$\text{Iron} + \text{Oxygen} \rightarrow \text{Iron(III) oxide}$$
$$4Fe(s) + 3O_2(g) \rightarrow 2Fe_2O_3(s)$$

Rusting is a nuisance. Cars, ships, bridges, machines and other costly items made from iron and steel rust. To prevent rusting, or at least to slow it down, saves a lot of money. Before you can prevent rusting, you first have to know what conditions speed up rusting. Figure 15.9A shows some experiments on the rusting of iron nails.

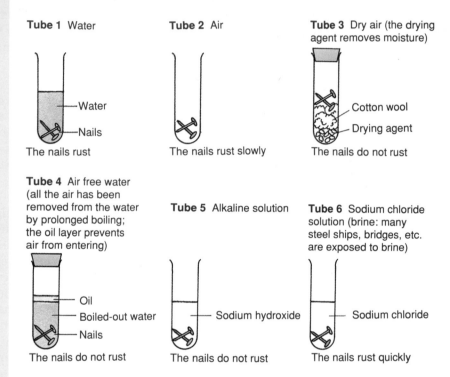

Tube 1 Water
—Water
—Nails
The nails rust

Tube 2 Air
The nails rust slowly

Tube 3 Dry air (the drying agent removes moisture)
Cotton wool
Drying agent
The nails do not rust

Tube 4 Air free water (all the air has been removed from the water by prolonged boiling; the oil layer prevents air from entering)
—Oil
—Boiled-out water
—Nails
The nails do not rust

Tube 5 Alkaline solution
—Sodium hydroxide
The nails do not rust

Tube 6 Sodium chloride solution (brine: many steel ships, bridges, etc. are exposed to brine)
—Sodium chloride
The nails rust quickly

Figure 15.9A ● Experiments on the rusting of iron nails

The experiments show that in order to rust, iron needs air and water and a trace of acid. The carbon dioxide in the air provides sufficient acidity. Rusting is accelerated by salt. Bridges and ships are exposed to brine, and cars are exposed to the salt that is spread on the roads in winter. It is obviously very important to find ways of rust-proofing these objects.

Devise experiments which you could do to answer the questions. If possible check your ideas with your teacher, and try them out. (Be careful to use iron nails, not rust-proofed nails, and don't forget to set up control experiments.)
(a) Does iron increase in mass when it rusts?
(b) Do salts other than sodium chloride speed up rusting?
(c) Does steel wool rust more quickly or more slowly than iron nails?

TOPIC 16 WATER

16.1  The water cycle

Where does all the rain come from? Why does the atmosphere never run out of water? Four-fifths of the world's surface is covered by water. From oceans, rivers and lakes, water evaporates into the atmosphere. Plants give out water vapour in **transpiration**. As it rises into a cooler part of the atmosphere, water vapour condenses to form clouds of tiny droplets. If the clouds are blown upward and cooled further, larger drops of water form and fall to the ground as rain (or snow). *Where does the rain go?* Rain water trickles through soil, where some is taken up by plants. The rest passes through porous rocks to become part of rivers, lakes, ground water and the sea. This chain of events is called the **water cycle** (see Figure 16.1A).

The first living things evolved in water. As more complex plants and animals evolved, water remained essential for life.

LOOK AT LINKS
for **transpiration**
See *KS: Biology*, Topic 14.1.

IT'S A FACT

A large tree can lose 300 litres of water vapour in an hour by transpiration.

LOOK AT LINKS
for **clouds**
See *KS: Physics*, Topics 3.4 and 3.5.

LOOK AT LINKS
for **acid rain**
See Topic 17.5.

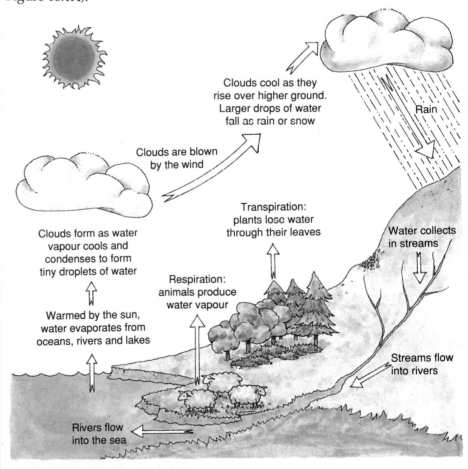

Clouds cool as they rise over higher ground. Larger drops of water fall as rain or snow

Rain

Clouds are blown by the wind

Transpiration: plants lose water through their leaves

Water collects in streams

Clouds form as water vapour cools and condenses to form tiny droplets of water

Respiration: animals produce water vapour

Streams flow into rivers

Warmed by the sun, water evaporates from oceans, rivers and lakes

Rivers flow into the sea

Figure 16.1A ● The water cycle

SUMMARY

The water cycle:
• Evaporation, transpiration and respiration send water vapour into the atmosphere.
• Condensation forms clouds which return water to the Earth as rain, hail or snow.

As rain water falls through the air, it dissolves oxygen, nitrogen and carbon dioxide. The dissolved carbon dioxide forms a solution of the weak acid, carbonic acid. Natural rain water is therefore weakly acidic. In regions where the air is polluted, rain water may dissolve sulphur dioxide and oxides of nitrogen, which make it strongly acidic; it is then called **acid rain**. As rain water trickles through porous rocks, it dissolves salts from the rocks. The dissolved salts are carried into the sea. When sea water evaporates, the salts remain behind.

16.2 Dissolved oxygen

Oxygen sensors connected to a computer are used to monitor the concentration of dissolved oxygen in commercial fish tanks. Fish farmers and breeders are warned when the concentration starts dropping and can take preventive action before any damage is done.

SUMMARY

Air dissolves in water. The dissolved oxygen in water keeps fish alive. If too much organic matter, e.g. sewage, is discharged into a river, the dissolved oxygen is used up in oxidising the organic matter, and the fish die.

The fact that oxygen dissolves in water is vitally important. The solubility is low: water can dissolve no more than 10 g oxygen per tonne of water, that is 10 p.p.m. (parts per million). This is high enough to sustain fish and other water-living animals and plants. When the level of dissolved oxygen falls below 5 p.p.m. aquatic plants and animals start to suffer.

Water is able to purify itself of many of the pollutants which we pour into it. Bacteria which are present in water feed on plant and animal debris. These bacteria are **aerobic** (they need oxygen). They use dissolved oxygen to oxidise organic material (material from plants and animals) to harmless products, such as carbon dioxide and water. This is how the bacteria obtain the energy which they need to sustain life. If a lot of untreated sewage is discharged into a river, the dissolved oxygen is used up more rapidly than it is replaced, and the aerobic bacteria die. Then **anaerobic** bacteria (which do not need oxygen) attack the organic matter. They produce unpleasant-smelling decay products.

Some synthetic (manufactured) materials, e.g. plastics, cannot be oxidised by bacteria. These materials are nonbiodegradable, and they last for a very long time in water.

16.3 Water treatment

The earliest human settlements were always beside rivers. The settlers needed water to drink and used the river to carry away their sewage and other waste. Obtaining clean water is more difficult now.

SCIENCE AT WORK

The water industry makes use of computers. Sensors detect the pH, oxygen concentration and other qualities of the water and relay the measurements to a computer. This constant monitoring enables the industry to control the quality of the water it provides.

IT — *Water Treatment* (program)

This program allows you to control and maintain a town's domestic water supply.

The water that we use is taken mainly from lakes and rivers. Water treatment plants purify the water to make it safe to drink. They do this by:
• filtration to remove solid matter followed by...
• bacterial oxidation to get rid of organic matter and...
• treatment with chlorine to kill germs.

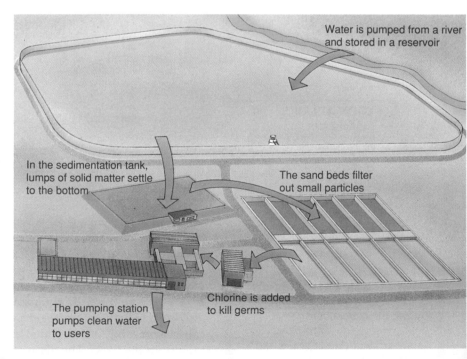

Water is pumped from a river and stored in a reservoir

In the sedimentation tank, lumps of solid matter settle to the bottom

The sand beds filter out small particles

The pumping station pumps clean water to users

Chlorine is added to kill germs

Figure 16.3A ● A water treatment works

SUMMARY

Water treatment works take water from lakes and rivers. After filtration, followed by chlorination, the water is safe to drink.

In some areas, the water supply comes from ground water (water held underground in porous layers of rock). As rain water trickles down from the surface through porous rocks, the solid matter suspended in it is filtered out. Ground water therefore does not need the complete treatment. It is pumped out of the ground and chlorinated before use.

Who's behind the science

In 1854, 50 000 people died of cholera in London. Dr John Snow did some scientific detective work to find the source of the disease. He marked the deaths from cholera and the positions of the street pumps from which people obtained their water on a map of London. Dr Snow came to the conclusion that one of the pumps was supplying contaminated water. *Which was it?* He tested his theory by removing the handle of the pump. There was a sudden fall in the number of people getting cholera. The pump water had been contaminated by sewage leaking into it.

FIRST THOUGHTS

16.4 Sewage works

Rivers carry away our sewage, our industrial grime, the waste chemicals from our factories and the waste heat from our power stations. Sewage works try to ensure that rivers are not overloaded with waste.

Homes, factories, businesses and schools all discharge their used water into sewers which take it to a sewage works. There, the dirty water is purified until it is fit to be discharged into a river (see Figure 16.4A). The river dilutes the remaining pollutants and oxidises some of them. The digested sludge obtained from a sewage works can be used as a fertiliser. Raw sewage cannot be used as fertiliser because it contains harmful bacteria.

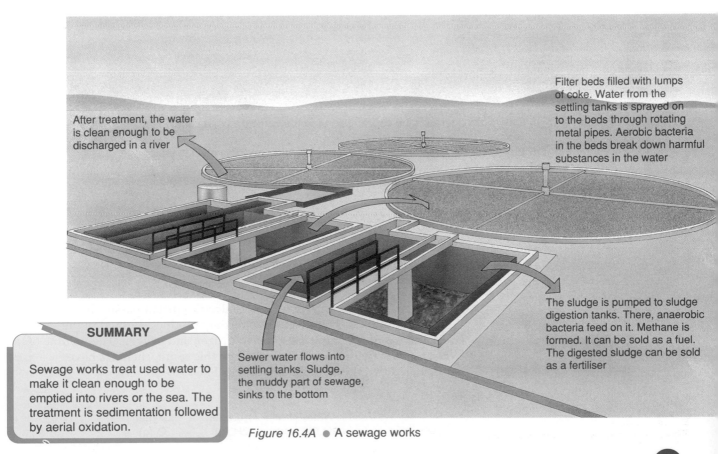

After treatment, the water is clean enough to be discharged in a river

Filter beds filled with lumps of coke. Water from the settling tanks is sprayed on to the beds through rotating metal pipes. Aerobic bacteria in the beds break down harmful substances in the water

The sludge is pumped to sludge digestion tanks. There, anaerobic bacteria feed on it. Methane is formed. It can be sold as a fuel. The digested sludge can be sold as a fertiliser

Sewer water flows into settling tanks. Sludge, the muddy part of sewage, sinks to the bottom

SUMMARY

Sewage works treat used water to make it clean enough to be emptied into rivers or the sea. The treatment is sedimentation followed by aerial oxidation.

Figure 16.4A ● A sewage works

16.5 Uses for water

| Washing and baths 50 litres | Lavatory 50 litres | Laundry 15 litres | Washing up 15 litres | Cooking 5 litres | Gardening 5 litres | Waste (dripping taps, leaking pipes) 20 litres |

Figure 16.5A ● Water: 160 litres a day

The uses shown in Figure 16.5A show only 10% of the total amount of water you use. The other 90% is used:
- to grow your food (agricultural use),
- to make your possessions (industrial use as a solvent, for cleaning and for cooling),
- to generate electricity (used as a coolant in power stations).

The total water consumption in an industrialised country amounts to around 80 000 litres (80 tonnes) a year per person. The manufacture of:
- 1 tonne of steel uses 45 tonnes of water,
- 1 tonne of paper uses 90 tonnes of water,
- 1 tonne of nylon uses 140 tonnes of water,
- 1 tonne of bread uses 4 tonnes of water,
- 1 motor car uses 450 tonnes of water,
- 1 litre of beer uses 10 litres of water.

Water used for many industrial purposes is purified and recycled.

> IT'S A FACT
>
> An industrial country uses about 80 tonnes of water per person per year.

CHECKPOINT

❶ The table shows the world consumption of water over the past 30 years.

Year	World consumption of water (millions of tonnes per day)
1960	10.0
1970	11.5
1975	13.0
1980	15.0
1985	17.0
1990	20.0

(a) On graph paper, plot the consumption (on the vertical axis) against the year (on the horizontal axis).
(b) Say what has happened to the demand for water over the past 30 years.
(c) Suggest three reasons for the change.
(d) From your graph, predict what the consumption of water will be in the year 2000.

16.6 Water: the compound

> LOOK AT LINKS
> for **electrolysis**
> See Topic 9.

Water is a compound. When a direct electric current passes through it, water splits up: it is electrolysed. The only products formed in the electrolysis of water are the gases hydrogen and oxygen. The volume of hydrogen is twice that of oxygen. From this result, chemists have calculated that the formula for water is H_2O.

$$\text{Water} \xrightarrow{\text{electrolyse}} \text{Hydrogen} + \text{Oxygen}$$
$$2H_2O(l) \rightarrow 2H_2(g) + O_2(g)$$

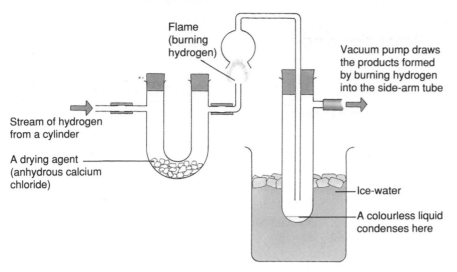

Flame (burning hydrogen)

Vacuum pump draws the products formed by burning hydrogen into the side-arm tube

Stream of hydrogen from a cylinder

A drying agent (anhydrous calcium chloride)

Ice-water

A colourless liquid condenses here

Water is the oxide of hydrogen. *Can it be made by the combination of hydrogen and oxygen?* Figure 16.6A shows an experiment to find out what forms when you burn hydrogen in air. The only product is a colourless liquid. You can test this liquid to see whether it is water.

Figure 16.6A ● What is formed when hydrogen burns in air?

LOOK AT LINKS
In Topic 3.5, Figures 3.5A, 3.5B and 3.5E show how to find melting point and boiling point.

SUMMARY

Tests for water:
• Turns anhydrous copper(II) sulphate from white to blue.
• Turns anhydrous cobalt(II) chloride from blue to pink.
Tests for pure water:
• Boiling point = 100 °C at 1 atm
• Freezing point = 0 °C at 1 atm
Water is formed when hydrogen burns in air.

● *Tests for water*
• Water turns white anhydrous copper(II) sulphate blue.

Copper(II) sulphate + Water → Copper(II) sulphate-5-water
$CuSO_4(s)$ + $5H_2O(l)$ → $CuSO_4.5H_2O(s)$
(white solid) (blue solid)

• Water turns blue anhydrous cobalt(II) chloride pink.

Cobalt(II) chloride + Water → Cobalt(II) chloride-6-water
$CoCl_2(s)$ + $6H_2O(l)$ → $CoCl_2.6H_2O(s)$
(blue solid) (pink solid)

Any liquid which contains water will give positive results in these tests. To find out whether a liquid is pure water, you can find its boiling point and freezing point. At 1 atm, pure water boils at 100 °C and freezes at 0 °C.

The tests show that the liquid formed when hydrogen burns in air is in fact water.

Hydrogen + Oxygen → Water
$2H_2(g)$ + $O_2(g)$ → $2H_2O(l)$

16.7  Pure water

LOOK AT LINKS
for **solubility**
See Topic 3.6.

SUMMARY

Water is a good solvent. The presence of a solute raises the boiling point and lowers the freezing point. Pure water is obtained by distillation.

Almost all substances dissolve in water to some extent: that is, water is a good **solvent**. Since water is such a good solvent, it is difficult to obtain pure water. Distillation is one method of purifying water. In some countries, distillation is used to obtain drinking water from sea water. The technique is called desalination (desalting). Hong Kong has a large desalination plant which has never been used because the cost of importing the oil needed to run it is so high. Saudi Arabia and Bahrain operate desalination plants. *Why do you think they need the plants and can afford to run them?*
When chemists describe water as pure, they mean that the water contains no dissolved material. This is different from what a water company means by pure water: they mean that the water contains no harmful substances. Safe drinking water contains dissolved salts. Water which contains substances that are bad for health is **polluted** water.

CHECKPOINT

These questions will enable you to revise solubility curves. Use the figure opposite to help you.

❶ One kilogram of water saturated with potassium chloride is cooled from 80 °C to 20 °C. What mass of potassium chloride crystallises out?

❷ One kilogram of water saturated with sodium chloride is cooled from 80 °C to 20 °C. What mass of sodium chloride crystallises out?

❸ Dissolved in 100 g of water at 100 °C are 30 g of sodium chloride and 50 g of potassium chloride. What will happen when the solution is cooled to 20 °C?

❹ Dissolved in 100 g of water at 80 °C are 15 g of potassium sulphate and 70 g of potassium bromide. What will happen when the solution is cooled to 20 °C?

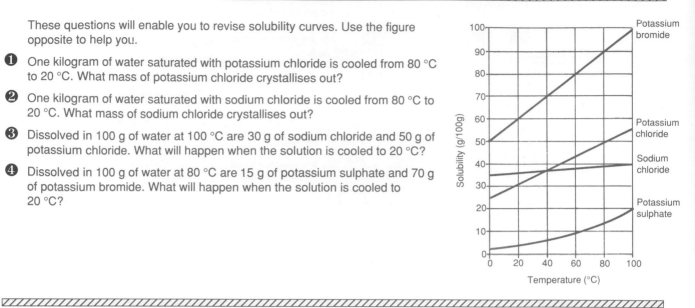

16.8 Underground caverns

LOOK AT LINKS
for **carbonates**
See Topic 12.1.

In limestone regions, rain water trickles over rocks composed of calcium carbonate (limestone) and magnesium carbonate. These carbonates do not dissolve in pure water, but they react with acids. The carbon dioxide dissolved in rain water makes it weakly acidic. It reacts with the carbonate rocks to form the soluble salts, calcium hydrogencarbonate and magnesium hydrogencarbonate.

Calcium carbonate + Water + Carbon dioxide → Calcium hydrogencarbonate
   (limestone)   solution
   $CaCO_3(s)$ + $H_2O(l)$ + $CO_2(g)$ → $Ca(HCO_3)_2(aq)$

LOOK AT LINKS
for **caves**
How the landscape was shaped is discussed in Topic 2.2.

Figure 16.8A ● A cavern at Wookey Hole (note the stalactites and stalagmites)

This chemical reaction is responsible for the formation of the underground **caves** and potholes which occur in limestone regions. Over thousands of years, large masses of carbonates have been dissolved out of the rock (see Figure 16.8A).

The reverse reaction can take place. Sometimes, in an underground cavern, a drop of water becomes isolated. With air all round it, water will evaporate. The dissolved calcium hydrogencarbonate turns into a grain of solid calcium carbonate.

$$\text{Calcium hydrogencarbonate} \rightarrow \text{Calcium carbonate} + \text{Water} + \text{Carbon dioxide}$$
$$Ca(HCO_3)_2(aq) \rightarrow CaCO_3(s) + H_2O(l) + CO_2(g)$$

Slowly, more grains of calcium carbonate are deposited. Eventually, a pillar of calcium carbonate may have built up from the floor of the cavern. This is called a **stalagmite**. The same process can lead to the formation of a **stalactite** on the roof of the cavern.

16.9 Soaps

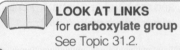
FIRST THOUGHTS

Have you ever tried to wash greasy hands without using soap? The problem is that grease and water do not mix. You can find out how soap solves the problem in this section.

LOOK AT LINKS
for **carboxylate group**
See Topic 31.2.

Soaps are able to form a bridge between grease and water. They are the sodium and potassium salts of organic acids. One soap is sodium hexadecanoate, $C_{15}H_{31}CO_2Na$. A model of the soap is shown in Figure 16.9A.
(Hexadecane means sixteen. Count up. *Are there 16 carbon atoms?*)
It consists of a sodium ion and a hexadecanoate ion, which we will call a *soap* ion for short. The *soap* ion has two parts (see Figure 16.9B). The head, which is attracted to water, is a —CO_2 group (a **carboxylate** group). The tail, which is repelled by water and attracted by grease, is a long chain of —CH_2— groups. Figure 16.9C shows how *soap* ions wash grease from your hands.

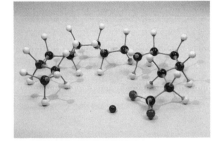
Figure 16.9A ● A model of the soap, sodium hexadecanoate

$$CH_3\ CH_2\ CH_2\ CH_2\ CH_2\ CH_2\ CH_2\ CH_2\ CH_2\ CH_2\ CH_2\ CH_2\ CH_2\ CH_2\ CH_2 \overset{O}{\underset{\|}{-C}}-O^- \ Na^+$$

Figure 16.9B ● A *soap* ion

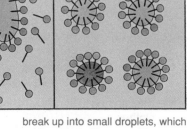

Figure 16.9C ● The cleansing action of soap

1 The tails of the *soap* ions begin to dissolve in the grease. The heads remain dissolved in the water
2 The negatively charged heads of the *soap* ions repel one another; this repulsion makes the grease

break up into small droplets, which are suspended in water. The soap has **emulsified** the grease and water (made them mix)
3 The emulsified grease is washed away by water

● *Manufacture of soap*

Soaps are made by boiling together animal fats or vegetable oils and a strong alkali, e.g. sodium hydroxide. The reaction is called **saponification** (soap-making).

Fat (or oil) + Sodium hydroxide → Soap + Glycerol

When sodium chloride is added to the mixture, soap solidifies. The product is purified to remove alkali from it. Perfume and colouring are added before the soap is formed into bars.

16.10 Soapless detergents

LOOK AT LINKS
Soapless detergents have some advantages over soaps.
See Topic 16.12.

Which is better for the condition of your hair, washing with soap or mild soapless detergent (shampoo)? Think up an experiment to find out. **Do not** experiment on your own hair: obtain samples of hair from a local hairdresser. Obtain your teacher's permission before you carry out your experiment.

Soaps are one type of detergent (cleaning agent). There is another type of detergent known as **soapless detergents**. Many washing powders and household cleaning fluids are soapless detergents. Often they are referred to simply as 'detergents'. Soapless detergents are made from petroleum oil. They are the sodium salts of sulphonic acids (see Figure acids (see Figure 16.10A).

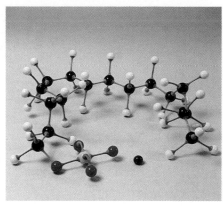

Figure 16.10A ● A model of a soapless detergent (note the tail, a chain of —CH_2— groups which dissolves in grease, and the head, a sulphate group, —SO_4^-, which dissolves in water)

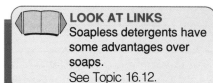

$CH_3\ CH_2\ CH_2\ CH_2\ CH_2\ CH_2\ CH_2\ CH_2\ CH_2\ CH_2\ CH_2\ CH_2\ —O—S—O^-\ Na^+$

Figure 16.10B ● A soapless detergent

SUMMARY

Washing powders contain soapless detergent.

Soapless detergents are very good at removing oil and grease. They are too powerful for use on the skin, and the gentler action of a soap is better. Shampoos are mild detergents.

CHECKPOINT

❶ Figure 22.10B shows the formula of a soapless detergent. Say which part of the ion will attach itself to grease and which part will remain in the water.

❷ Explain how a soapless detergent is able to dislodge grease from dirty clothes.

❸ Why is it important that the water in a washing machine is agitated?

❹ Why is it important to rinse clothes well after washing them?

16.11 Bleaches and alkaline cleaners

Bleaches

Many household bleaches contain chlorine compounds, e.g. sodium chlorate(I), NaClO. They are powerful oxidising agents, killing germs and oxidising dirt. You should not use bleaches together with other cleaning agents. An acid will react with sodium chlorate(I) to liberate the poisonous gas chlorine.

Alkaline cleaners

Many household cleaners are alkalis. They react with grease and oil to form an emulsion of glycerol and soap, which can be washed away. The reaction is saponification. Sodium hydroxide, NaOH, is used in oven-cleaners; sodium carbonate, Na_2CO_3, is used in washing powders, and ammonia is used in solution as a household cleaner. You should not use ammonia together with a bleach because they can react to form poisonous chloroamines.

SUMMARY

Household bleaches are chlorine compounds. They work by oxidising dirt and germs. It is not safe to use a bleach together with an acid. Alkaline cleaners work by saponifying grease and oil. It is not safe to use ammonia and a bleach together.

CHECKPOINT

Oven-cleaners contain sodium hydroxide. A greasy oven is wiped with a pad of oven-cleaner and left for a few minutes. The grease can then be washed off with water.

❶ Explain how sodium hydroxide makes it easier to remove grease.

❷ Explain why you should wear rubber gloves when you use this kind of oven-cleaner.

❸ What effect does it have on the cleaning job if you warm the oven first?

16.12 Hard water and soft water

In some parts of the country, the tap water is described as **hard water**. This means that it is hard to get a lather with soap. Instead of forming a lather, soap forms an insoluble scum. Water in which soap lathers easily is **soft water**. Hard water contains soluble calcium and magnesium salts. They combine with *soap* ions to form insoluble calcium and magnesium compounds. These compounds are the insoluble scum that floats on the water.

Soap ions	+	Calcium ions	➔	Scum
(in solution)		(in solution)	➔	(insoluble solid)

If you go on adding soap, eventually all the calcium ions and magnesium ions will be precipitated as scum. After that, the soap will be able to work as a cleaning agent.

Soapless detergents are able to work in hard water because their calcium and magnesium salts are soluble. For many purposes, people prefer soapless detergents to soaps. Sales of soapless detergents are four times as high as those of soaps.

16.13 Methods of softening hard water

Temporary hardness

Figure 16.13A ● Scale on a kettle element

Hardness which can be removed by boiling is called temporary hardness. Temporarily hard water contains dissolved calcium hydrogencarbonate and magnesium hydrogencarbonate, and these compounds decompose when the water is boiled. The resulting water is soft water.

Calcium hydrogencarbonate → Calcium carbonate + Carbon dioxide + Water
$$Ca(HCO_3)_2(aq) \rightarrow CaCO_3(s) + CO_2(g) + H_2O(l)$$

A deposit of calcium carbonate and magnesium carbonate forms. This is the **scale** which is deposited in kettles and water pipes.

Permanent hardness

Hardness which cannot be removed by boiling is called permanent hardness. It is caused by dissolved chlorides and sulphates of calcium and magnesium. These compounds are not decomposed by heat.

Washing soda

Washing soda is sodium carbonate-10-water. It can soften both temporary and permanent hardness. Washing soda precipitates calcium ions and magnesium ions as insoluble carbonates.

Calcium ions + Carbonate ions → Calcium carbonate
$$Ca^{2+}(aq) + CO_3^{2-}(aq) \rightarrow CaCO_3(s)$$

Exchange resins

Ion exchange resins are substances which take ions of one kind out of aqueous solution and replace them with ions of a different kind. Permutits are manufactured ion exchange resins. They replace calcium and magnesium ions in water by sodium ions.

Calcium ions + Sodium permutit → Sodium ions + Calcium permutit

16.14 Advantages of hard water

SUMMARY

- Temporary hardness is removed by boiling.
- Permanent hardness is removed by adding sodium carbonate (washing soda) or by running water through an exchange resin.
- Hard water is better than soft water for drinking.

Hard water has some advantages over soft water for health reasons. The **calcium** compounds in hard water strengthen bones and teeth. The calcium content is also beneficial to people with a tendency to develop heart disease.

Some industries prefer hard water. The leather industry prefers to cure leather in hard water. The brewing industry likes hard water for the taste which the dissolved salts give to the beer.

CHECKPOINT

❶ Gwen washes her hair in hard water. Which kind of shampoo would you advise her to choose: a mild soapless detergent or a soap? Explain your advice.

❷ (a) Explain the difference between hard and soft water.
  (b) Why is drinking hard water better for health than drinking soft water?
  (c) Which solutes make water hard? Explain how the substances you mention get into tap water.
  (d) Name a use for which soft water is preferred to hard water. Explain why.
  (e) Describe one method of softening hard water. Explain how it works.
  (f) Why are detergents preferred to soaps for use in hard water?
  (g) Why is it better to use distilled water rather than tap water in a steam iron?

❸ The table gives some information on three brands of shampoo.

Brand	Price of bottle (p)	Volume (cm³)
Soffen	40	204
Sheeno	50	350
Silken	60	480

  (a) Which brand is sold in the smallest bottle?
  (b) Calculate what volume of shampoo (in cm³) you get for 1p if you buy
    (i) Soffen (ii) Sheeno and (iii) Silken. Say which shampoo is the cheapest.
  (c) Suggest three things which a person might consider, other than price, when choosing a shampoo.
  (d) Describe an experiment you could do to find out which of the shampoos is best at producing a lather. Mention any steps you would take to make sure the test was fair.

❹ Some washing powders contain enzymes. Zenab decides to test whether the washing powder Biolwash, which contains an enzyme, washes better than Britewash, which does not. Zenab decides to use 1 g of washing powder in 100 cm³ of warm water and to do her tests on squares of cotton fabric.
  (a) Suggest some everyday substances which stain cloth and which would be interesting to experiment on.
  (b) Describe how Zenab could do a fair test to compare the washing action of Biolwash and Britewash. What factors must be kept the same in the two experiments?
  (c) Zenab finds that Biolwash washes better than Britewash on many stains. Another student, Ahmed, did his tests at 80 °C and found that Britewash gave a cleaner result than Biolwash. Can you explain the difference between Zenab's and Ahmed's results? For enzymes, see Topic 31.6.

16.15 Colloids

The bluish haze of tobacco smoke and the brilliant sunsets in deserts – what do they have in common? The answer is the scattering of light by **colloidal particles** suspended in air. In this topic you will find out what colloidal particles are.

LOOK AT LINKS
For **filtration** and **centrifugation**
See Topics 11.4 and 11.5.

LOOK AT LINKS
Colloid chemistry is important in the study of biological systems.
For colloidal dispersions in cells
See *KS: Biology*, Topic 9;
for the formation of colloids in the digestion of food
See *KS: Biology*, Topic 11.6.

Sand is largely silica, SiO_2. Silica does not dissolve in water. Why is it then that, when an aqueous solution of a silicate is acidified, silica is not precipitated? Instead there forms either a clear solution with a slight pearly sheen or a jelly-like precipitate or a solid-like gel in which liquid is trapped. The form of the product depends on the pH. The solution and the jelly-like precipitate and the gel are all **colloids**, also called **colloidal dispersions** and **colloidal suspensions**. The particles of silica are **suspended** or **dispersed** (spread) through the liquid. The solid (silica) is the **disperse phase**, and the liquid (water) is the **dispersion phase** or **dispersion medium**. The silica content may be as high as 30% by mass. Measurements show that the silica particles contain between 2000 and 20 000 SiO_2 units.

How do colloids differ from solutions and suspensions?

Solutions are homogeneous (the same all through) mixtures of two or more substances. The particles of the solute (the dissolved substance) are of atomic or molecular size (about one nanometre long, 1 nm; $1\,nm = 10^{-9}\,m$).

Suspensions are heterogeneous (not the same all through) mixtures. They contain relatively large particles (over 1000 nm long) of insoluble solid or liquid suspended in (scattered throughout) a liquid. In time, the particles settle out. Suspensions of solid particles in a liquid can be separated by filtration through filter paper and by centrifugation. For example, in blood the red and white blood cells are in suspension in the plasma. If a sample of blood is left for a time, the cells settle out. The separation is accelerated by centrifuging the sample.

Colloids are heterogeneous mixtures whose particles are larger than the molecules and ions which form solutions but smaller than the particles which form suspensions (between 1 nm and 1000 nm long). The particles cannot be separated by filtration through filter paper. The sizes of some colloidal particles are illustrated in Figure 16.15A.

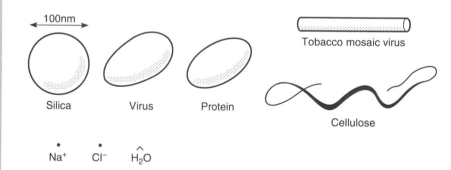

Figure 16.15A ● The size of colloidal particles (nm = 10^{-9} m)

Optical properties

The particles of a colloid are too small to be visible, but they are large enough to scatter light. The effect resembles the scattering of light in dust-laden air. It is named the Tyndall effect after its discoverer (see Figure 16.15B).

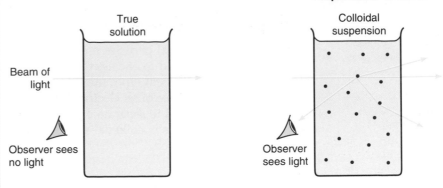

Figure 16.5B ● The scattering of light by a colloid

Charge

The reason colloidal particles do not clump together to form a precipitate is that they are charged. Colloidal silica particles are negatively charged. Repulsion between the charged particles keeps them apart and in suspension.

In order to precipitate a colloid, the charge on the particles must be neutralised. This can be done by adding ions of opposite charge. When water is purified for drinking, clay particles and other colloidal suspensions must be removed. The impure water is treated with, for example, aluminium sulphate. Aluminium ions, Al^{3+}, are small in size and highly charged. They neutralise the negative charges on the clay particles, which are then able to clump together and settle out of solution.

The proteins in blood are colloidal and are negatively charged. Small cuts can be treated with styptic pencils. These contain Al^{3+} or Fe^{3+} ions which neutralise the charges on the colloidal particles of protein and help the blood to clot.

Electrostatic precipitation

When gases and air are fed into an industrial process, they often contain colloidal particles. To remove these particles, **electrostatic precipitation** is used. The charge on the particles is used to attract the particles to charged metal plates (see Figure 16.15C).

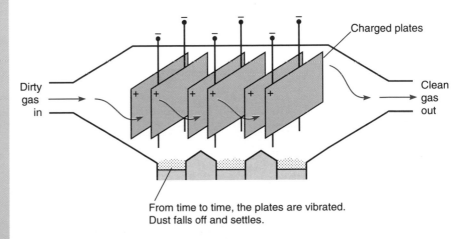

From time to time, the plates are vibrated.
Dust falls off and settles.

Figure 16.15C ● Electrostatic precipitation

Electrophoresis

Colloidal particles migrate in an electric field as ions do. They are attracted to one electrode and repelled by the other. Colloidal particles of different types migrate at different rates. This movement under the influence of an electric field is called **electrophoresis**. Electrophoresis can be used to separate different substances. Blood contains a number of proteins in colloidal suspension. Figure 16.15D shows how these can be separated.

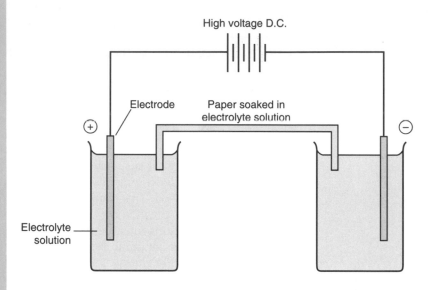

(a) Apparatus for electrophoresis

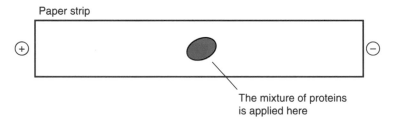

(b) Applying a liquid containing proteins, e.g. blood, to the paper strip

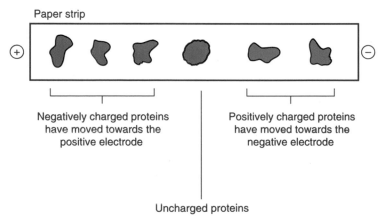

(c) Electrophoresis pattern obtained

Figure 15.15D ● Electrophoresis

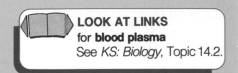

LOOK AT LINKS
for **blood plasma**
See *KS: Biology*, Topic 14.2.

Classification

Some colloids of different kinds are tabulated in Table 16.1.

Table 16.1 ● Some types of colloids

Dispersed phase	Dispersion medium	Type	Example
Liquid	Gas	Aerosol	Fog, mist from aerosol spray can, clouds
Solid	Gas	Aerosol	Smoke, dust-laden air
Gas	Liquid	Foam	Soap suds, whipped cream
Liquid	Liquid	Emulsion	Oil in water, milk, mayonnaise, protoplasm
Solid	Liquid	Sol	Clay, starch in water, protein in water, gelatin in water
Gas	Solid	Solid foam	Lava, pumice, styrofoam, marshmallows
Liquid	Solid	Solid emulsion	Pearl, opal, jellies, butter, cheese
Solid	Solid	Solid sol	Some gems, e.g. black diamond, coloured glass, some alloys

Emulsifiers

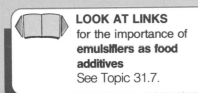

LOOK AT LINKS
for the importance of
**emulsifiers as food
additives**
See Topic 31.7.

An emulsion is a colloidal mixture of oil and water. The oil is dispersed through the water as a suspension of tiny drops. Examples are milk, cream, mayonnaise and many sauces. The natural tendency is for oil and water to separate. An emulsifier keeps the two together as a colloidal dispersion. An emulsifier ion has two parts. One is a polar group, with a negative charge, which is attracted to water (the water-loving group). The other is a hydrocarbon chain which is attracted to fat and oil (the fat-loving group) (see Figure 16.15E(a)). When the emulsifier is added to a mixture of oil and water, the emulsifier ions orient themselves so that the water-loving group dissolves in the water and the fat-loving group dissolves in the oil. Emulsifier ions arrange themselves round each droplet of oil (Figure 16.15E(b)). As the surface of each droplet is negatively charged, the drops repel one another and do not run together.

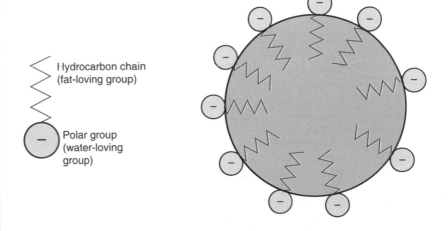

Hydrocarbon chain
(fat-loving group)

Polar group
(water-loving
group)

Figure 16.15E ● (a) An emulsifier ion; (b) A drop of oil surrounded by emulsifier ions

187

 **LOOK AT LINKS**
for **dialysis**
See *KS: Biology*, Topic 13.1.

Dialysis

Some membranes allow ions and small molecules to pass through them, but not large molecules or colloidal particles. They are called **dialysing membranes**. The movement of ions and small molecules through dialysing membranes is called **dialysis**. Most cell membranes are dialysing membranes. Dialysis can be used to separate colloids from solutions that contain both colloids and solutes, e.g. small molecules and ions (see Figure 16.15F).

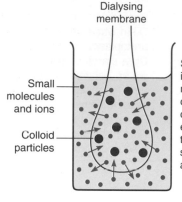

Dialysing membrane

Small molecules and ions

Colloid particles

Small molecules and ions pass through the membrane, while colloidal particles cannot. Equilibrium is established between the concentrations of small particles inside and outside the bag

(a)

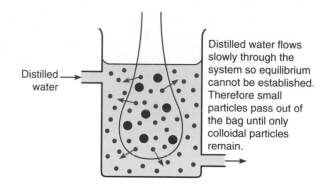

Distilled water

Distilled water flows slowly through the system so equilibrium cannot be established. Therefore small particles pass out of the bag until only colloidal particles remain.

(b)

Figure 16.15F ● Dialysis (● = colloidal particle; (• = small molecule or ion)

SUMMARY

Colloidal particles are larger than those in solutions but smaller than those in suspensions. They scatter light. They are charged, and if they lose their charge they settle out of the colloidal dispersion. Colloidal particles move under the influence of an electric field in electrophoresis. Colloidal particles do not pass through a dialysing membrane; this is the basis of artificial kidney machines.

The function of the kidneys is to remove waste products from the blood. If a person's kidneys do not work properly, waste substances build up in the blood. These waste products have small molecules. A dialysis machine called an **artificial kidney** can be used to send blood from one of the patient's arteries through a coil of dialysis tubing immersed in a solution of salts and glucose, called the dialysis fluid. Small molecules, e.g. molecules of the waste substance, urea, pass out of the blood into the dialysis fluid. Blood cells and large molecules, e.g. protein molecules, are too large to pass through and are retained in the blood. Dialysis fluid has sugars and salts dissolved in it so that while wastes diffuse out of the blood, sugars and salts do not. After dialysis, purified blood is returned to the patient's body through a vein.

CHECKPOINT

❶ How can you tell the difference between:
(a) a solution and a suspension,
(b) a suspension and a colloid,
(c) a solution and a colloid?

❷ How can solid colloidal particles be made to separate:
(a) from a sol,
(b) from an aerosol?

❸ Explain:
(a) why a salad dressing contains an emulsifier,
(b) how the emulsifier works.

❹ Briefly explain how an artificial kidney helps a person whose kidneys do not function properly.

TOPIC 17 AIR POLLUTION

17.1 Smog

Four thousand people died in the great London smog of December 1952. Smog is a combination of smoke and fog. Fog consists of small water droplets. It forms when warm air containing water vapour is suddenly cooled. The cool air cannot hold as much water vapour as it held when it was warm, and water condenses. When smoke combines with fog, fog prevents smoke escaping into the upper atmosphere. Smoke stays around, and we inhale it. Smoke contains particles which irritate our lungs and make us cough. Smoke also contains the gas sulphur dioxide. This gas reacts with water and oxygen to form sulphuric acid, H_2SO_4. This strong acid irritates our lungs, and they produce a lot of mucus which we cough up.

The Government did very little about the cause of smog until 1956. Then there was another killer smog. A private bill brought by a Member of Parliament (the late Mr Robert Maxwell, the newspaper owner) gained such widespread support that the Government was forced to act. The Government introduced its own bill, which became the Clean Air Act of 1956. The Act allowed local authorities to declare smokeless zones. In these zones, only low-smoke and low-sulphur fuels can be burned. The Act banned dark smoke from domestic chimneys and industrial chimneys.

17.2 The problem

All the dust and pollutants in the air pass over the sensitive tissues of our lungs. Any substance which is bad for health is called a **pollutant**. The lung diseases of cancer, bronchitis and emphysema are common illnesses in regions where air is highly polluted. From our lungs, pollutants enter our bloodstream to reach every part of our bodies. The main air pollutants are shown in Table 17.1.

Hyperbook
(data base)

Use *Hyperbook* to read about:
- acid rain,
- the greenhouse effect,
- pollution generally.

With the computer, try moving from one article to another and perhaps printing out any particularly interesting documents.

Table 17.1 ● The main pollutants in air (Emissions are given in millions of tonnes per year in the UK.)

Pollutant	Emission	Source
Carbon monoxide, CO	100	Vehicle engines and industrial processes
Sulphur dioxide, SO_2	33	Combustion of fuels in power stations and factories
Hydrocarbons	32	Combustion of fuels in factories and vehicles
Dust	28	Combustion of fuels; mining; factories
Oxides of nitrogen, NO and NO_2	21	Vehicle engines and fuel combustion
Lead compounds	0.5	Vehicle engines

In this topic, we shall look at where these pollutants come from, what harm they do and what can be done about them.

17.3 Dispersing air pollutants

LOOK AT LINKS
for **convection currents**
Why does warm air rise?
See Topic 6.5 of *KS: Physics*.

The surface of the Earth absorbs energy from the Sun and warms up. The Earth warms the lower atmosphere. The air in the upper atmosphere is cooler than the air near the Earth. **Convection currents** carry warm air upwards. Cold air descends to take its place (see Figure 17.3A). In this way, the warm dirty air from factories and motor vehicles is carried upwards and spread through the vast upper atmosphere.

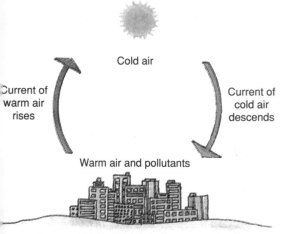

Figure 17.3A ● Convection currents disperse pollutants

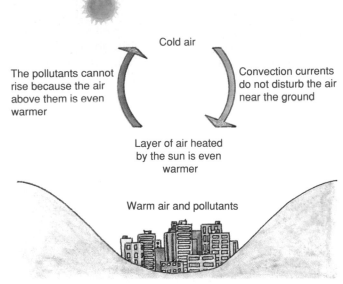

Figure 17.3B ● A temperature inversion traps pollutants

SUMMARY

Pollutants are carried upwards by rising currents of warm air. A temperature inversion stops the dispersal of pollutants. Temperature inversions occur in places with a hot climate and still air.

A low-lying area surrounded by higher ground tends to have still air. If an area like this has a hot climate, it is possible for the Sun to warm a layer of air in the upper atmosphere (Figure 17.3B). If the Sun is very hot, this layer of air may become warmer than that near the ground. There is a **temperature inversion**. The air near the ground is no longer carried upwards and dispersed. Pollutants accumulate in the layer of still air at ground level, and the city dwellers are forced to breathe them.

17.4 Sulphur dioxide

IT   *Pollution*

Use a word-processing system or better still a desktop publishing package to design and print a poster or news-sheet on the topic of pollution.

● *Where does sulphur dioxide come from?*

Worldwide, 150 million tonnes of sulphur dioxide a year are emitted. Almost all the sulphur dioxide in the air comes from industrial sources. The emission is growing as countries become more industrialised. Half of the output of sulphur dioxide comes from the burning of coal. Most of the coal is burned in power stations. All coal contains between 0.5 and 5 per cent sulphur.

$$\text{Sulphur} \quad + \quad \text{Oxygen} \quad \rightarrow \quad \text{Sulphur dioxide}$$
$$\text{(coal)} \qquad\qquad \text{(air)}$$
$$S(s) \quad + \quad O_2(g) \quad \rightarrow \quad SO_2(g)$$

Industrial smelters, which obtain metals from sulphide ores, also produce tonnes of sulphur dioxide daily.

In Canada, the countryside surrounding the nickel smelter in Sudbury, Ontario, was so devastated by sulphur dioxide emissions that in 1968 US astronauts practised moon-walking there before attempting the first lunar landing.

SUMMARY

Sulphur dioxide causes bronchitis and lung diseases. The Clean Air Acts have reduced the emission of sulphur dioxide from low chimneys. Factories, power stations and metal smelters send sulphur dioxide into the air. In the upper atmosphere, sulphur dioxide reacts with water to form acid rain.

● **What harm does sulphur dioxide do?**

Sulphur dioxide is a colourless gas with a very irritating smell. Inhaling sulphur dioxide causes coughing, chest pains and shortness of breath. It is poisonous; at a level of 0.5%, it will kill. Sulphur dioxide is thought to be one of the causes of bronchitis and lung diseases.

● **What can be done about it?**

After the Clean Air Acts of 1956 and 1968, the emission of sulphur dioxide and smoke from the chimneys of houses decreased. At the same time, the emission of sulphur dioxide and smoke from tall chimneys increased. Tall chimneys carry sulphur dioxide away from the power stations and factories which produce it (see Figure 17.4A). Unfortunately it comes down to earth again as acid rain (see Figure 17.5B).

Figure 17.4A ● Smoking chimneys

CHECKPOINT

❶ The emission of sulphur dioxide from low chimneys decreased between 1955 and 1975 from 1.7 million tonnes a year to 0.6 million tonnes a year. In the same period, the emission of sulphur dioxide from tall chimneys increased from 1.4 million tonnes to 3.0 million tonnes.
(a) Explain why there was a decrease in sulphur dioxide emission from low chimneys.
(b) Who benefited from the decrease in emission from low chimneys?
(c) Explain why tall chimneys are not a complete answer to the problem of sulphur dioxide emission.

17.5 Acid rain

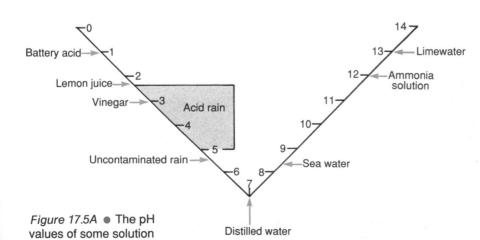

Figure 17.5A ● The pH values of some solution

Rain water is naturally weakly acidic. It has a pH of 5.4. Carbon dioxide from the air dissolves in it to form the weak acid, carbonic acid, H_2CO_3. What we mean by acid rain is rain which contains the strong acids, sulphuric acid and nitric acid. Acid rain has a pH between 2.4 and 5.0 (see Figure 17.5A).

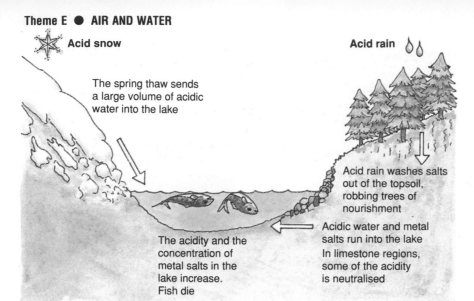

Acid snow

The spring thaw sends a large volume of acidic water into the lake

Acid rain

Acid rain washes salts out of the topsoil, robbing trees of nourishment

Acidic water and metal salts run into the lake
In limestone regions, some of the acidity is neutralised

The acidity and the concentration of metal salts in the lake increase. Fish die

Figure 17.5B ● Where acid rain comes from

How do sulphuric acid and nitric acid get into rain water? Tall chimneys emit sulphur dioxide and other pollutant gases, such as oxides of nitrogen. Air currents carry the gases away. Before long, the gases react with water vapour and oxygen in the air. Sulphuric acid, H_2SO_4, and nitric acid, HNO_3, are formed. The water vapour with its acid content becomes part of a cloud. Eventually it falls to earth as acid rain or acid snow which may turn up hundreds of miles away from the source of pollution (see Figure 17.5B).

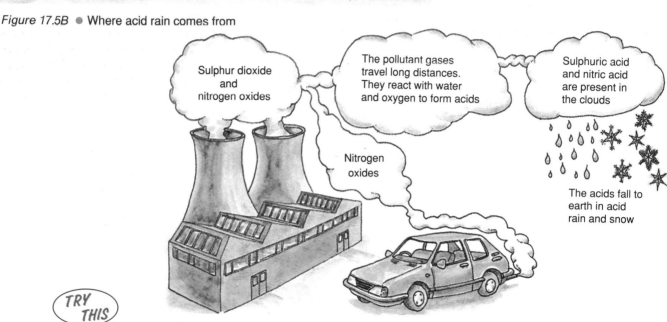

Sulphur dioxide and nitrogen oxides

The pollutant gases travel long distances. They react with water and oxygen to form acids

Sulphuric acid and nitric acid are present in the clouds

Nitrogen oxides

The acids fall to earth in acid rain and snow

Figure 17.5C ● Where acid rain goes

TRY THIS

- Collect samples of rainwater from drain pipes, garden water-butts, freshwater ponds.
- Find a tree with rainwater channels running down its bark. Dam a water channel with plasticine, and wait for rainwater to collect behind the dam.
- Collect some soil from the base of the tree. Pour a little distilled water onto the soil, filter and collect the filtrate.
- Dip a strip of universal indicator paper into each water sample, and record its pH.
- Comment on the differences in pH between the different samples of water.

Acid rain which falls on land is absorbed by the soil. At first, the nitrates in the acid rain fertilise the soil and encourage the growth of plants. But acid rain reacts with minerals, converting the metals in them into soluble salts. The rain water containing these soluble salts of calcium, potassium, aluminium and other metals trickles down through the soil into the subsoil where plant roots cannot reach them. In this way, salts are leached out of the topsoil, and crops are robbed of nutrients. One of the salts formed by acid rain is aluminium sulphate. This salt damages the roots of trees. The damaged roots are easily attacked by viruses and bacteria, and the trees die of a combination of malnutrition and disease. The Black Forest is a famous beauty spot in Germany which makes money from tourism. About half the trees there are now damaged or dead. Pollution is an important issue in Germany. The Green Party is a major political party. It is campaigning for the reduction of pollution. There are dead forests also in Czechoslovakia and Poland. In 1987, the UK Forestry Commission reported that damage to spruce and pine trees is as widespread in the UK as in Germany. One third of British trees are damaged, but not everyone agrees that the cause of the damage is acid rain.

The acidic rain water trickles through the soil until it meets rock. Then it travels along the layer of rock to emerge in lakes and rivers. Lakes are more affected by acid rain than rivers are. They become more and more acidic, and the concentrations of metal salts increase. Fish cannot live in acidic water. Aluminium compounds, e.g. aluminium hydroxide, come out of solution and are deposited on the gills. The fish secrete mucus to try to get rid of the deposit. The gills become clogged with mucus, and the fish die. An acid lake is perfectly transparent because plants, plankton, insects and other living things have perished.

Thousands of lakes in Norway, Sweden and Canada are now 'dead' lakes. One reason why these countries suffer badly is that acidic snow piles up during the winter months. In the spring thaw, the accumulated snow melts suddenly, and a large volume of acidic water flows into the lakes. Acid rain is partially neutralised as it trickles slowly through soil and over rock. Limestone, in particular, keeps damage to a minimum by neutralising some of the acidity. There is not time for this partial neutralisation to occur when acid snow melts and tonnes of water flow rapidly down the hills and into the lakes.

The UK is affected too. In 1982, lakes and rivers in south-west Scotland had become so acidic that the water companies started treating the lakes with calcium hydroxide (lime). The aim is to neutralise the acidic water and revive stocks of fish. In Wales, the water company has for some years poured tonnes of powdered limestone into acidic lakes. A number of lakes are 'dead' and the fish in many others are threatened.

Figure 17.5D ● The effects of acid rain

● ***What can be done about acid rain?***

There are three main methods of attacking the problem of acid rain. They all cost money, but then the damage done by acid rain costs money too.

❶ Low-sulphur fuels can be used. Crushing coal and washing it with a suitable solvent reduces the sulphur content by 10 to 40 per cent. The dirty solvent must be disposed of without creating pollution on land or in rivers. Oil refineries could refine the oil which they sell to power stations. The cost of the purified oil would be higher, and the price of electricity would increase.

❷ Flue gas desulphurisation, FGD, is the removal of sulphur from power station chimneys after the coal has been burnt and before the waste gases leave the chimneys. As the combustion products pass up the chimney, they are bombarded by jets of wet powdered limestone. The acid gases are neutralised to form a sludge. The method will remove 95 per cent of the acid combustion products. FGD can be fitted to existing power stations. One of the products is calcium sulphate, which can be sold to the plaster board industry and to cement manufacturers.

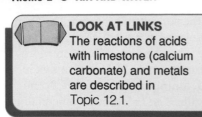

LOOK AT LINKS
The reactions of acids with limestone (calcium carbonate) and metals are described in Topic 12.1.

❸ Pulverised fluidised bed combustion, PFBC, uses a new type of furnace. The furnace burns pulverised coal (small particles) in a bed of powdered limestone. An upward flow of air keeps the whole bed in motion. The sulphur is removed during burning. The PFBC uses much more limestone than the FGD method: one power station needs 1 million tonnes of limestone a year (4 times as much as the FGD method). The PFBC method also produces a lot more waste material, which has to be dumped.

CHECKPOINT

❶ What is the advantage of building a power station close to a densely populated area? What is the disadvantage?

❷ Why do power stations and factories have tall chimneys? Are tall chimneys a solution to the problem of pollution? Explain your answer.

❸ Why does acid rain attack (a) iron railings (b) marble statues and (c) stone buildings?

❹ (a) Why does Sweden suffer badly from acid rain?
(b) Why do lakes suffer more than rivers from the effects of acid rain?

❺ A country decides to increase the price of electricity so that the power stations can afford to use refined low-sulphur fuel oil. In what ways will the country actually *save* money by reducing the emission of sulphur dioxide?

17.6  Carbon monoxide

● *Where does it come from?*

Worldwide, the emission of carbon monoxide is 350 million tonnes a year. Most of it comes from the exhaust gases of motor vehicles. Vehicle engines are designed to give maximum power. This is achieved by arranging for the mixture in the cylinders to have a high fuel to air ratio. This design leads to incomplete combustion. The result is the discharge of carbon monoxide, carbon and unburnt **hydrocarbons**.

LOOK AT LINKS
for **hydrocarbons**
See Theme E, Topic 16.5.

LOOK AT LINKS
for **haemoglobin**
See *KS: Biology*, Topic 14.2.

● *What harm does carbon monoxide do?*

Oxygen combines with **haemoglobin**, a substance in red blood cells. Carbon monoxide is 200 times better at combining with haemoglobin than oxygen is. Carbon monoxide is therefore able to tie up haemoglobin and prevent it combining with oxygen. A shortage of oxygen causes headache and dizziness, and makes a person feel sluggish. If the level of carbon monoxide reaches 0.1% of the air, it will kill. Carbon monoxide is especially dangerous in that, being colourless and odourless, it gives no warning of its presence. Since carbon monoxide is produced by motor vehicles, it is likely to affect people when they are driving in heavy traffic. This is when people need to feel alert and to have quick reflexes.

● *What can be done?*

Soil contains organisms which can convert carbon monoxide into carbon dioxide or methane. This natural mechanism for dealing with carbon monoxide cannot cope in cities, where the concentration of carbon monoxide is high and there is little soil to remove it. People are trying out a number of solutions to the problem.

SUMMARY

Carbon monoxide is emitted by vehicle engines. It is poisonous. Catalytic converters fitted in the exhaust pipes of cars reduce the emission of carbon monoxide.

- Vehicle engines can be tuned to take in more air and produce only carbon dioxide and water. Unfortunately, this increases the formation of oxides of nitrogen (see Topic 17.7).
- Catalytic converters are fitted to the exhausts of many cars. The catalyst helps to oxidise carbon monoxide in the exhaust gases to carbon dioxide (see Topic 17.6).
- New fuels may be used in the future. Some fuels, e.g. alcohol, burn more cleanly than hydrocarbons (see Topic 30.1).

CHECKPOINT

1 (a) What are the products of complete combustion of petrol?
(b) What harm do these products do?
(c) What conditions lead to the formation of carbon monoxide?
(d) What harm does it do?

2 How does carbon monoxide act on the body?

3 Which types of people are likely to breathe in carbon monoxide? Is there anything they can do to avoid it?

4 A family was spending the weekend in their caravan. At night, the weather turned cold, so they shut the windows and turned up the paraffin heater. In the morning, they were all dead. What had gone wrong? Why did they have no warning that something was wrong?

17.7 Oxides of nitrogen

When fuels are burned in air, nitrogen is present. Combustion temperatures are high enough to make some nitrogen combine with oxygen. As a result, the gases nitrogen monoxide, NO and nitrogen dioxide, NO_2, are formed. These gases enter the air from the chimneys of power stations and factories and from the exhausts of motor vehicles. This mixture of gases is sometimes shown as NO_x or even NOX.

● What harm do they do?
Nitrogen monoxide, NO, is not a very dangerous gas. However, it quickly reacts with air to form nitrogen dioxide, NO_2. Nitrogen dioxide is highly toxic and irritates the breathing passages. It reacts with oxygen and water to form nitric acid, an ingredient of acid rain.

● What can be done?
A reaction which can be used to reduce the quantity of nitrogen monoxide in exhaust gases is

Nitrogen monoxide	+	Carbon monoxide	→	Nitrogen	+	Carbon dioxide
$2NO(g)$	+	$2CO(g)$	→	$N_2(g)$	+	$2CO_2(g)$

This reaction takes place in the presence of a catalyst. A metal cylinder containing the catalyst is fitted in the vehicle exhaust (see Figure 23.8B). All German, Japanese and US cars are now fitted with these **catalytic converters**.

SUMMARY

Oxides of nitrogen, NO and NO_2, (NO_x) enter the air through the exhausts of motor vehicles. They react with air and water to form nitric acid. This strong acid is very toxic and corrosive. Catalytic converters can be fitted in cars to convert nitrogen oxides into nitrogen.

17.8 Hydrocarbons

● *Where do they come from?*

Hydrocarbons are present naturally in air. Methane, CH_4, is one of the products of decay of plant material. Only 15 per cent of the hydrocarbons in the air come from human activities. They affect our health because they are concentrated in city air.

● *What harm do they do?*

Hydrocarbons by themselves cause little damage. In intense sunlight, **photochemical reactions** occur ('photo' means light). Hydrocarbons react with oxygen and oxides of nitrogen to form irritating and toxic compounds.

● *What can be done?*

The hydrocarbons in the exhausts of petrol engines can be reduced by increasing the oxygen supply so as to burn the petrol completely. This also decreases the formation of carbon monoxide. There is a snag, however. Increasing the oxygen supply increases the formation of oxides of nitrogen in the engine. The problem may have a solution. Research workers are trying the idea of running the engine at a lower temperature (to reduce the combination of oxygen and nitrogen) with a catalyst (to assist complete combustion of hydrocarbons at the lower temperature).

SUMMARY

Hydrocarbons from vehicle exhausts take part in photochemical reactions to form irritating and toxic compounds. The emission of hydrocarbons may be reduced by running the engine at a lower temperature with a catalyst.

Figure 17.8A ● Testing the emission from a vehicle exhaust

● *Catalytic converters*

Pollution by carbon monoxide, oxides of nitrogen and hydrocarbons is reduced when catalytic converters are fitted to vehicle exhausts. Figure 17.8B shows the structure of a catalytic converter.

SUMMARY

Catalytic converters reduce carbon monoxide, oxides of nitrogen and hydrocarbons in vehicle exhausts.

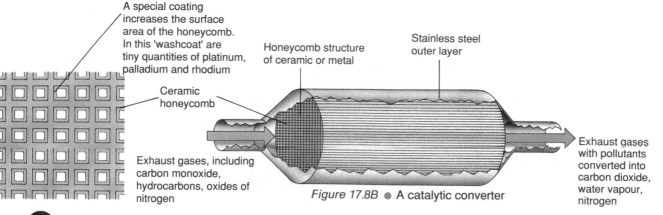

A special coating increases the surface area of the honeycomb. In this 'washcoat' are tiny quantities of platinum, palladium and rhodium

Ceramic honeycomb

Honeycomb structure of ceramic or metal

Stainless steel outer layer

Exhaust gases, including carbon monoxide, hydrocarbons, oxides of nitrogen

Exhaust gases with pollutants converted into carbon dioxide, water vapour, nitrogen

Figure 17.8B ● A catalytic converter

❶ What products are formed by the combustion of hydrocarbons in petrol engines?

❷ How does the supply of air affect the course of combustion?

❸ What is the advantage of increasing the air to fuel ratio in the combustion chamber?

❹ What are the pollutants that form when the air to fuel ratio is high? What can be done about them?

❺ Copy and complete this summary.
In internal combustion engines, a high air to fuel ratio:
decreases the emission of unburnt _____ A
decreases the emission of _____ B
increases the emission of _____ C
A way of reducing the emission of C would be to run the engine at a lower temperature. A _____ would be needed to promote _____ combustion and reduce the emission of A and B.

❻ The figures opposite show approximately how the emissions of carbon monoxide, oxides of nitrogen and hydrocarbons change with the speed of a vehicle. (Note that the scale for carbon monoxide goes up to 30 g/l, while that of the other pollutants goes up to 3 g/l.)
(a) Say what speed is best for reducing the emission of (i) carbon monoxide (ii) oxides of nitrogen (iii) hydrocarbons.
(b) (i) What speed would you recommend as the best to reduce overall pollution? (ii) What is this speed in miles per hour (5 mile = 8 km)?

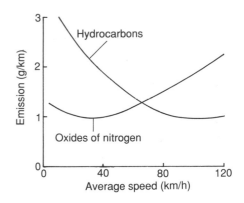

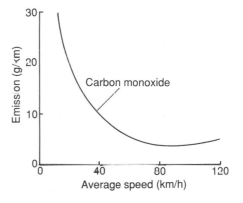

17.9 Smoke, dust and grit

LOOK AT LINKS
for **smog**
See Topic 17.1.

LOOK AT LINKS
for **electrostatic attraction**
How do electrostatic precipitators work?
See Topic 16.15.

SUMMARY

Particles of smoke and dust and grit are sent into the air by factories, power stations and motor vehicles. Dirt damages buildings and plants. It pollutes the air we breathe; mixed with fog, it forms smog.

Millions of tonnes of smoke, dust and grit are present in the atmosphere. Dust storms, forest fires and volcanic eruptions send matter into the air. Human activities such as mining, land-clearing and burning coal and oil add to the solid matter in the air.

● **What harm do particles do?**
Particles darken city air by scattering light. Smoke increases the danger of smog. Solid particles fall as grime on people, clothing, buildings and plants.
Sunlight which meets dust particles is reflected back into space and prevented from reaching the Earth. Some scientists believe that the increasing amount of dust in the atmosphere is serious. A fourfold increase in the amount of dust would make the Earth's temperature fall by about 3 °C. This would affect food production.

● **How can particles be removed?**
Industries use a number of methods. These include:
• using sprays of water to wash out particles from their waste gases
• passing waste gases through filters,
• electrostatic precipitators, which remove dust particles from waste gases by **electrostatic attraction**

17.10  Metals

Many heavy metals and their compounds are serious air pollutants. 'Heavy' metals are metals with a density greater than 5 g/cm³.

Mercury

Earth-moving activities, such as mining and road-making, disturb soil and rock and allow the mercury which they contain to escape into the air. Mercury vapour is also released into the air during the smelting of many metal ores and the combustion of coal and oil. Both mercury and its compounds cause kidney damage, nerve damage and death.

Lead

● **Where does it come from?**

The lead compounds in the air all come from human activity. Vehicle engines, the combustion of coal and the roasting of metal ores send lead and its compounds into the air. Unlike the other pollutants in exhaust gases, lead compounds have been purposely added to the fuel. Tetraethyl lead, TEL, is added to improve the performance of the engine.

● **What harm does it do?**

Lead compounds settle out of the air on to plant crops, and contaminate our food. The level of lead in our environment is high: some areas still have lead plumbing; old houses may have peeling lead-based paint. City dwellers take in lead from many sources. Many people have blood levels of lead which are nearly high enough to produce the symptoms of lead poisoning. Symptoms of mild lead poisoning are headache, irritability, tiredness and depression. Higher levels of lead cause damage to the brain, liver and kidneys. Scientists have suggested that behaviour disorders such as hooliganism and vandalism may be due in part to lead poisoning.

● **What can be done?**

This type of pollution can be remedied. We can stop adding lead compounds to petrol. Research chemists have found other compounds which can be used to improve engine performance. Vehicles made in the UK after autumn 1990 are adjusted to run on lead-free petrol. Most petrol stations now stock lead-free petrol. The USA, Germany and Japan use lead-free petrol because the catalytic converters fitted to their cars are 'poisoned' by lead compounds.

Figure 17.10A ● City dwellers breathe exhaust gases

SUMMARY

Heavy metals are serious air pollutants. Levels of mercury and lead and their compounds in the air are increasing.

CHECKPOINT

❶ Name the pollutants which come from motor vehicles.

❷ Name the pollutants which can be reduced by fitting catalytic converters into vehicle exhausts. What effect will this modification have on the price of cars?

❸ Catalytic converters will only work with lead-free petrol. When TEL is no longer added to petrol, motorists will have to use high octane (4 star) fuel. What effect will this have on the cost of motoring?

④ What effect does the use of TEL have on the air, apart from its effect on catalytic converters?

⑤ In which ways will the control of pollution from vehicles cost money? In which ways will a reduction in the level of pollutants in the air save money? (Consider the effects of pollution on people and materials.) Will the expense be worthwhile?

17.11 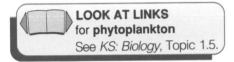 Chlorofluorohydrocarbons

The ozone layer

There is a layer of ozone, O_3, surrounding the Earth. It is 5 km thick at a distance of 25–30 km from the Earth's surface. The ozone layer cuts out some of the ultraviolet light coming from the Sun. Ultraviolet light is bad for us and for crops. Long exposure to ultraviolet light can cause skin cancer. This complaint is common in Australia among people who spend a lot of time out of doors. If anything happens to decrease the ozone layer, the incidence of skin cancer from exposure to ultraviolet light will increase. An excess of ultraviolet light kills **phytoplankton**, the minute plant life of the oceans which are the primary food on which the life of an ocean depends.

> **LOOK AT LINKS**
> for **phytoplankton**
> See *KS: Biology*, Topic 1.5.

● ***The problem***
Ozone is a very reactive element. If the upper atmosphere becomes polluted, ozone will oxidise the pollutants. Two pollutants are accumulating in the upper atmosphere. One is the **propellant** from aerosol cans (see Figure 17.11A).

When the pressure is released, the propellant liquid vaporises and forces the polish out of the can

Mixture of propellant and useful liquid, e.g. polish or insecticide, under pressure

Figure 17.11A ● An aerosol can

Many of the propellants are chlorofluorohydrocarbons (CFCs). They are very unreactive compounds. They spread through the atmosphere without reacting with other substances and drift into the upper atmosphere. There they meet ozone, which oxidises CFCs and in doing so is converted into oxygen.

Ozone + CFC → Oxygen + Oxidation products

Another pollutant found at this height is nitrogen monoxide, NO. It comes from the exhausts of high-altitude aircraft, such as Concorde. Ozone oxidises nitrogen monoxide to nitrogen dioxide:

Ozone + Nitrogen monoxide → Oxygen + Nitrogen dioxide
$O_3(g)$ + $NO(g)$ → $O_2(g)$ + $NO_2(g)$

> ══ *SCIENCE AT WORK* ══
> Microcomputers and sensors are taken up in aircraft to keep watch on the ozone layer. The sensor detects the thickness of the ozone layer and the microcomputer records the measurements.

199

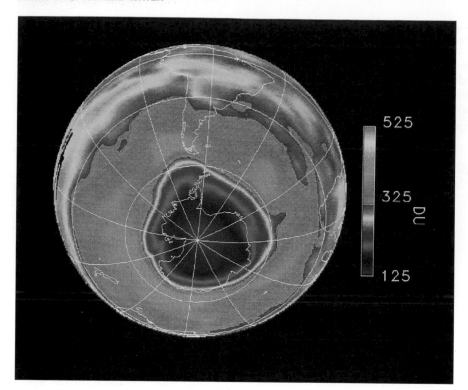

Figure 17.11B ● The ozone hole

● What should be done?

Is it happening? Is the ozone layer becoming thinner? In June 1980 the British Antarctic Expedition discovered that there was a gap in the ozone layer over Antarctica during certain months. In 1987, research workers in the US confirmed that there was a thinning of the ozone layer which was 'large, sudden and unexpected... far worse than we thought'. In 1988 a team of scientists working in the Arctic Ocean discovered that the ozone layer over Northern Europe was thinner than it had been.

Knowing that the danger had appeared over more populated regions of the globe, spurred many countries to take action. At a meeting in Montreal in 1987 many countries agreed to reduce their use of CFCs by 50% by the year 2000. Since that date, many countries have agreed to speed up their programme of phasing out CFCs. Aerosols containing CFCs have been banned in the USA since 1988. In 1988 many makers of toiletries in the UK agreed to stop using CFCs by the end of 1989. They are now using spray cans with different propellants, which they label 'ozone-friendly', or pump-action cans. There is more of a problem with the CFCs used as refrigerants, in air conditioners, in the manufacture of polyurethane foam and as solvents. Chemists are now finding stable compounds to replace CFCs. In the USA, Du Pont Chemicals have agreed to stop using CFCs after the year 2000. In the UK, ICI chemists are working hard to find substitutes to enable ICI to do the same.

SUMMARY

The ozone layer protects animals and plants from ultraviolet radiation. As it reacts with pollutants in the upper atmosphere, the ozone layer is becoming thinner. CFCs and nitrogen monoxide from high altitude planes are the culprits. The use of CFCs is being reduced.

CHECKPOINT

❶ Look round your kitchen, bathroom and garage. How many products in aerosol cans do you buy? How convenient is it to have each of these products in an aerosol can? What inconvenience would you suffer if aerosol cans were banned? How many of the aerosol cans are labelled 'ozone-friendly'? What does this mean?

❷ Speaking on 23 February 1988, the Prince of Wales announced that he had banned aerosols from his household. He said that some members of his household had difficulty in finding a suitable alternative hairspray.
(a) What concern led the Prince of Wales to take this step?
(b) What properties must the propellant in the hairspray possess to work effectively and to be safe in use?
(c) What substitute can you suggest for an aerosol hairspray?

❸ How does their lack of chemical reactivity make CFCs (a) useful and (b) dangerous?

17.12 Cigarette smoke

Another pollutant in the air is cigarette smoke. There are more than 1000 compounds in tobacco smoke (see Figure 17.12A). One is carbon monoxide. This is why sitting in a smoke-filled room can give you a headache. Nicotine is another. Nicotine is a poison. It increases the pulse rate and the blood pressure, depresses breathing and causes nausea and diarrhoea. People become able to tolerate nicotine when they take small quantities regularly. This is why, although people feel sick when they start to smoke, after a while these symptoms do not occur. People smoke because it produces a calming effect. The tar that condenses in the lungs of smokers is a mixture of carcinogenic (cancer-producing) hydrocarbons. The World Health Organisation estimates that 30 000 people die each year as a result of smoking.

FIRST THOUGHTS

Can you name a poisonous oily liquid with an unpleasant smell, a burning taste and an alkaline reaction?

LOOK AT LINKS
for evidence of the harmful effects of smoking
See *KS: Biology*, Topic 12.2.

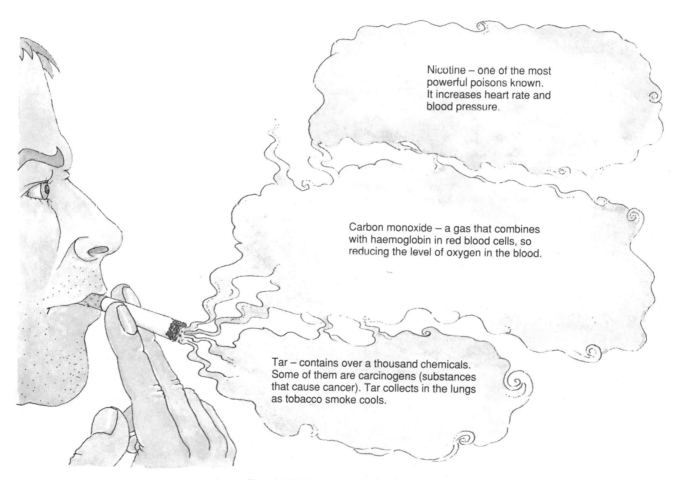

Nicotine – one of the most powerful poisons known. It increases heart rate and blood pressure.

Carbon monoxide – a gas that combines with haemoglobin in red blood cells, so reducing the level of oxygen in the blood.

Tar – contains over a thousand chemicals. Some of them are carcinogens (substances that cause cancer). Tar collects in the lungs as tobacco smoke cools.

Figure 17.12A ● Smoking

TOPIC 18 WATER POLLUTION

18.1 Pollution by industry

FIRST THOUGHTS

What's the problem? We need clean drinking water – nothing could be more important. We need industry, and industry needs to dispose of waste products; in the process our water is polluted.

SUMMARY

The National Rivers Authority controls pollution of inland rivers. It does not regulate the discharge of pollutants into tidal rivers, estuaries and the sea. The estuaries in the UK are heavily polluted by industry and by sewage.

IT'S A FACT

B & N Chemicals in 1981 filled the town of Haverhill in Suffolk with chemical smells. The River Stow was so polluted that the water works was unable to draw water from the river. The company was prosecuted and convicted. The fine was a mere £325.

British Tissues was prosecuted in 1983 after discharging far more waste than it was allowed into the River Don. The fine was £750. It was obviously cheaper to pay fines than to treat the waste.

SUMMARY

Many industrial firms do not keep their discharges of wastes within the limits set by law.

Controls

You will notice that many industrial firms are on river banks. These firms can get rid of waste products by discharging them into rivers. Until 1989, the quantities of waste which industries were allowed to discharge into rivers were controlled by the water authority of each region. Under the 1974 Control of Pollution Act, the water authorities had power to control pollution in inland rivers but not in tidal rivers, estuaries and the sea (except for the discharge of radioactive waste: see Topic 8.8. In spite of the Act, more than 2800 km of Britain's largest rivers are too dirty and lacking in oxygen to keep fish alive.

In 1989 the UK Water Privatisation Bill became law. The water authorities were sold to private companies, and are now run for profit as other industries are. The Government set up a National Rivers Authority to watch over the quality of water and prosecute polluters.

● Estuaries

Many of the worst polluters discharge into coastal waters and estuaries. The oil refineries, chemical works, steel plants and paper mills on coasts and estuaries can pour all the waste they want into estuaries and the sea. In the 1930s, fishermen could make a living in the Mersey. Now, it is too foul to keep fish alive. One reason is the discharge of raw sewage into the Mersey. The other is that too many firms pour waste into the estuary. There is unemployment in Merseyside, and the Government does not want to make life difficult for industry in the area. The industries on the banks of the Mersey have been given permission to fall below the standards of the Control of Pollution Act.

Other estuaries, such as the Humber, the Tees, the Tyne and the Clyde, are also polluted by industry.

Figure 18.1A ● The Mersey

Mercury and its compounds

FIRST THOUGHTS

Why has it taken so long for industry to react to the tragedy of Minamata?

LOOK AT LINKS
for **food chains**
See Topic 3.6 of *KS: Biology*.

A well-known case of industrial pollution is the tragedy of Minamata, a fishing village on the shore of Minamata Bay in Japan. A plastics factory started discharging waste into the bay in 1951. By 1953, a thousand people in Minamata were seriously ill. Some were crippled, some were paralysed, some went blind, some became mentally deranged, and some died. The cause of the disease was found to be the mercury compounds which the plastics factory discharged into the Bay. Although the level of mercury compounds in the Bay water was low, mercury was concentrated by a **food chain** (see Figure 18.1B). The level of mercury in the fish in the Bay was high, and fishers and their families became ill through eating the fish.

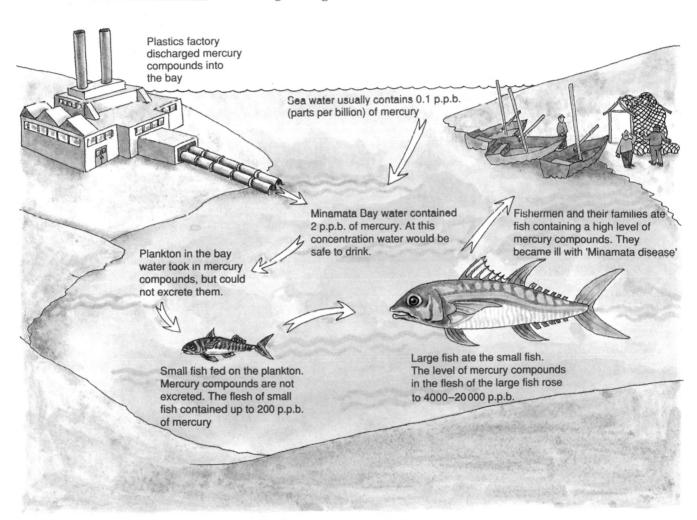

Figure 18.1B ● The food chain which led to the Minamata disease

SUMMARY

Mercury and its compounds are poisonous. If mercury gets into a lake or river, it is converted slowly into soluble compounds. These are likely to accumulate in fish and may be eaten by humans.

Other countries have experienced the results of mercury pollution. In 1967, many lakes and rivers in Sweden were found to be so contaminated by mercury that fishing had to stop. In 1970, high mercury levels were found in hundreds of lakes in Canada and the USA. As late as 1988, the ICI plant on Merseyside discharged more mercury than the permitted level. Now that the danger is known, the polluting plants have taken care to reduce spillage of mercury. The danger is still there, however. Mercury from years of pollution lies in the sediment at the bottom of lakes. Slowly it is converted by bacteria into soluble mercury compounds. These may get into a food chain.

CHECKPOINT

❶ When a car engine has an oil change, the waste oil is sometimes poured down the drain. What is wrong with doing this?

❷ Does it matter whether rivers are clean and stocked with fish or foul and devoid of life? Explain your answer.

❸ The Minamata tragedy happened when Japan was building up its industry after the war. In spite of Japan's experience, Sweden, Canada and the USA found an excess of mercury in their lakes twenty years later. Why had they not learned from Japan's mistake? (You will not find the answer in the back!)

FIRST THOUGHTS

18.2 Thermal pollution

What's wrong with warming up the water?

SUMMARY

Thermal pollution means warming rivers and lakes. It reduces the concentration of oxygen dissolved in the water.

Industries use water as a coolant. A large nuclear power station uses 4000 tonnes of water a minute for cooling. River water is circulated round the power station, where its temperature increases by 10 °C, and is returned to the river. If the temperature of the river rises by many degrees, the river is **thermally polluted**. As the temperature rises, the solubility of oxygen decreases. At the same time, the **biochemical oxygen demand** increases. Fish become more active at the higher temperature, and need more oxygen. The bacteria which feed on decaying organic matter become more active and use more oxygen.

FIRST THOUGHTS

18.3 Pollution by sewage

The population of the UK is increasing. One result is the need for more sewage works. What happens when a country does not keep up with this need?

Inside Science
(program)

Use this software to deal with an accidental pollution spillage.

Life and Death of a River
(data base)

Explore a case of river pollution and see what can be done to clean it up. Use this with either the *Key* or *Keysheet* package.

Figure 18.3A ● British beaches and the EC standard

In Topic 16.4, you read how sewage is treated before it is discharged into rivers or the sea. Unfortunately, as some water companies do not have enough plants to treat all their area's sewage, they discharge some raw sewage into rivers and estuaries. The Mersey receives raw sewage from Liverpool and other towns. In Sussex, sewage treatment works are inadequate and sewage is discharged into the sea. This creates some nasty results at several bathing beaches in the country.

SUMMARY

During the 1980s, the United Nations set a target of safe water and sanitation for all by 1990. The aim was to provide wells and pumps, kits for disinfecting water and hygienic toilets. The sum needed was £25 billion, slightly more than the world spends on its armies in one month. The target was not reached by 1990, but the work is continuing.

The quality of the water at dozens of Britain's bathing beaches fails to meet standards set by the European Community (EC). Many British beaches have more coliform bacteria and faecal bacteria in the water than the EC standard.

The Third World

Of the four billion people in the world, two billion have no toilets, and one billion have unsafe drinking water. In Third World countries (the developing countries) three out of five people have difficulty in obtaining clean water. Some Third World communities have to use a river as a source of drinking water as well as for disposal of their sewage. Bacteria are present in faeces, and they infect the water. Many diseases are spread by contaminated water. They include cholera, typhoid, river blindness, diarrhoea and schistosomiasis. Four-fifths of the diseases in the Third World are linked to dirty water and lack of sanitation. Five million people each year are killed by water-borne diseases.

Figure 18.3B ● Their water supplies

18.4 Pollution by agriculture

FIRST THOUGHTS

Farmers need to use fertilisers. What happens when a crop does not use all the fertiliser applied to it? There can be pollution, as this section explains.

Fertilisers

A lake has a natural cycle. In summer, algae grow on the surface, fed by nutrients which are washed into the lake. In autumn the algae die and sink to the bottom. Bacteria break down the algae into nutrients. Plants need the elements carbon, hydrogen, oxygen, nitrogen and phosphorus. Water always provides enough carbon, hydrogen and oxygen; plant growth is limited by the supply of nitrogen and phosphorus. Sometimes farm land surrounding a lake receives more fertiliser than the crops can absorb. Then the unabsorbed nitrates and phosphates in the fertiliser wash out of the soil into the lake water. When fertilisers wash into a lake, they upset the natural cycle. The algae multiply rapidly to produce an **algal bloom**. The lake water comes to resemble a cloudy greenish soup. When the algae die, bacteria feed on the dead material and multiply. The increased bacterial activity consumes much of the dissolved oxygen. There is little oxygen left in the water, and fish die from lack of oxygen. The lake becomes difficult for boating because masses of algae snag the propellers. The name given to this accidental fertilisation of lakes and rivers is **eutrophication**.

Many parts of the Norfolk Broads are now covered with algal bloom. The tourist industry centred on the Broads would like to see them restored to their former condition.

IT

Use an oxygen sensor and probe to measure the percentage of oxygen in air. The same probe can be used to measure dissolved oxygen in water. You'll need to connect the sensor to a computer or a datalogger.

Lough Neagh in Northern Ireland is the UK's biggest inland lake. It supplies Belfast's water and it also supports eel-fishing. Algae now block the filters through which water flows to the water treatment plant. Eels and other fish are in danger as the level of dissolved oxygen falls. The problem is being tackled by removing phosphates from the treated sewage which enters the lough. Treatment with a solution containing aluminium ions and iron(III) ions precipitates phosphates. Each year, this stops 60 tonnes of phosphorus in the form of phosphates from entering Lough Neagh. This pollution is unnecessary. Detergents without phosphates would leave laundry only a little less sparkling white, but would not pollute our rivers and lakes.

Fertiliser which is not absorbed by crops can be carried into the ground water (the water in porous underground rock). Ground water provides one third of Britain's drinking water. The EC has set a maximum level of nitrates in drinking water at 50 mg/l (12 p.p.m. of nitrogen in the form of nitrate). Four out of the ten water companies in England and Wales have drinking water which exceeds this nitrate level. In 1989, the EC decided to prosecute the UK for falling below EC water standards.

Figure 18.4A ● Algal bloom

SUMMARY

When a crop receives more fertiliser than it can use, nitrates and phosphates wash into lakes and rivers. There, they stimulate the growth of weeds and algae. When the plants die, bacterial decay of the dead material uses oxygen. The resulting shortage of dissolved oxygen kills fish.

The level of nitrates in ground water, from which we obtain much of our drinking water, is rising.

Pollution can be reduced by reducing the application of fertilisers and by omitting phosphates from detergents.

There are two health worries over nitrates. Nitrates are converted into nitrites (salts containing the NO_2^- ion). Some chemists think that nitrites are converted in the body into nitrosoamines. These compounds cause cancer. The other worry is that nitrites oxidise the iron in haemoglobin. The oxidised form of haemoglobin can no longer combine with oxygen. The extreme form of nitrite poisoning is 'blue baby' syndrome, in which the baby turns blue from lack of oxygen. Babies are more at risk than adults because babies' stomachs are less acidic and assist the conversion of nitrates into nitrites.

The level of nitrites in drinking water permitted by the EC is 0.1 mg/l. Some parts of London have nitrite levels which are higher than this. The UK Government has agreed to bring the UK into line with the rest of Europe. To install nitrate-stripping equipment would cost £200 million. *Should the Government reduce the use of fertilisers? How? Should the Government introduce a tax on fertilisers or ration fertilisers?*

Pesticides

Other pollutants which must worry us are the pesticides dieldrin, endrin and aldrin (sometimes called the 'drins'). They cause liver cancer and affect the central nervous system. The EC sets a maximum level of 5×10^{-9} g/l for 'drins'. Half the water in the UK exceeds this level. The danger with the 'drins' is that fish take them in and do not excrete them. The level of 'drins' in fish may build up to 6000 times the level in water.

FIRST THOUGHTS

What are the 'drins'? Why is the EC worried about the level of drins in UK water?

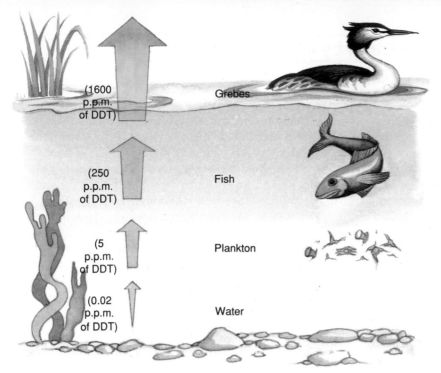

(1600 p.p.m. of DDT)

Grebes

(250 p.p.m. of DDT)

Fish

(5 p.p.m. of DDT)

Plankton

(0.02 p.p.m. of DDT)

Water

Figure 18.4B ● A food chain in Clear Lake, California

Figure 18.4B shows what happened when DDT, another powerful insecticide, was used to spray Clear Lake in California to get rid of mosquitoes. It is another example of pollutants being concentrated by a food chain.

SUMMARY

Pesticides are serious pollutants, especially when they can be concentrated through a food chain.

CHECKPOINT

❶ Groups of settlers in North America always built their villages on river banks, and discharged their sewage into the river. How did the river dispose of the sewage? Why can this method of sewage disposal not be used for larger settlements?

❷ British Tissues make toilet paper, paper towels, paper handkerchiefs, etc. They use a lot of bleach on the paper, and this bleach is one of the chemicals which the firm has to dispose of. Can you suggest how the firm could reduce the problem of bleach disposal?

❸ (a) Why do some lakes develop a thick layer of algal bloom?
(b) Why is algal bloom less likely to occur in a river?
(c) What harm does algal bloom do to a lake that is used as (i) a reservoir (ii) a fishing lake (iii) a boating lake?

❹ The concentration of nitrates in ground water is rising. Explain:
(a) why this is happening,
(b) why some people are worried about the increase.

❺ Water companies can tackle the problem of high nitrate levels by:
• blending water from high-nitrate sources with water from low-nitrate sources
• closing some sources of water
• treating the water with chemicals
• ion exchange
• microbiological methods

(a) Say what you think are the advantages and disadvantages of each of these methods.
(b) Which do you think would be the most expensive treatments? How will water companies be able to pay for the treatment?
(c) Suggest a different method of reducing the level of nitrates in ground water.
(d) Say who would pay for the method which you mention in (c) and how they would find the money.

18.5 Pollution by lead

FIRST THOUGHTS

Lead water pipes have been used for centuries. What is wrong with them?

LOOK AT LINKS
for **carbonates**
See Topic 16.2.

SUMMARY

Some parts of the UK have lead water pipes. Lead compounds are toxic. In hard water areas, the deposition of insoluble scale in the pipes stops lead dissolving. In other areas, the solution of lead is reduced by keeping the water alkaline.

Lead was used to make water pipes in Roman times. Until the 1950s, lead water pipes were used widely in the UK. Slowly, lead dissolves in water. Lead and its compounds are poisonous. Since 1950, copper has been used for water pipes. In cold water systems, plastic pipes are used. In hard water areas, lead pipes are safe because a layer of insoluble lead carbonate, calcium carbonate and magnesium carbonate builds up and acts as a protective barrier which stops lead dissolving. In soft water areas, the pH of the water may be less than 5, and lead dissolves more rapidly. Water companies make the water more alkaline by adding calcium hydroxide. During the 1983 strike of UK water workers, the treatment of many water supplies with calcium hydroxide stopped. Within two weeks, the level of lead in the tap water of some houses with lead pipes had risen from 40 µg/l to over 800 µg/l. In the North-West, 600 000 houses have lead pipes. Some parts of Scotland have water which contains more lead than the EC standard.

CHECKPOINT

❶ (a) How does lead get into our drinking water?
  (b) What harm does it do?
  (c) In what ways can the amount of lead in drinking water be reduced?
  (d) Pollution of air by lead is another problem. How does lead get into the air? (See Topic 17 if you need to revise.)

18.6  Pollution by oil

FIRST THOUGHTS

Huge oil tankers sail the sea. Sometimes one has an accident – not often, but often enough to cause a pollution disaster.
The first big oil spill off the UK coast was when the oil tanker, the *Torrey Canyon*, sank off Cornwall in 1967. The pollution fouled beaches in Cornwall and killed thousands of sea birds. Now, tankers of up to 550 000 tonnes are in use, and accidents can happen on a much larger scale.

Prince William Sound was a beautiful unspoiled bay in Alaska. It was home to a huge variety of sea animals. Death struck over an area of 1300 square kilometres in 1989. The supertanker *Exxon Valdez* left Valdez with a cargo of oil from Alaska. Only 40 km out of port, the tanker hit a submerged reef and 60 million litres of crude oil leaked from her tanks. Fish, sea mammals and migrating birds from all parts of the American continent perished in the giant oil slick.

Sea birds dive to obtain food. If the sea is polluted by a discharge of oil, sea birds may find themselves in an oil slick when they surface. Then oil sticks to their feathers and they cannot fly. They drift on the surface, becoming more and more waterlogged until they die of hunger and exhaustion. Thirty five thousand sea birds died in the *Exxon Valdez* disaster.

Oil spills at sea are the results of capsizings, collisions and accidental spills during loading and unloading at oil terminals. There is another source. After a large tanker has unloaded, it may have 200 tonnes of oil left in its tanks. While in port, the tanker is flushed out with water sprays, and the cleaning water is collected in a special tank, where the oil separates. Some captains save time by flushing out their tanks at sea and pumping the wash water overboard. This is illegal. Maritime nations have tried to set up standards to stop pollution of the seas, but several

Figure 18.6A ● The *Torrey Canyon*

Figure 18.6B ● The *Exxon Valdez* accident

Figure 18.6C ● Ten thousand sea otters perished in the *Exxon Valdez* spill. Some ate poisoned fish. Others drowned when their fur became clogged with oil

Figure 18.6D ● Burning the spilled oil from a smaller scale accident

Figure 18.6E ● Another victim

SCIENCE AT WORK

Bacteria can be used to clean out a tanker's storage compartment. The empty tank is filled with sea water, nutrient, air and bacteria. When the tanker reaches its destination, the tank contains clean water, a small amount of recoverable oil and an increased number of bacteria. The bacteria can be used as animal feed.

RESOURCE
ACTIVITY
PACK

SUMMARY

Spillage of oil from large tankers is a source of pollution at sea. It kills marine animals and washes ashore to pollute beaches.

nations have not signed the agreements. Enforcing agreements is very difficult as it is impossible to detect everything that happens at sea.

Various methods have been tried for the removal of oil from the surface of the sea.

- **Dispersal**   Chemicals are added to emulsify the oil. The danger is that they may be toxic to marine life.

- **Sinking**   Oil may be treated with sand and other fine materials to make it sink. A danger is that the sunken oil may cover and destroy the feeding areas of marine creatures.

- **Burning**   Burning oil is dangerous as a fire can spread rapidly over the sea. Research has been done on safe methods of burning oil, but they leave 15 per cent of the oil behind as lumps of tar.

- **Absorbing**   Absorbents do not work well in the open sea. They provide the best way of cleaning a beach or preventing an oil spill from reaching the shore.

- **Skimming off**   The method of surrounding an oil spill with a line of booms to prevent it spreading and then pumping oil off the surface has been used with some success.

- **Solidifying**   Scientists at British Petroleum have discovered chemicals which will solidify oil spills. The chemicals must be sprayed on to the oil slick from the air. They convert the oil into a rubber-like solid which can be skimmed off the surface in nets.

- **Bacteria**   There are bacteria which will decompose petroleum. A mixture of bacteria (of the correct strain) and nutrients is sprinkled on to the spill from the air.

CHECKPOINT

 (a) What are the causes of oil spills at sea?
   (b) What damage do they do?
   (c) Who pays to clean up the mess?
   (d) Suggest what can be done to stop pollution of the sea by oil.

? THEME QUESTIONS

1 (a) What happens to the temperature of a gas if the gas is compressed suddenly?
  (b) What happens to the temperature of a gas if the gas is allowed to expand suddenly? How is this effect used to liquefy air? Why is this method chosen for the liquefaction of air?
  (c) Boiling points are nitrogen, −196 °C; oxygen, −183 °C.
     Explain how the difference in boiling points makes it possible to separate oxygen and nitrogen from liquid air.
  (d) Give one-large scale industrial use for (i) oxygen (ii) nitrogen and (iii) another gas which is obtained from liquid air. Say why each gas is chosen for that particular use.

2 Three of the gases in air dissolve in water.
  (a) Which of them dissolves to give an acidic solution? What use is made of this solution?
  (b) Which of the three gases is a nuisance to deep-sea divers? Explain why, and say how the problem has been solved.
  (c) The life processes of plants and animals depend on the solubility of two of these gases. Explain why.

3 The table shows the composition of inhaled air and exhaled air, excluding water vapour, and a comment on the content of water vapour.

	Percentage by volume	
	Inhaled air	Exhaled air
Oxygen	21	17
Nitrogen	78	78
Carbon dioxide	0.03	4
Noble gases	1	1
Water vapour	Variable	Saturated

  (a) Describe the differences between inhaled air and exhaled air.
  (b) Briefly give the cause of each of these differences.

4 A classroom contains 36 pupils. The doors and windows are closed for half an hour. Answer these questions about the air at the end of the half hour.
  (a) Will the air temperature be higher or lower? Explain your answer.
  (b) Will the air be more or less humid (moist)? Explain your answer.
  (c) Will the percentage of carbon dioxide in the air be higher or lower? Explain your answer. Say how the change in carbon dioxide content will affect the class.

5 You are given four gas jars. One contains oxygen, one nitrogen, one carbon dioxide and one hydrogen. Describe how you would find out which is which.

6 The diagram shows an apparatus which is being used to pass a sample of air slowly over heated copper.

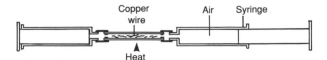

  (a) Describe how the appearance of the copper changes.
  (b) Which gas is removed from the air by copper? Name the solid product formed.
  (c) If 250 cm³ of air are treated in this way, what volume of gas will remain?
  (d) Name the chief component of the gas that remains.
  (e) Name two other gases that are present in air.

7 When petrol burns in a car engine, carbon dioxide and carbon monoxide are two of the products.
  (a) Write the formula of (i) carbon dioxide (ii) carbon monoxide.
  (b) Explain the statement 'Carbon monoxide is a product of incomplete combustion.'
  (c) Red blood cells contain haemoglobin. What vital job does haemoglobin do in the body?
  (d) If people breathe in too much carbon monoxide, it may kill them. How does carbon monoxide cause death?
  (e) Explain how people can be poisoned accidentally by carbon monoxide.
  (f) What precautions can people take to make sure that carbon monoxide is not formed in their homes?
  (g) The blood of people who smoke contains more carbon monoxide than the blood of non-smokers. Can you explain why?

8 Ruth carried out an experiment to compare the hardness of the water from three towns. She measured 50 cm³ of each water sample into separate conical flasks. She added soap solution gradually to each flask, shaking them until a lather was formed. Her results are shown in the table.

Water sample	Volume of soap solution (cm³)
Distilled water	2.0
Johnstown water	7.5
Mansville water	10.0
Rumchester water	4.0

  (a) Say what piece of apparatus Ruth could use for measuring (i) the 50 cm³ samples of water and (ii) the volume of soap solution added.
  (b) Explain why she did a test on distilled water.
  (c) Why does distilled water require the smallest volume of soap solution to form a lather?
  (d) Which town has the hardest water? Explain your answer.

(e) When Ruth boiled a 50 cm³ of Rumchester water before testing it, she found that the volume of soap solution needed to produce a lather was 2.0 cm³. Explain why she got a different result with boiled water.

(f) Recommend two measures that the hard water towns could take to cut down on their soap consumption.

9 The diagram shows rain falling on the ground and trickling over underground rocks.

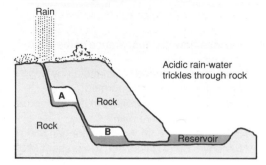

(a) Explain why natural rain water is weakly acidic.
(b) Name a type of rock which will be attacked by acidic rain water.
(c) Name the type of cavity, A, that is formed as a result of the action of rain water on the rock.
(d) Name the type of water that accumulates at B.
(e) Water flows from B into a reservoir. What treatment does the water need before it is fit to drink? Explain why this water receives a different treatment from river water.
(f) Will the water in the reservoir be hard water or soft water? Explain your answer.
(g) State one advantage and one disadvantage of hard water compared with soft water.

10 (a) What is the difference between raw sewage and digested sludge?
(b) Why can raw sewage not be used as a fertiliser?
(c) The digested sludge which sewage works produce is sold to farmers. Who benefits from this sale?
(d) Why do water treatment works treat drinking water with chlorine?
(e) How and why does the treatment of ground water differ from that of river water?

11 (a) Name three pollutants that are produced by power stations.
(b) For one of the pollutants, describe the kind of cleaning system that can be used to stop the pollutant being discharged into the air.
(c) How will the cost of electricity be affected by (i) installing the cleaning system and (ii) stocking the chemicals consumed in running the cleaning system?

12 (a) What is the difference between oxygen and ozone?
(b) What converts oxygen into ozone?
(c) What converts ozone into oxygen?
(d) What Is the ozone layer? Where is it?
(e) Why is the ozone layer becoming thinner?
(f) Why does the decrease in the ozone layer make people worry?

13 The nitrogen monoxide content of the atmosphere is 5 x 10⁶ tonnes. One supersonic transport (SST) flies on average 2500 hours per year and emits 3 tonnes of nitrogen monoxide per hour.
(a) How much nitrogen monoxide will be emitted in one year by a fleet of 50 SSTs?
(b) Calculate the ratio

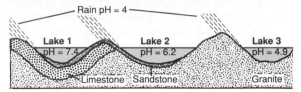

14 Cleopatra's needle has corroded more in London since 1878 than it did in 30 centuries in the Egyptian desert. Can you explain this?

15 The diagram shows acid rain falling on the shores of three lakes.

Rain pH = 4

Lake 1 pH = 7.4   Lake 2 pH = 6.2   Lake 3 pH = 4.9

Limestone Sandstone Granite

(a) Unpolluted rain water has a pH of 6.8. What gives it this weak acidity?
(b) By acid rain, we mean rain with a pH below 5.6. Name two substances that react with rain to make it strongly acidic.
(c) Explain why Lake 3 is more acidic than Lake 2.
(d) Explain why Lake 2 is more acidic than Lake 1.
(e) Lakes in Sweden become more acidic in the spring. Suggest an explanation.
(f) Acidic lakes in Sweden are treated with crushed limestone. Explain how this reduces the acidity. Give two disadvantages of this solution to the problem.

16 Many industrial plants take water from a river and then return it at a higher temperature. What harm can this do? What name is given to this practice?

17 Bacteria in river water are able to convert many pollutants into harmless products. What, then, is the harm in dumping waste into rivers?

18 What happens when plastics are dumped in lakes and rivers?

19 It takes 60 g of oxygen per day to oxidise the sewage from one person.
Water contains 10 p.p.m. of dissolved oxygen. What mass of water is robbed of its dissolved oxygen by oxidising the sewage from a village of 400 people in one day? (1 tonne = 1000 kg)

20 Water normally contains 10 p.p.m. of oxygen. It takes 4 g of oxygen to oxidise completely 1 g of oil. A garage does an oil change on a car, and pours the 4 kg of dirty oil down the drain. What mass of water will be stripped of its dissolved oxygen by oxidising the oil from this car?

21 Why is the pollution of estuaries so common? Who gains from being able to pollute estuaries? Who loses from this pollution?

THEME F
Using Earth's Resources

The history of the human race is divided into three ages: the Stone Age, the Bronze Age and the Iron Age. Two of the three ages are named after metals. In this theme, we look at the variety of metals found on the surface of the Earth. We review their structure and their chemical reactions to find out what gives them their unique role in the operation of a civilised society.

We look at the chemical cells which provide us with power. We study the chemicals which give us safe drinking water and those which preserve our food.

We could not make good use of Earth's resources without being able to find out what elements and compounds are present in the substances which we find in the Earth's crust. Chemical analysis enables us to do this.

TOPIC 19 METALS

19.1 Metals and alloys

Metals and alloys have played an important part in history. The discovery of bronze made it possible for the human race to advance out of the Stone Age into the Bronze Age. Centuries later, smiths found out how to extract iron from rocks, and the Iron Age was born. In the nineteenth century, the invention of steel made the Industrial Revolution possible.

Metals are strong materials. They are used for purposes where strength is required. Metals can be worked into complicated shapes. They can be ground to take a cutting edge. They conduct heat and electricity. As science and technology advance, metals are put to work for more and more purposes.

Alloys

An alloy is a mixture of metallic elements and in some cases non-metallic elements also. Many metallic elements are not strong enough to be used for the manufacture of machines and vehicles which will have to withstand stress. Alloying a metallic element with another element is a way of increasing its strength. Steel is an alloy of iron with carbon and often other metallic elements. Duralumin is an alloy of aluminium with copper and magnesium. This alloy is much stronger than pure aluminium and is used for aircraft manufacture (Figure 19.1A).

Alloys have different properties from the elements of which they are composed. Brass is made from copper and zinc. It has a lower melting point than either of these metals. This makes it easier to cast, that is, to melt and pour into moulds. Brass musical instruments have a more pleasant and sonorous sound than instruments made from either of the two elements copper or zinc.

Figure 19.1A ● Concorde

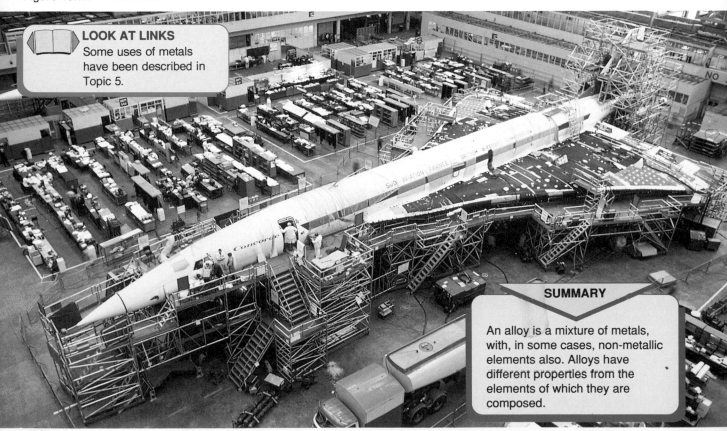

LOOK AT LINKS
Some uses of metals have been described in Topic 5.

SUMMARY

An alloy is a mixture of metals, with, in some cases, non-metallic elements also. Alloys have different properties from the elements of which they are composed.

FIRST THOUGHTS

Metals possess their remarkable and useful properties because of the type of chemical bond between the atoms: the metallic bond.

LOOK AT LINKS
for **the conduction of electricity**
See Topic 20.3 of KS·
Physics.

SUMMARY

The metallic bond is a strong bond. It gives metals their strength and allows them to conduct electricity.

The metallic bond

A piece of metal consists of positive metal ions and free electrons (see Figure 19.2A). The free electrons are the outermost electrons which break free when the metal atoms form ions. Free electrons move about between the metal ions. This is what prevents the metal ions from being driven apart by repulsion between their positive charges.

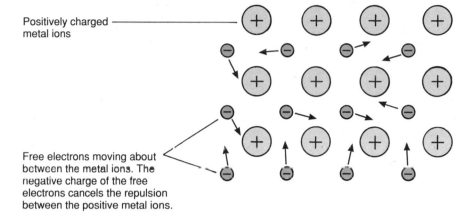

Positively charged metal ions

Free electrons moving about between the metal ions. The negative charge of the free electrons cancels the repulsion between the positive metal ions.

Figure 19.2A ● The metallic bond

The metallic bond explains how metals **conduct electricity**. It also explains how metals can change their shape without breaking. When a metal is bent, the shape changes but the free electrons continue to hold the metal ions together.

The crystalline structure of metals

● *Grains*

When a metal solidifies from the molten state, many crystals start to grow at the same time in the liquid, and they produce a solid mass of small crystals (see Figure 19.2B). These individual crystals are called **'grains'** by metallurgists. Each grain grows independently, and the directions in which the atoms are arranged are not related to those in neighbouring grains (Figure 19.2B). Between the grains are atoms which have not fitted into the crystal structure. These are often atoms of impurities. The crystal structure of the metal grains is imperfect. There are faults in the orderly arrangement of atoms called **dislocations** (see Figure 19.2C).

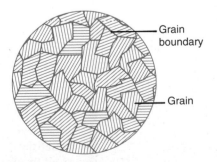

Grain boundary

Grain

Figure 19.2B ● Metal grains

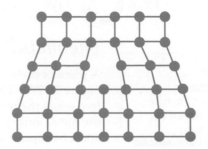

Figure 19.2C ● A dislocation

The presence of dislocations causes weakness. When the structure is stressed, the dislocation may be forced to take up another position. As the process is repeated, the dislocation may travel along to the crystal boundary (Figure 19.2D). The mobility of dislocations explains the malleability of metals.

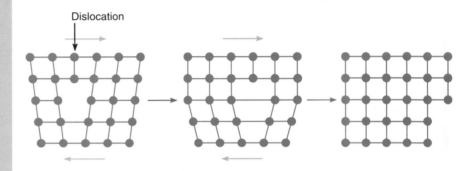

Figure 19.2D ● Dislocations tend to move towards a grain boundary

Other imperfections in the structure are caused by the presence of extra atoms and by atoms being missing from a particular position (see Figure 19.2E).

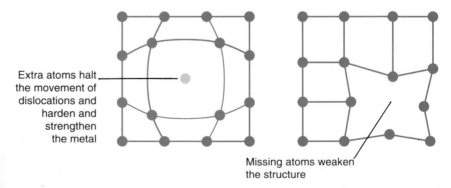

Extra atoms halt the movement of dislocations and harden and strengthen the metal

Missing atoms weaken the structure

Figure 19.2E ● Imperfections in the structure of grains

Alloying is used to harden a metal by providing additional atoms which will be trapped between the grains. The greater the number of grains, the stronger is the metal. Therefore 'grain refining agents' are sometimes added to molten metal to provide nuclei from which crystals grow to produce 'fine-grained' metal, which contains a large number of small grains.

SUMMARY

A metal consists of a mass of tiny crystals called grains. Faults in the crystal structure of a grain are called dislocations. The presence of dislocations weakens a metal. The mobility of dislocations under stress gives metals their ability to change shape without breaking. A piece of metal with a large number of small grains is stronger than a piece with a smaller number of larger grains.

Work-hardening

Metals can be **cold-worked**, i.e. forced into new shapes without being heated. Eventually they become **work-hardened**. This happens because, as forces are applied, large numbers of dislocations are produced. At first dislocations move easily as the shape changes. After a time, movement becomes difficult as each dislocation becomes entangled with others trying to move in different directions. A point is reached where dislocations cannot move and the grains can deform no further without risk of fracture.

Recovery

If the work-hardened metal is heated, its atoms gain enough energy to move by diffusion and can rearrange within the grains. The heating time is kept short enough and the temperature is kept low enough to prevent grains from growing. This heat treatment is called **recovery** or **stress relief**.

Recrystallisation

If the temperature of the work-hardened metal is raised further, it will recrystallise to form numerous fine grains. The temperature at which this happens is called the **recrystallisation temperature**.

Annealing

Work-hardened metals can have their ductility restored by a treatment called **annealing**. The metal object is put into a furnace and left until both recovery and recrystallisation are complete. When a metal is shaped by cold-working, it may need to be annealed several times during the process, that is, each time work-hardening makes the metal too hard for further cold-working.

Hot-working

Hot-working or **forging** means working a metal above its recrystallisation temperature. The metal is being annealed at the same time as it is worked. Forging requires less mechanical energy than cold-working because the metal is softer and there are no interruptions for annealing.

Quenching and tempering

Quenching is a sudden immersion of a hot metal in oil or water at room temperature. Quenching is important in the treatment of cutting tools. Many crystalline materials, including metals, can change from one crystalline form into another on heating or cooling. The **austenite** form of steel, which exists at higher temperatures, can hold more carbon than the **ferrite** form, which exists at lower temperatures. Rapid quenching does not give the carbon atoms enough time to move, and the austenite form persists at the lower temperature and retains its carbon content. The carbon atoms prevent the movement of dislocations, and the metal becomes harder, less malleable and more brittle.

To reduce brittleness, some hardness must be sacrificed. A compromise is reached; the steel is **tempered**. The metal object is heated a second time to a selected temperature and then quenched again. The temperature at which the steel is worked must not be above the **tempering temperature**.

Age-hardening

The solute may gradually precipitate from the solid solution. Small particles of solute form inside each grain. They harden the material still further by blocking the movement of dislocations. The process is called **age-hardening**.

IT'S A FACT

Age-hardening takes place in a few days in Duralumin (aluminium containing 3–4% of copper). This discovery led to the development of the alloy used for airships in the First World War. Copper atoms migrate to form small regions of $CuAl_2$ which impede the movement of dislocations.

SUMMARY

Improvements in the hardness and strength of metals can be made by
• cold-working
• quenching and tempering
• alloying
• a combination of these factors.
The improvements are due to restricting the movement of dislocations.

CHECKPOINT

❶ (a) Why is it appropriate to describe a metal as a polycrystalline material?
(b) How does it come about that a molten metal crystallises in this way?

❷ Explain what a dislocation is and how the malleability of a metal depends on the presence of dislocations.

❸ Recovery, recrystallisation and annealing are three methods of enabling a work-hardened metal to regain its malleability. Explain the differences between them.

❹ What is the advantage of hot-working a metal over cold-working?

❺ What is quenching? Why does it make a metal harder?

19.3 The chemical reactions of metals

> **FIRST THOUGHTS**
>
> You have already met the reaction between metals and acids in Topic 12.1 and the reaction between metals and air in Topic 15.5. In this section, you will find out how to classify metals on the basis of the vigour of their chemical reactions.

Reactions of metals with air

Many metals react with the oxygen in the air (see Table 19.1).

Table 19.1 ● The reactions between metals and oxygen

Metal	Symbol	Reaction when heated in air	Reaction with cold air
Potassium	K		
Sodium	Na		
Calcium	Ca		
Magnesium	Mg	Burn in air to form oxides	React slowly with air to form a surface film of the metal oxide. This reaction is called **tarnishing**
Aluminium	Al		
Zinc	Zn		
Iron	Fe		
Tin	Sn		
Lead	Pb	When heated in air, these metals form oxides without burning	
Copper	Cu		
Silver	Ag		Silver tarnishes in air
Gold	Au	Do not react	
Platinum	Pt		Do not react

Reactions of metals with water

Table 19.2 ● Reactions of metals with cold water and steam

Metal	*Reaction with water*
Potassium 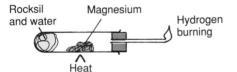 The hydrogen that is formed burns — Potassium reacts violently with water	A violent reaction occurs. Hydrogen and potassium hydroxide solution, a strong alkali, are formed. The reaction is so exothermic that the hydrogen burns. The flame is coloured lilac by potassium vapour. Potassium + Water → Hydrogen + Potassium hydroxide 2K (s) + 2H$_2$O(l) → H$_2$(g) + 2KOH(aq) Potassium is kept under oil to prevent water vapour and oxygen in the air from attacking it.
Sodium Lighted taper — Pyrex tube — Sodium — Water	Reacts slightly less violently than potassium does. Hydrogen and sodium hydroxide solution, a strong alkali, are formed. The hydrogen formed burns with a yellow flame. The flame colour is due to the presence of sodium vapour. Sodium + Water → Hydrogen + Sodium hydroxide 2Na(s) + 2H$_2$O(l) → H$_2$(g) + 2NaOH(aq) Sodium is kept under oil to prevent water vapour and oxygen in the air attacking it.
Calcium Hydrogen — Water — Calcium reacts steadily with water	Reacts readily but not violently with cold water to form hydrogen and calcium hydroxide solution, the alkali limewater. Calcium + Water → Hydrogen + Calcium hydroxide Ca(s) + 2H$_2$O(l) → H$_2$(g) + Ca(OH)$_2$(aq)
Magnesium Test tube — Hydrogen Filter tunnel — Magnesium	Reacts slowly with cold water to form hydrogen and magnesium hydroxide. Magnesium + Water → Hydrogen + Magnesium hydroxide Mg(s) + 2H$_2$O(l) → H$_2$(g) + Mg(OH)$_2$(aq)
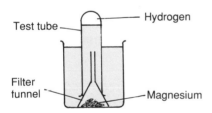 Rocksil and water — Magnesium — Hydrogen burning — Heat	Burns in steam to form hydrogen and magnesium oxide. Magnesium + Steam → Hydrogen + Magnesium oxide Mg(s) + H$_2$O(g) → H$_2$(g) + MgO(s)
Aluminium	Aluminium has a surface layer of aluminium oxide which is unreactive. When the oxide layer is removed, aluminium reacts readily with water to form hydrogen and aluminium oxide.
Zinc 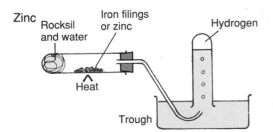 Rocksil and water — Iron filings or zinc — Hydrogen — Heat — Trough	Reacts with steam to form hydrogen and zinc oxide. Zinc + Steam → Hydrogen + Zinc oxide Zn(s) + H$_2$O(g) → H$_2$(g) + ZnO(s)

(continued overleaf)

Table 19.2 ● Reactions of metals with cold water and steam (continued)

Metal	Reaction with water
Iron	Reacts with steam to form hydrogen and the oxide, Fe_3O_4, tri-iron tetraoxide, which is blue-black in colour.

$$\text{Iron} + \text{Steam} \rightarrow \text{Hydrogen} + \text{Tri-iron tetraoxide}$$
$$3Fe(s) + 4H_2O(g) \rightarrow 4H_2(g) + Fe_3O_4(s)$$

Metal	Reaction with water
Tin Lead Copper Silver Gold Platinum	Do not react

Reactions of metals with dilute acids

Table 19.3 ● The reactions of metals with dilute acids

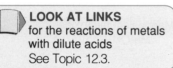

LOOK AT LINKS
for the reactions of metals
with dilute acids
See Topic 12.3.

Metal	Reaction with dilute acid
Potassium Sodium Lithium	The reaction is dangerously violent **Do not try it**
Calcium Magnesium Aluminium Zinc Iron Tin Lead	These metals react with dilute hydrochloric acid to give hydrogen and a solution of the metal chloride, e.g. $\text{Zinc} + \text{Hydrochloric acid} \rightarrow \text{Hydrogen} + \text{Zinc chloride}$ $Zn(s) + 2HCl(aq) \rightarrow H_2(g) + ZnCl_2(aq)$ The vigour of the reaction decreases from calcium to lead. Lead reacts very slowly. With dilute sulphuric acid, the metals give hydrogen and sulphates.
Copper Silver Gold Platinum	These metals do not react with dilute hydrochloric acid and dilute sulphuric acid. (Copper reacts with dilute nitric acid to form copper nitrate. Nitric acid is an oxidising agent as well as an acid.)

CHECKPOINT

See Theme E, Topic 15.6 if you need to revise.

❶ State whether the oxides of the following elements are acidic or basic or neutral.
   (a) iron   (d) sulphur
   (b) carbon   (e) zinc
   (c) copper

❷ Write (a) word equations (b) balanced chemical equations for the combustion of the following elements in oxygen to form oxides.
   (i) zinc   (iii) sodium
   (ii) magnesium   (iv) carbon

❸ (a) Which are attacked more by acid rain: lead gutters or iron fall pipes?
   (b) Food cans made of iron are coated with tin. How does this help them to resist attack by the acids in foods?

❹ Write word equations and symbol equations for the reactions between:
   (a) magnesium and hydrochloric acid,
   (b) iron and hydrochloric acid to form iron(II) chloride,
   (c) zinc and sulphuric acid,
   (d) iron and sulphuric acid to form iron(II) sulphate.

19.4 The reactivity series

There are over seventy metallic elements. One way of classifying them is to arrange them in a sort of league table, with the most reactive metals at the top of the league and the least reactive metals at the bottom. This section shows how it is done.

Structure and Bonding (program)

Find the section of the program on the reactivity series. Use it to investigate the properties of different metals in the series.

Chemistry Super Stars (program)

Test your chemistry skill using the game in this program on the reactivity of metals. Try to identify the metals involved.

SUMMARY

- Many metals react with oxygen; some metals burn in oxygen. The oxides of metals are bases.
- A few metals (for example sodium) react with cold water to give a metal hydroxide and hydrogen. Some metals react with steam to give a metal oxide and hydrogen.
- Many metals react with dilute acids to give hydrogen and the salt of the metal.
Metals fall into the same order of reactivity in all these reactions. This order is called the reactivity series of the metals.

There are over 70 metals in the Earth's crust. Table 19.4 summarises the reactions of metals with oxygen, water and acids. You will see that the same metals are the most reactive in the different reactions.

Table 19.4 ● Reactions of metals

Metal	Reaction when heated in oxygen	Reaction with cold water	Reaction with dilute hydrochloric acid
Potassium Sodium Lithium Calcium	Burn to form oxides	Displace hydrogen; form alkaline hydroxides	React dangerously fast
Magnesium		Slow reaction	Displace hydrogen; form metal chlorides
Aluminium Zinc Iron		No reaction, except for slow rusting of iron; all react with steam	
Tin Lead	Oxides form slowly without burning	Do not react, even with steam	React very slowly
Copper Silver			Do not react
Gold Platinum	Do not react		

The metals can be listed in order of **reactivity**, in order of their readiness to take part in chemical reactions. This list is called the **reactivity series** of the metals. Table 19.5 shows part of the reactivity series.

Table 19.5 ● Part of the reactivity series

Metal	Symbol	
Potassium	K	
Sodium	Na	
Calcium	Ca	
Magnesium	Mg	
Aluminium	Al	
Zinc	Zn	Reactivity
Iron	Fe	decreases
Tin	Sn	from top
Lead	Pb	to bottom
Copper	Cu	
Silver	Ag	
Gold	Au	
Platinum	Pt	

If you have used aluminium saucepans, you may be surprised to see aluminium placed with the reactive metals high in the reactivity series. In fact, aluminium is so reactive that, as soon as aluminium is exposed to the air, the surface immediately reacts with oxygen to form aluminium oxide. The surface layer of aluminium oxide is unreactive and prevents the metal from showing its true reactivity.

CHECKPOINT

❶ In some parts of the world, copper is found 'native'. Why is zinc never found native?

❷ The ancient Egyptians put gold and silver objects into tombs. Explain why people opening the tombs thousands of years later find the objects still in good condition. Why are no iron objects found in the tombs?

❸ The following metals are listed in order of reactivity:

sodium > magnesium > zinc > copper

Describe the reactions of these metals with (a) water, (b) dilute hydrochloric acid. Point out how the reactions illustrate the change in reactivity.

19.5 Metals in the Periodic Table

FIRST THOUGHTS

The reactivity series is one way of classifying metals. The Periodic Table of the elements, which you met in Topic 7.5, is another.

LOOK AT LINKS

Silicon and germanium are on the borderline between metals and non-metals. They are semiconductors. For the importance of semiconductors in computers, see Topic 5.1 and *KS: Physics*, Topic 23.

Table 19.6 ● The Periodic Table

In Table 19.6 you see how a line has been drawn across the Periodic Table to separate the metallic elements (on the left) from the non-metallic elements (on the right). On the far left are the very reactive metals lithium, sodium and potassium. These metals are called the **alkali metals** because their hydroxides are strong alkalis. They form ions of formula M^+ and occupy Group 1 of the Periodic Table. The metals of Group 1 are at the top of the reactivity series. The metals in Group 2 form ions of formula M^{2+}. They are called the **alkaline earths**. They are lower in the reactivity series than Group 1 metals. Aluminium is in Group 3. The less reactive metals, tin and lead, are in Group 4.

The metals in between Group 2 and Group 3 are called the **transition metals**. They include iron, copper, silver, gold and zinc. Transition metals form more than one kind of ion, for example iron forms the ions Fe^{2+} and Fe^{3+}. The ions of transition metals are often coloured. Transition metals and their ions often act as catalysts. Vanadium(V) oxide is used as a catalyst in the manufacture of sulphuric acid. Iron and molybdenum catalyse the manufacture of ammonia.

SUMMARY

The Periodic Table is a well-known method of classifying the elements. Group 1 contains the most reactive metals, the alkali metals. Group 2 contains other reactive metals, the alkaline earths. Aluminium is in Group 3. Less reactive metals are in Group 4 and in the block of transition metals. Groups 5, 6, 7 and 0 contain the non-metallic elements.

LOOK AT LINKS
The differences between the physical properties of metallic and non-metallic elements are summarised in Table 5.1.

SUMMARY

Metallic and non-metallic elements differ in:
• their reactions with acids,
• the acid–base nature of their oxides,
• the nature of their chlorides,
• the type of ions they form.

Table 19.7 summarises the differences between the chemical properties of metallic and non-metallic elements.

Table 19.7 ● Chemical properties of metallic and non-metallic elements

Metallic elements	Non-metallic elements
Metals which are high in the reactivity series react with dilute acids to give hydrogen and a salt of the metal	Non-metallic elements do not react with dilute acids
Metallic elements form positive ions, e.g. Na^+, Zn^{2+}, Fe^{3+}	Non-metallic elements form negative ions, e.g. Cl^-, O^{2-}
Metal oxides and hydroxides are bases, e.g. Na_2O, CaO, $NaOH$. If they dissolve in water they give alkaline solutions, e.g. $NaOH$	Many oxides are acids and dissolve in water to give acidic solutions, e.g. CO_2, SO_2. Some oxides are neutral and insoluble, e.g. CO
The chlorides of metals are ionic crystalline solids, e.g. $NaCl$	The chlorides of the non-metals are covalent liquids or gases, e.g. $HCl(g)$, $CCl_4(l)$

LOOK AT LINKS
for the reactions of metals
See Topics 19.3 and 19.4.

A closer look at metals

Looking more closely, what is the difference between the alkali metals in Group 1, the alkaline earths in Group 2 and the transition metals? Table 19.8 summarises some of their reactions.

Table 19.8 ● Some reactions of metals (M stands for the symbol of a metallic element.)

	Reaction with air	Reaction with water	Reaction with dilute acids	Trend
Group 1 Lithium Sodium Potassium Rubidium Caesium	Burn vigorously to form the strongly basic oxide, M_2O	React vigorously to form hydrogen and a solution of the strong alkali, MOH	The reaction is dangerously violent	The vigour of all these reactions increases down the group
Group 2 Beryllium Magnesium Calcium Strontium Barium	Burn to form the strongly basic oxide, MO	Reacts slowly. Burns in steam. React readily to form hydrogen and the alkali $M(OH)_2$	React readily to give hydrogen and a salt, e.g. MCl_2	The vigour of all these reactions increases down the group
Transition metals Iron	When heated, form oxides, without burning	Rusts slowly, reacts with steam to form hydrogen and iron oxide.	Reacts to give hydrogen and a salt	
Copper		Does not react	Does not react	

The alkali metals

From Table 19.8, you can see how the Periodic Table makes it easier to learn about all the elements. Look at the elements in Group 1: lithium, sodium, potassium, rubidium and caesium. They are all very reactive metals. Their oxides and hydroxides have the general formulas M_2O and MOH (where M is the symbol of the metallic element) and are strongly basic. The oxides and hydroxides dissolve in water to give strongly alkaline solutions. The reactivity of the alkali metals increases as you pass down the group. If you know these facts, you do not need to learn the properties of all the metals separately. If you know the properties of sodium, you can predict those of potassium and lithium. The Periodic Table saves you from having to learn the properties of 106 elements separately!

The alkaline earths

The metals in Group 2 are less reactive than those in Group 1. They form basic oxides and hydroxides with the general formulas MO and $M(OH)_2$. Their oxides and hydroxides are either sparingly soluble or insoluble. The reactivity of the alkaline earths increases as you pass down the group. Again, if you know the chemical reactions of one element, you can predict the reactions of other elements in the group.

The transition metals

The transition metals in the block between Group 2 and Group 3 are a set of similar metallic elements. They are less reactive than those in Groups 1 and 2. Their oxides and hydroxides are less strongly basic and are insoluble.

● Physical properties

Transition metals are hard and dense. They are good conductors of heat and electricity (e.g. copper wire in electrical circuits). Their melting points, boiling points and heats of melting are all higher than those for Group 1 and Group 2 metals. These are all a measure of the strength of the metallic bond. Iron, cobalt and nickel are strongly magnetic. Some of the other transition metals are weakly magnetic.

● Extraction

LOOK AT LINKS
for more detail on the extraction of zinc
See Topic 19.9.

Transition metals are less reactive than those in Groups 1 and 2. They can be extracted from their ores by means of chemical reducing agents. The method of extraction follows the steps listed below.
1 The ore is concentrated, e.g. by flotation.
2 Sulphide ores are roasted to convert them into oxides.
3 The oxide is reduced by heating with coke to form the metal and carbon monoxide.
4 The metal is purified.

● Valency

LOOK AT LINKS
for the uses of transition metals
See Tables 19.11, 19.14 and 19.15.

Transition metals characteristically use more than one valency.

Table 19.9 ● Some transition metals, ions and compounds

Metal	Ions	Compounds
Copper	Cu^+, Cu^{2+}	Cu_2O, $CuSO_4$
Iron	Fe^{2+}, Fe^{3+}	$FeSO_4$, $Fe_2(SO_4)_3$
Cobalt	Co^{2+}, Co^{3+}	$CoCl_2$, $CoCl_3$
Chromium	Cr^{2+}, Cr^{3+}	$CrCl_2$, $CrCl_3$
	$Cr_2O_7^{2-}$	$K_2Cr_2O_7$
Manganese	Mn^{2+}, MnO_4^-	$MnSO_4$, MnO_2, $KMnO_4$

LOOK AT LINKS
The colour of the hydroxide can be used to identify the metal present See Topic 22.2.

Table 19.10 ● The colours of some ions

Ion	Colour
Cu^{2+}(aq)	blue
Fe^{2+}(aq)	green
Fe^{3+}(aq)	rust
Ni^{2+}(aq)	green
Mn^{2+}(aq)	pink
Cr^{3+}(aq)	blue
Zn^{2+}(aq)	colourless

● *Colour*

Transition metals have coloured ions, except for zinc (see Table 19.10).

In addition to forming simple cations, transitions metals form oxo-ions in combination with oxygen. Examples are:

chromate(VI), CrO_4^{2-} (yellow), dichromate(VI), $Cr_2O_7^{2-}$ (orange)
manganate(VII), MnO_4^- (purple), zincate ZnO_2^{2-} (colourless)

● *Catalysis*

Many transition metals and their compounds are important catalysts. Examples are:

● iron and iron(III) oxide in the Haber process for making ammonia (Topic 29.2)
● platinum in the oxidation of ammonia during the manufacture of nitric acid (Topic 29.3)
● vanadium(V) oxide in the Contact process for making sulphuric acid (Topic 29.4)
● nickel in the hydrogenation of oils to form fats (Topic 30.1).

FIRST THOUGHTS

19.6 **Predictions from the reactivity series**

The classifications you have been learning about are useful: they enable you to make predictions about chemical reactions.

Competition between metals for oxygen

Aluminium is higher in the reactivity series than iron. When aluminium is heated with iron(III) oxide, a very vigorous, exothermic reaction occurs

Aluminium + Iron(III) oxide → Iron + Aluminium oxide
$$2Al(s) + Fe_2O_3(s) → 2Fe(s) + Al_2O_3(s)$$

This reaction is called the **thermit reaction** (therm = heat). It is used to mend railway lines because the iron formed is molten and can weld the broken lines together (see Figure 19.6A).

Figure 19.6A ● Using the thermit reaction

SUMMARY

Reactive metals displace metals lower down the reactivity series from their compounds.

Displacement reactions

Metals can displace other metals from their salts. A metal which is higher in the reactivity series will displace a metal which is lower in the reactivity series from a salt. Zinc will displace lead from a solution of lead nitrate.

CHECKPOINT

❶ Which metal is best for making saucepans: zinc, iron or copper? Explain your choice.

❷ Gold is used for making electrical contacts in space capsules. Explain (a) why gold is a good choice and (b) why it is not more widely used.

❸ A metal, romin, is displaced from a solution of one of its salts by a metal, sarin, A metal, tonin, displaces sarin from a solution of one of its salts. Place the metals in order of reactivity.

❹ The following metals are listed in order of reactivity, with the most reactive first:

Mg, Zn, Fe, Pb, Cu, Hg, Au

List the metals which will:
(a) occur 'native',   (b) react with cold water,
(c) react with steam,   (d) react with dilute acids,
(e) displace copper from copper(II) nitrate solution.

19.7 **Uses of metals and alloys**

FIRST THOUGHTS

As you read through this section, think about the reasons which lead to the choice of a metal or alloy for a particular purpose.

Metals and alloys have thousands of uses. The purposes for which a metal is used are determined by its physical and chemical properties. Sometimes a manufacturer wants a material for a particular purpose and there is no metal or alloy which fits the bill. Then metallurgists have to invent a new alloy with the right characteristics. Table 19.11 gives some examples.

Table 19.11 ● What are metals and alloys used for?

Metal/Alloy	Characteristics	Uses
Aluminium (Duralumin is an important alloy)	Low density Never corroded Good electrical conductor Good thermal conductor Reflector of light	Aircraft manufacture (Duralumin) Food wrapping Electrical cable Saucepans Car headlamps
Brass (alloy of copper and zinc)	Not corroded Easy to work with Sonorous Yellow colour	Ships' propellers Taps, screws Trumpets Ornaments
Bronze (alloy of copper and tin)	Harder than copper Not corroded Sonorous	Coins, medals Statues, springs Church bells

TRY THIS

To see some beautiful crystals, stand a piece of iron wire in a solution of lead nitrate. **Take care** – remember lead salts are poisonous.

Table 19.11 ● What are metals and alloys used for? (continued)

Metal/Alloy	Characteristics	Uses
Copper	Good electrical conductor Not corroded	Electrical circuits Water pipes and tanks
Gold	Beautiful colour Never tarnishes Easily worked	Jewellery Electrical contacts Filling teeth
Iron	Hard, strong, inexpensive, rusts	Motor vehicles, trains, ships, buildings
Lead	Dense Unreactive	Protection from radioactivity Was used for all plumbing (Lead is no longer used for water pipes as it reacts very slowly with water)
Magnesium	Bright flame	Distress flares, flash bulbs
Mercury	Liquid at room temperature	Thermometers Electrical contacts Dental amalgam for filling teeth
Silver	Good electrical conductor Good reflector of light Beautiful colour and shine (tarnishes in city air)	Electrical contacts Mirrors Jewellery
Sodium	High thermal capacity	Coolant in nuclear reactors
Solder (alloy of tin and lead)	Low melting point	Joining metals, e.g. in an electrical circuit
Steel (alloy of iron)	Strong	Construction, tools, ball bearings, magnets, cutlery, etc.
Tin	Low in reactivity series	Coating 'tin cans'
Titanium	Low in density Stays strong at high and low temperatures	Supersonic aircraft
Zinc	High in reactivity series	Protection of iron and steel; see Table 19.16.

SUMMARY

Metals and alloys are essential for many different purposes. Metals and alloys are chosen for particular uses because of their physical properties and their chemical reactions.

CHECKPOINT

❶ Name the metal which is used for each of these purposes. Explain why that metal is chosen.
(a) Thermometers
(b) Window frames
(c) Sinks and draining boards
(d) Radiators
(e) Water pipes
(f) Household electrical wiring
(g) Scissor blades

❷ Explain the following:
(a) Some mirrors have aluminium sprayed on to the back of the glass instead of silver, which was used previously.
(b) Although brass is a colourful, shiny material, it is not used for jewellery.
(c) Titanium oxide has replaced lead carbonate as the pigment in white paint.

19.8  Compounds and the reactivity series

The stability of metal oxides

- Reactive metals form compounds readily.
- The compounds of reactive metals are difficult to split up.

Magnesium is a reactive metal, and magnesium oxide is difficult to reduce to magnesium.

Copper is an unreactive metal, and copper(II) oxide is easily reduced to copper. Hydrogen will reduce hot copper(II) oxide to copper.

LOOK AT LINKS
for **reduction**
See Topic 15.6 and Topic 21.

$$\begin{array}{ccccccc} & & & & \text{Heat} & & \\ \text{Copper(II) oxide} & + & \text{Hydrogen} & \rightarrow & \text{Copper} & + & \text{Water} \\ \text{CuO(s)} & + & \text{H}_2\text{(g)} & \rightarrow & \text{Cu(s)} & + & \text{H}_2\text{O(l)} \end{array}$$

LOOK AT LINKS
for the extraction of
aluminium
See Topic 19.9.

Carbon is another reducing agent. When heated, it will reduce the oxides of metals which are fairly low in the reactivity series. Reduction by carbon is often the method employed to obtain metals from their ores.

$$\begin{array}{ccccccc} & & & & \text{Heat} & & \\ \text{Zinc oxide} & + & \text{Carbon} & \rightarrow & \text{Zinc} & + & \text{Carbon monoxide} \\ \text{ZnO(s)} & + & \text{C(s)} & \rightarrow & \text{Zn(s)} & + & \text{CO(g)} \end{array}$$

Neither hydrogen nor carbon will reduce the oxides of metals which are high in the reactivity series. Aluminium is high in the reactivity series; its oxide is difficult to reduce. The method used to obtain aluminium from aluminium oxide is electrolysis.

If a metal is very low in the reactivity series, its oxide will decompose when heated. The oxides of silver and mercury decompose when heated.

The stability of other compounds

When compounds are described as stable it means that they are difficult to decompose (split up) by heat. Table 19.12 shows how the stability of the compounds of a metal is related to the position of the metal in the reactivity series.

The sulphates, carbonates and hydroxides of the most reactive metals are not decomposed by heat. Those of other metals decompose to give oxides.

SUMMARY

The oxides of metals low in the reactivity series (e.g. Cu) are easily reduced by carbon and hydrogen. The oxides of metals high in the reactivity series are difficult to reduce.
Compounds of metals low in the reactivity series are decomposed by heat.
Compounds of metals high in the reactivity series are stable to heat.

Table 19.12 ● Action of heat on compounds

Cation	Anion				
	Oxide	Chloride	Sulphate	Carbonate	Hydroxide
Potassium Sodium			No decomposition		
Calcium Magnesium Aluminium Zinc Iron Lead Copper	No decomposition	No decomposition	Oxide and sulphur trioxide $MO + SO_3$ Some also give SO_2	Oxide and carbon dioxide $MO + CO_2$	Oxide and water $MO + H_2O$
Silver Gold	Metal + oxygen Not formed		Metal + $O_2 + SO_3$	Metal + $O_2 + CO_2$	Do not form hydroxides

19.9 Extraction of metals from their ores

One of the important jobs which chemists do is to find ways of extracting metals from the rocks of the Earth's crust.

Metals are found in rocks in the Earth's crust. A few metals, such as gold and copper, occur as the free metal, uncombined. They are said to occur 'native'. Only metals which are very unreactive can withstand the action of air and water for thousands of years without being converted into compounds. Most metals occur as compounds.

Rock containing the metal compound is mined. Then machines are used to crush and grind the rock. Next a chemical method must be found for extracting the metal. All these stages cost money. If the rock contains enough of the metal compound to make it profitable to extract the metal, the rock is called an ore.

Figure 19.9A ● 'Native' copper

The method used to extract a metal from its ore depends on the position of the metal in the reactivity series (see Table 19.13).

Table 19.13 ● Methods used for the extraction of metals

Metal	Method
Potassium Sodium Calcium Magnesium	The anhydrous chloride is melted and electrolysed
Aluminium	The anhydrous oxide is melted and electrolysed
Zinc Iron Copper Lead	Found as sulphides and oxides. The sulphides are roasted to give oxides; the oxides are reduced with carbon
Silver Gold	Found 'native' (as the free metal)

Sodium

Sodium is obtained by the electrolysis of molten dry sodium chloride. The same method is used for potassium, calcium and magnesium. This is an expensive method of obtaining metals because of the cost of the electricity consumed.

Aluminium – the 'newcomer' among metals ■

Aluminium is mined as **bauxite**, an ore which contains aluminium oxide, $Al_2O_3.2H_2O$. This ore is very plentiful, yet aluminium was not extracted from it until 1825. Another 60 years passed before a commercial method of extracting the metal was invented. *What was the problem?*

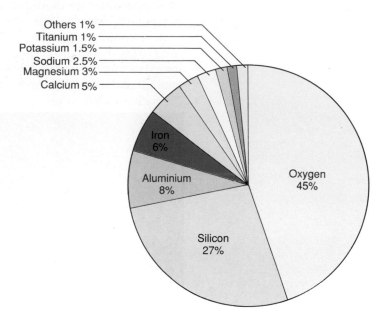

Others 1%
Titanium 1%
Potassium 1.5%
Sodium 2.5%
Magnesium 3%
Calcium 5%

Iron 6%

Aluminium 8%

Oxygen 45%

Silicon 27%

Figure 19.9B ● Elements in the Earth's crust

A Danish scientist called Hans Christian Oersted succeeded in obtaining aluminium from aluminium oxide in 1825. First he made aluminium chloride from aluminium oxide. Then he used potassium amalgam (an alloy of potassium and mercury) to displace aluminium from aluminium chloride.

The German chemist Friedrich Wöhler altered the method somewhat. He used potassium (instead of potassium amalgam).

A French chemist, Henri Sainte-Claire Déville, tackled the problem of scaling up the reaction. He used sodium in the displacement reaction. In 1860 he succeeded in making the reaction yield aluminium on a large scale. The price of aluminium tumbled. Instead of being an expensive curiosity, it became a useful commodity. With its exceptional properties, aluminium soon found new uses, and the demand for the new metal increased.

Aluminium was expensive because of the cost of the sodium used in its extraction. Chemists were keen to find a less costly method of extracting the new metal. Many thought it should be possible to obtain aluminium by electrolysing molten aluminium oxide. The difficulty was the high melting point, 2050 °C, which made it impossible to keep the compound molten while a current was passed through it.

The big breakthrough came in 1886. Two young chemists working thousands of miles apart made the discovery at the same time. An American called Charles Martin Hall, aged 21, and a Frenchman called Paul Héroult, aged 23, discovered that they could obtain a solution of aluminium oxide by dissolving it in molten cryolite, Na_3AlF_6, at 700 °C. On passing electricity through the melt, the two men succeeded in obtaining aluminium. Their method is still used. The Hall-Héroult cell is shown in Figure 19.9C. Anhydrous pure aluminium oxide must be obtained from the ore and then electrolysed in the cell.

=== *SCIENCE AT WORK* ===

In 1985, a firm in Northampton had the idea of buying empty aluminium drink cans for recycling. They installed machines which pay 1p for every two aluminium cans fed in but not for steel cans. *How can the machines tell the difference between aluminium cans and steel cans?*

LOOK AT LINKS
for **electrolysis**
See Topic 9.

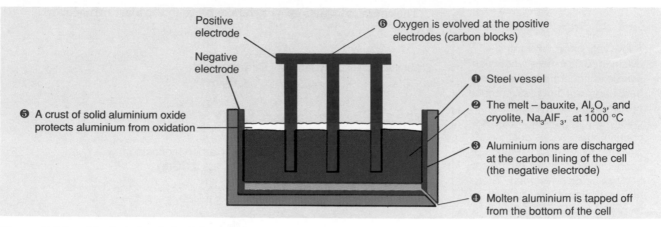

⑥ Oxygen is evolved at the positive electrodes (carbon blocks)

Positive electrode

Negative electrode

❶ Steel vessel

❷ The melt – bauxite, Al_2O_3, and cryolite, Na_3AlF_3, at 1000 °C

❸ Aluminium ions are discharged at the carbon lining of the cell (the negative electrode)

❺ A crust of solid aluminium oxide protects aluminium from oxidation

❹ Molten aluminium is tapped off from the bottom of the cell

Figure 19.9C ● Electrolysis of aluminium oxide

Iron

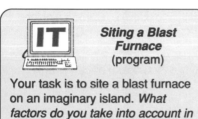

Siting a Blast Furnace
(program)

Your task is to site a blast furnace on an imaginary island. *What factors do you take into account in siting the furnace?*

Many countries have plentiful resources of iron ores, **haematite**, Fe_2O_3, **magnetite**, Fe_3O_4 and **iron pyrites**, FeS_2. The sulphide ore is roasted in air to convert it to an oxide. The oxide ores are reduced to iron in a **blast furnace** (Figure 19.9D). Iron ore and coke and limestone are fed into the furnace. Iron ores and limestone are plentiful resources. Coke is made by heating coal.

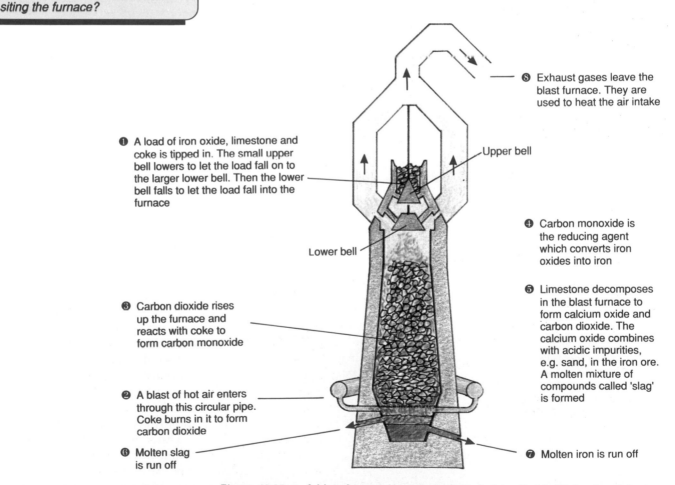

❶ A load of iron oxide, limestone and coke is tipped in. The small upper bell lowers to let the load fall on to the larger lower bell. Then the lower bell falls to let the load fall into the furnace

❽ Exhaust gases leave the blast furnace. They are used to heat the air intake

Upper bell

Lower bell

❹ Carbon monoxide is the reducing agent which converts iron oxides into iron

❺ Limestone decomposes in the blast furnace to form calcium oxide and carbon dioxide. The calcium oxide combines with acidic impurities, e.g. sand, in the iron ore. A molten mixture of compounds called 'slag' is formed

❸ Carbon dioxide rises up the furnace and reacts with coke to form carbon monoxide

❷ A blast of hot air enters through this circular pipe. Coke burns in it to form carbon dioxide

❻ Molten slag is run off

❼ Molten iron is run off

Figure 19.9D ● A blast furnace is a tower of steel plates lined with heat-resistant bricks. It is about 50 m high.

The chemical reactions which take place in the blast furnace are:

Step 2

Carbon (coke) + Oxygen $\xrightarrow{\text{Heat}}$ Carbon dioxide

$C(s)$ + $O_2(g)$ → $CO_2(g)$

Step 3

Carbon dioxide + Carbon (coke) $\xrightarrow{\text{Heat}}$ Carbon monoxide

$CO_2(g)$ + $C(s)$ → $2CO(g)$

Step 4

Iron(III) oxide + Carbon monoxide $\xrightarrow{\text{Heat}}$ Iron + Carbon dioxide

$Fe_2O_3(s)$ + $3CO(g)$ → $2Fe(s)$ + $3CO_2(g)$

Step 5

(a) Calcium carbonate (limestone) $\xrightarrow{\text{Heat}}$ Calcium oxide + Carbon dioxide

$CaCO_3(s)$ → $CaO(s)$ + $CO_2(g)$

(b) Calcium oxide + Silicon(IV) oxide (sand) $\xrightarrow{\text{Heat}}$ Calcium silicate(slag)

$CaO(s)$ + $SiO_2(s)$ → $CaSiO_3(l)$

The blast furnace runs continuously. The raw materials are fed in at the top, and molten iron and molten slag are run off separately at the bottom. The slag is used in foundations by builders and road-makers. The process is much cheaper to run than an electrolytic method. With the raw materials readily available and the cost of extraction low, iron is cheaper than other metals.

❶ The electrolyte is copper(II) sulphate solution

❷ The negative electrode is a strip of pure copper. Copper ions are discharged and copper atoms are deposited on the electrode. The strip of pure copper becomes thicker
$Cu^{2+}(aq) + 2e^- \rightarrow Cu(s)$

❸ The positive electrode is a lump of impure copper. Copper atoms supply electrons to this electrode and become copper ions which enter the solution. The lump of impure copper becomes smaller
$Cu(s) \rightarrow Cu^{2+}(aq) + 2e^-$

❹ Anode sludge, this is the undissolved remains of the lump of impure copper

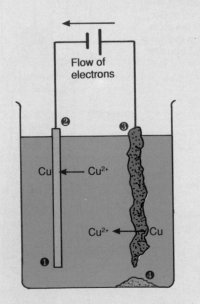

Electrons flow through the external circuit from the positive electrode to the negative electrode

Flow of electrons

Figure 19.9E ● Purification of copper

Copper

Copper is low in the reactivity series. It is found 'native' (uncombined) in some parts of the world. More often, it is mined as the sulphide. This is roasted in air to give impure copper. Pure copper is obtained from this by the **electrolytic** method shown in Figure 19.9E.

After the cell has been running for a week, the negative electrode becomes very thick. It is lifted out of the cell, and replaced by a new thin sheet of copper. When all the copper has dissolved out of the positive electrode, a new piece of impure copper is substituted.

Other metals are present as impurities in copper ores. Iron and zinc are more reactive than copper, and their ions therefore stay in solution while copper ions are

SUMMARY

The method used for extracting a metal from its ores depends on the position of the metal in the reactivity series. Very reactive metals, such as sodium, are obtained by electrolysis of a molten compound. Less reactive metals, such as iron, are obtained by reducing the oxide with elements such as carbon or hydrogen. The metals at the bottom of the reactivity series occur 'native'.

discharged. Silver and gold are less reactive than copper. They do not dissolve and are therefore present in the anode sludge. They can be extracted from this residue.

Silver and Gold

Silver and gold, are found 'native'. A new deposit of gold was discovered in the Sperrin mountains Northern Ireland in 1982.

CHECKPOINT

❶ The Emperor Napoleon invested in the French research on new methods of obtaining aluminium. He was interested in the possibility of aluminium suits of armour for his soldiers. What advantage would aluminium armour have had over iron or steel?

❷ (a) Arrange in order of the reactivity series the metals copper, iron, zinc, gold and silver.
   (b) Arrange the same metals in order of their readiness to form ions, that is, the ease with which the reaction $M(s) \rightarrow M^{n+}(aq) + ne^-$ takes place.
   (c) Arrange the same metals in order of the ease of discharging their ions, that is, the ease with which the reaction $M^{n+}(aq) + ne^- \rightarrow M(s)$ takes place.
   (d) Refer to Figure 19.9E showing the electrolytic purification of copper. Use your answers to (b) and (c) to explain why copper is deposited on the negative electrode but iron, zinc, gold and silver are not.

❸ Explain why (a) the human race started using copper, silver and gold long before iron was known and (b) iron tools were a big improvement on tools made from other metals.

❹ Give the names of metals which fit the descriptions A, B, C, D and E.
   A Reacts immediately with air to form a layer of oxide and then reacts no further.
   B Reacts violently with water to form an alkaline solution.
   C Reacts slowly with cold water and rapidly with steam.
   D Is a reddish-gold coloured metal which does not react with dilute hydrochloric acid.
   E Can be obtained by heating its oxide with carbon.

IT'S A FACT

All the magnesium that has been produced so far could have been extracted from 4 cubic kilometres of sea water. The sea is an almost inexhaustible reserve of magnesium.

Magnesium

Magnesium has a low density ($1.7\,g/cm^3$) and its alloys are widely used where lightweight components are required, e.g. the aircraft industry. The alloy Dowmetal, 89% Mg, 9% Al, 2% Zn, has a tensile strength which is close to steel. Magnesium is present to the extent of 0.14% in sea water. Figure 19.9F (overleaf) illustrates the extraction of magnesium by the Dow process.

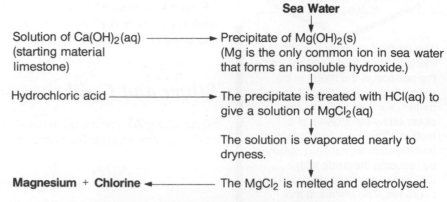

Figure 19.9F ● The Dow process: magnesium from sea water

Zinc

Zinc is found chiefly as the sulphide, ZnS, *zinc blende*, and as the carbonate, $ZnCO_3$, *calamine*. Extraction is as follows:

Step 1 Concentration: The ore is concentrated by **flotation** (see Figure 19.9G).

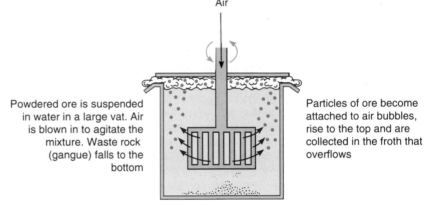

Figure 19.9G ● Froth flotation

The success of the method depends on the use of a suitable foam-stabiliser and a substance that will coat the particles without wetting them.

Step 2 Roasting: The zinc ore is roasted in air to convert the carbonate and the sulphide into zinc oxide.

Zinc sulphide + Oxygen → Zinc oxide + Sulphur dioxide
$$2ZnS(s) + 3O_2(g) → 2ZnO(s) + 2SO_2(g)$$

Step 3 Reduction: The oxide is reduced to zinc by heating with powdered carbon (as coke or coal).

Zinc oxide + Carbon → Zinc + Carbon monoxide
$$ZnO(s) + C(s) → Zn(s) + CO(g)$$

Step 4 Purification: Zinc is usually purified by distillation. Alternatively it is converted into zinc sulphate which is electrolysed.

Tin

Tin is mined as the ore SnO_2, *cassiterite* or *tinstone*. The extraction of tin is as follows:

Step 1 Washing: The ore is pulverised and washed with water so that particles of less dense materials float away.

Step 2 Roasting: This removes sulphur and arsenic as volatile oxides.

Step 3 Reduction: The ore is heated with anthracite (carbon). Limestone is added to produce a slag with the impurities.

Tin(IV) oxide + Carbon → Tin + Carbon monoxide
$$SnO_2(s) + 2C(s) → Sn(s) + 2CO(g)$$

Step 4 Purification: The impure tin is heated just enough to melt it. The solid residue contains iron and other metals which have higher melting points.

Lead

Lead is found chiefly as the sulphide PbS, *galena*. The metal is extracted as follows:

Step 1 Roasting of galena

Lead(II) sulphide + Oxygen → Lead(II) oxide + Sulphur dioxide
$$2PbS(s) + 3O_2(g) → 2PbO(s) + 2SO_2(g)$$

Step 2 Reduction of lead(II) oxide by coke

Lead(II) oxide + Coke → Lead + Carbon monoxide
$$PbO(s) + C(s) → Pb(s) + CO(g)$$

Alternatively *galena* can be used as the reducing agent.

Step 3 Purification: The lead is treated by processes which remove copper, silver and other metals.

SUMMARY

Magnesium can be extracted from sea water because it forms an insoluble hydroxide. Zinc is mined as the sulphide and the carbonate. These are roasted to form the oxide, which is reduced by carbon. Tin is mined as the oxide which is reduced by carbon. Lead is mined as lead sulphide. This is roasted to form lead oxide, which is reduced by carbon.

CHECKPOINT

❶ (a) Suggest a use for the sulphur dioxide produced by roasting zinc blende.
   (b) When zinc sulphate is electrolysed, what is discharged at the anode? What change takes place in the nature of the electrolyte? How does this change benefit the process?

❷ Write an equation for the reduction of lead(II) oxide by galena, PbS to lead and sulphur dioxide.

❸ Limestone is added in the reduction of tin(IV) oxide.
   (a) What basic product is formed when limestone is heated?
   (b) Present in tin ores is the oxide of another Group 4 element. This oxide combines with product (a). Name this oxide and the slag which is formed.

19.10 Iron and steel

FIRST THOUGHTS

The machines that manufacture our possessions, our means of transport and the frameworks of our buildings: all these depend on the strength of steel.

The iron that comes out of the blast furnace is called **cast iron** (or pig iron). It contains three to four per cent carbon. The carbon content makes it brittle, and cast iron cannot be bent without snapping. The impurity makes the melting point lower than that of pure iron so that cast iron is easier to cast – to melt and mould – than pure iron. Cast iron expands slightly as it solidifies. This helps it to flow into all the corners of a mould and reproduce the shape exactly. By casting, objects with complicated shapes can be made, for example the cylinder block of a car engine (which contains the cylinders in which the combustion of petrol vapour in air takes place).

Figure 19.10A ● Wrought iron

Iron which contains less than 0.25% carbon is called wrought iron. Wrought iron is strong and easily worked (Figure 19.10A). Nowadays, mild steel has replaced wrought iron.

Steel

Steel is made by reducing the carbon content of cast iron, which makes it brittle, to less than one per cent. Carbon is burnt off as its oxides, the gases carbon monoxide, CO, and carbon dioxide, CO_2. Iron is less easily oxidised. The sulphur, phosphorus and silicon in the iron are also converted into acidic oxides. These are not gases, and a base, such as calcium oxide, must be added to remove them. The base and the acidic oxides combine to form a slag (a mixture of compounds of low melting point).

Figure 19.10B shows a basic oxygen furnace. In it, cast iron is converted into steel. One converter can produce 150–300 tonnes of steel in an hour.

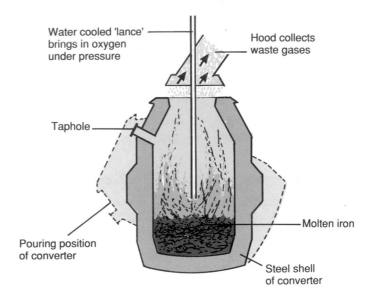

Water cooled 'lance' brings in oxygen under pressure

Hood collects waste gases

Taphole

Molten iron

Pouring position of converter

Steel shell of converter

Figure 19.10B ● A basic oxygen furnace holding 150–300 tonnes of steel

There are various types of steel. They differ in their carbon content and are used for different purposes. Low-carbon steel (mild steel) is pliable; high-carbon steel is hard.

Alloy steels

Many elements are alloyed with iron and carbon to give alloy steels with different properties. They all have different uses, for example nickel and chromium give stainless steel, manganese and molybdenum increase strength, and vanadium increases springiness.

SUMMARY

Cast iron (up to 4% carbon) is easy to mould, but is brittle. Wrought iron (the purest form of iron) is easily worked without breaking. Low-carbon steel (mild steel) is pliable; high-carbon steel is hard. Alloy steels contain other elements in addition to iron and carbon. Different steels are suited to different uses.

CHECKPOINT

❶ What is the difference between cast iron and wrought iron in (a) composition (b) strength and (c) ease of moulding? Name two objects made from cast iron and two objects made from wrought iron.

The uses of alloy steels

Some steels are alloys of iron and carbon. They have more uses than cast iron (see Table 19.14).

Table 19.14 ● Cast iron and steel and their uses

Type of steel	Description	Uses
Mild steel (<0.25% carbon)	Pliable (can be bent without breaking)	Chains and pylons
Medium steel (0.25–0.45% carbon)	Tougher than mild steel, more springy than high carbon steel	Nuts and bolts Car springs and axles Bridges
High carbon steel (0.45–1.5% carbon)	The carbon content makes it both hard and brittle	Chisels, files, razor blades, saws, cutting tools
Cast iron (2.5–4.5% carbon)	Cheaper than steel, easily moulded into complicated shapes	Fire grates, gear boxes, brake discs, engine blocks. These articles will break if they are hammered or dropped

Many elements are used for alloying with steel. The different alloy steels have different properties which fit them for different uses (see Table 19.15).

Table 19.15 ● Some alloy steels

Element	Properties of alloy	Uses
Chromium	Prevents rusting if >10%	Stainless steels, acid-resisting steels, cutlery, car accessories, tools
Cobalt	Takes a sharp cutting edge Can be strongly magnetised	High-speed cutting tools Permanent magnets
Manganese	Increases strength and toughness	Some is used in all steels; steel in railway points and safes contains a high percentage of manganese
Molybdenum	Strong even at high temperatures	Rifle barrels, propeller shafts
Nickel	Resists heat and acids	Stainless steel cutlery, industrial plants which must withstand acidic conditions
Tungsten	Stays hard and tough at high temperatures	High-speed cutting tools
Vanadium	Increases springiness	Springs, machinery

LOOK AT LINKS
for quenching and tempering of steel
See Topic 19.2.

SUMMARY

Cast iron contains up to 4% carbon. It is easy to mould, but is brittle. Steels contain carbon. Low-carbon steel is pliable; high-carbon steel is hard. The steels are suited to different uses. Alloy steels contain added metals. The different elements give a range of steels with different characteristics suited to different uses.

CHECKPOINT

❶ Think what is the most important characteristic of the steel that must be used for making the articles listed below. Should the steel be pliable or springy or hard? Then say whether you would use mild steel or medium steel or high-carbon steel or cast iron for making the following articles.
(a) car axles (b) axes (c) car springs (d) ornamental gates (e) drain pipes (f) chisels (g) sewing needles (h) picks (i) saws (j) food cans

❷ Suggest which type of alloy steel could be chosen for: (a) an electric saw (b) a car radiator (c) the suspension of a car (d) a gun barrel (e) a set of steak knives.

19.11 Rusting of iron and steel

FIRST
THOUGHTS

The rusting of iron is an expensive nuisance. Replacing rusted iron and steel structures costs the UK £500 million a year. In this section we look at some remedies.

Iron and most kinds of steel rust. Rust is hydrated iron(III) oxide, $Fe_2O_3.nH_2O$ (n, the number of water molecules in the formula, varies).

Some of the methods which are used to protect iron and steel against rusting are listed in Table 19.16. Some methods use a coating of some substance which excludes water and air. Other methods work by sacrificing a metal which is more reactive than iron.

Figure 19.11A ● Rust protection

(a) Paint protects the car

(b) Chromium plating protects the bicycle handlebars

(c) Galvanised steel girders

(d) Zinc bars protect the ship's hull

Table 19.16 ● Rust prevention

Method	Where it is used	Comment
A coat of paint	Ships, bridges, cars, other large objects (see Figure 19.11A(a))	If the paint is scratched, the exposed iron starts to rust. Corrosion can spread to the iron underneath the paintwork which is still sound.
A film of oil or grease	Moving parts of machinery, e.g. car engines	The film of oil or grease must be renewed frequently.
A coat of plastic	Kitchenware, e.g. draining rack	If the plastic is torn, the iron starts to rust.
Chromium plating	Kettles, cycle handlebars (see Figure 19.11A(b))	The layer of chromium protects the iron beneath it and also gives a decorative finish. It is applied by electroplating.
Galvanising (zinc plating)	Galvanised steel girders are used in the construction of buildings and bridges (see Figure 19.11A(c))	Zinc is above iron in the reactivity series: zinc will corrode in preference to iron. Even if the layer of zinc is scratched, as long as some zinc remains, the iron underneath does not rust. Zinc cannot be used for food cans because zinc and its compounds are poisonous.
Tin plating	Food cans	Tin is below iron in the reactivity series. If the layer of tin is scratched, the iron beneath it starts to rust.
Stainless steel	Cutlery, car accessories, e.g. radiator grille	Steel containing chromium (10–25%) or nickel (10–20%) does not rust.
Sacrificial protection	Ships (see Figure 19.11A(d))	Blocks of zinc are attached to the hulls of ships below the waterline. Being above iron in the reactivity series, zinc corrodes in preference to iron. The zinc blocks are sacrificed to protect the iron. As long as there is some zinc left, it protects the hull from rusting. The zinc blocks must be replaced.

LOOK AT LINKS
The conditions which make iron rust are water and air and acidity. If salts are present, rusting is accelerated.
See Topic 15.9.

SUMMARY

Exposure to air and water in slightly acidic conditions makes iron and steel rust. Some of the treatments for protecting iron and steel from rusting are:
• oil or grease,
• a protective coat of another metal, e.g. chromium, zinc, tin,
• alloying with nickel and chromium,
• attaching a more reactive metal, e.g. magnesium or zinc, to be sacrificed.

The Earth contains huge deposits of iron ores. We need iron for our machinery and for our means of transport. During the twentieth century, we have used more metal than in all the previous centuries put together. If we keep on mining iron ores at the present rate, the Earth's resources may one day be exhausted. We allow tonnes of iron and steel to rust every year. We throw tonnes of used iron and steel objects on the scrap heap. The Earth's iron deposits will last longer if we take the trouble to collect scrap iron and steel and recycle it, that is, melt it down and reuse it.

Figure 19.11B ● Scrap iron dump

CHECKPOINT

❶ Say how the rusting of iron is prevented:
  (a) in a bicycle chain,
  (b) in a food can,
  (c) in parts of a ship above the water line,
  (d) in parts of a ship below the water line,
  (e) in a galvanised iron roof.

❷ 'The Industrial Revolution would not have been possible without steel.' Say whether or not you agree with this statement. Give your reasons.

❸ The map opposite shows possible sites, A, B, C and D, for a steelworks. Say which site you think is the best, and explain your choice. You will have to consider the need for:
  • iron ore, coke (from coal), limestone,
  • a work-force,
  • transporting iron and steel to customers,
  • removing slag.

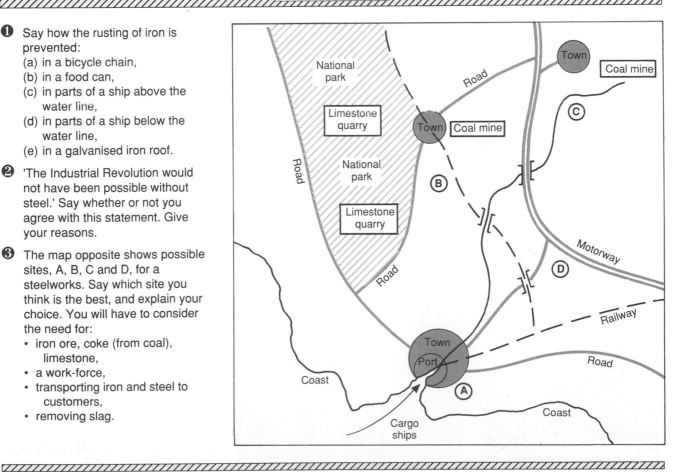

Aluminium is the most plentiful metal in the Earth's crust, yet aluminium was not manufactured until the nineteenth century. The twentieth century has seen aluminium finding more and more vitally important uses.

Uses of aluminium

Aluminium is a metal with thousands of applications. Some of these are listed in Table 19.17. Pure aluminium is not a very strong metal. Its alloys, such as duralumin (which contains copper and magnesium), are used when strength is needed.

Table 19.17 ● Some uses of aluminium and its alloys

Property	Use for aluminium which depends on this property
Never corroded (except by bases)	Door frames and window frames are often made of aluminium. 'Anodised aluminium' is used. The thickness of the protective layer of aluminium oxide has been increased by anodising (making it the positive electrode in an electrolytic cell).
Low density	Packaging food: milk bottle tops, food containers, baking foil. The low density and resistance to corrosion make aluminium ideal for aircraft manufacture. Alloys such as duralumin are used because they are stronger than aluminium.
Good electrical conductor	Used for overhead cables. The advantage over copper is that aluminium cables are lighter and need less massive pylons to carry them.
Reflects light when polished	Car headlamp reflectors
Good thermal conductor	Saucepans, etc.
Reflects heat when highly polished	Highly polished aluminium reflects heat and can be used as a thermal blanket. Aluminium blankets are used to wrap premature babies. They keep the baby warm by reflecting heat back to the body. Firefighters wear aluminium fabric suits to reflect heat away from their bodies.

The cost to the environment

Bauxite is found near the surface in Australia, Jamaica, Brazil and other countries. The ore is obtained by open cast mining (Figure 19.12A). A layer of earth 1 m to 60 m thick is excavated, and the landscape is devastated. In some places, mining companies have spent money on restoring the landscape after they have finished working a deposit.

Figure 19.12A ● An open cast bauxite mine

Pure aluminium oxide must be extracted from bauxite before it can be electrolysed. Iron(III) oxide, which is red, is one of the impurities in bauxite. The waste produced in the extraction process is an alkaline liquid containing a suspension of iron(III) oxide. It is pumped into vast red mud ponds.

Jamaica has a land area of 11 000 km^2. Every year, 12 km^2 of red mud are created. The Jamaicans are worried about the loss of land. Even when it dries out, red mud is not firm enough to build on. They also worry about the danger of alkali seeping into the water supply. The waste cannot be pumped into the sea because it would harm the fish.

Purified aluminium oxide is shipped to an aluminium plant. The electrolytic method of extracting aluminium is expensive to run because of the electricity it consumes. Aluminium plants are often built in areas which have hydroelectric power (electricity from water-driven generators). This is relatively cheap electricity. The waterfalls and fast-flowing rivers which provide hydroelectric power are found in areas of natural beauty. Local residents often object to the siting of aluminium plants in such beauty spots.

There may be other difficulties over hydroelectric power. Purified aluminium oxide has to be transported to a remote area and aluminium has to be transported away. If there are not enough local workers, a workforce may have to be brought into the area and provided with housing. Often it pays to build an aluminium plant in an area which has a big population and good transport, even if the cost of electricity is higher.

The exhaust gases from aluminium plants contain fluorides from the electrolyte. Before leaving the chimney, the exhaust gases are 'scrubbed' with water. The waste water, which contains fluoride, is discharged into rivers. The remaining exhaust gases are discharged into the atmosphere through tall chimneys. In the past, fluoride emissions have been known to pollute agricultural land, killing grass and causing lameness in cattle. Farmers sued for the damage to their cattle. Aluminium plants now take more care to control fluoride emission.

Industrial Chemistry
(program)

Use the program to investigate:
• The making of sulphuric acid.
• The factors involved in siting an aluminium plant.
Make a list of the factors involved in both production processes.

SUMMARY

Aluminium has brought us many benefits. There is, however, a cost to the environment. The open cast mining of bauxite spoils the landscape. The purification creates unsightly red mud ponds. The extraction can cause pollution through fluoride emission.

Factors which decide the siting of aluminium plants include
• the cost of electricity,
• the cost of transporting the raw material and the product,
• the availability of a workforce.

Figure 19.12B ● Recycling aluminium

//////// CHECKPOINT ////////

❶ The percentages of some metals recycled in the UK are shown below.
   (a) Explain what 'recycled' means.
   (b) Plot the figures as a bar chart or as a pie chart.
   (c) Why is lead easy to recycle?
   (d) How can iron be separated from other metals for recycling?
   (e) What resources are saved by recycling aluminium?

Metal	Aluminium	Zinc	Iron	Tin	Lead	Copper
Percentage recycled	28	30	50	30	56	19

19.13 Recycling

Few materials are destroyed during their use. Recycling is a possible method of saving resources and energy. Metals and alloys are prime candidates for recycling (see Table 19.18). Other materials are wood, paper, plastics and textiles. Some objects are easier to recycle than others. An iron machine has a high scrap value, but iron bars embedded in concrete are more difficult to reuse. In many cases a number of different metals are used in the manufacture of an item, e.g. a motor vehicle. A motor vehicle contains about 1% by mass of copper and this must be reduced to 0.1% before the scrap can go to the steelworks.

Table 19.18 ● Recovery of metals

Metal	Recovery
Aluminium	About 50% is recycled. Recycling uses only 5% of the energy need to make aluminium from its ore
Iron	About 50% of the feedstock for steel furnaces is scrap
Copper	Pipes, vehicle radiators, etc are recycled
Lead	Batteries, pipes, sheet metal, type metal are recycled
Zinc	Recovery is from alloys
Tin	Recovery is from alloys and from tin-plate
Mercury	A large percentage is recovered from mercury cells, instruments and apparatus
Gold, silver and platinum	Recovery is high from jewellery, watches and chemical plants

//////// CHECKPOINT ////////

❶ Why is the recovery of gold and silver high?

❷ Why do people find it more convenient to recycle mercury than to dump it?

❸ Why is copper an easy metal to recycle?

❹ How can other metals be separated from scrap iron?

CHEMICAL CELLS

20.1 Simple chemical cells

A car needs a battery to start the motor and to power the instruments. A battery is a series of chemical cells. A chemical cell is a system for converting the energy of a chemical reaction into electricity. This topic will tell you more about how chemical cells work.

You have studied metals in this theme. You have learned that metals are made up of positive ions and a cloud of moving electrons (Figure 19.2A). What happens when a strip of metal is put into water or a solution of an electrolyte? Figure 20.1A(a) shows what happens when the metal is a reactive metal such as zinc. Some zinc ions pass into solution, leaving their electrons behind on the metal:

$$Zn(s) \rightarrow Zn^{2+}(aq) + 2e^-$$

As a result, the strip of zinc becomes negatively charged. After a while the negative charge on the zinc builds up to a level which prevents any more Zn^{2+} ions from leaving the metal.

LOOK AT LINKS
for the **metal bond**
See Topic 19.2;
for the **reactivity series of metals**
See Topic 19.4.

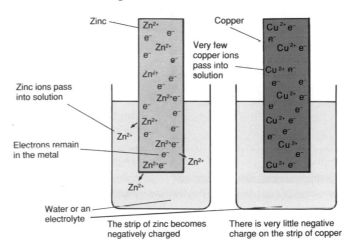
Figure 20.1A ● (a) Zinc (b) Copper

The reactivity series of metals (see Topic 19.4) is based on the observation that some metals are more ready to form ions (and are therefore more reactive) than others. Copper is one of the less reactive metals. A strip of copper placed in water or a solution of an electrolyte has very little tendency to form ions and to acquire a negative charge (see Figure 20.1A(b)).

Perhaps you can predict what will happen if you immerse a strip of zinc and a strip of copper in a solution and then connect the two metals. Electrons flow through the external circuit from zinc, which is negatively charged, to copper, which has very little charge (see Figure 20.1B).

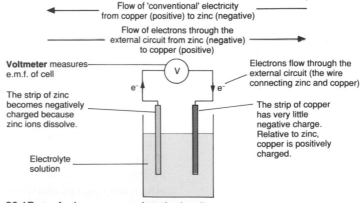

LOOK AT LINKS
for the current convention,
See *KS: Physics*, Topic 20.3.

Figure 20.1B ● A zinc–copper chemical cell

In Figure 20.1B, electrons flow from zinc to copper through the external circuit. Before the nature of electricity was understood, the flow of electricity was regarded as taking place from the positive electrode to the negative electrode, in this case from copper to zinc. The flow of 'conventional' electricity is from copper to zinc (right to left in Figure 20.1B). The direction of flow of electrons is from a metal higher in the reactivity series to a metal lower in the series. The difference in electric potential between the two electrodes is called the electromotive force (e.m.f.) of the cell. The e.m.f. of the cell depends on the difference between the positions of the two metals in the reactivity series (and the electrochemical series; see Table 20.1).

In electrolytic cells (see Topic 9), an electric current causes a chemical reaction to take place. Electrical energy is converted into chemical energy. In the zinc–copper cell described above, the chemical reaction taking place inside the cell causes a current to flow through the external circuit. Chemical energy is converted into electrical energy. This kind of cell is called a **chemical cell** (or a **galvanic cell** or a **voltaic cell**).

The voltage obtained from the simple cell shown in Figure 20.1B falls quickly because of changes at the electrodes. For example, the strip of zinc becomes coated with copper due to the displacement reaction:

Zinc + Copper ions → Zinc ions + Copper
$$Zn(s) + Cu^{2+}(aq) \rightarrow Zn^{2+}(aq) + Cu(s)$$

Changes of this kind which occur at the electrodes are called **polarisation**. In the zinc–copper cell, they can be stopped by dividing the cell into two half-cells. One way of separating the two half-cells is by means of a porous pot. The Daniell cell, which uses this method, can supply an e.m.f. of 1.1 V for along time (see Figure 20.1C on p. 246).

SUMMARY

When a metal is placed in a solution of an electrolyte, it acquires an electric charge. When two metals which dip into an electrolyte are connected, there is a difference in electric potential between them, called the e.m.f. of the cell. An electric current flows from one metal to the other. The difference in electric potential between the two metals is called the e.m.f. of the cell. The further apart the metals are in the reactivity series, the greater is the e.m.f. of the cell.

Arranging metals in order of their tendency to lose electrons

In the zinc–copper cell (Figure 20.1A), zinc is the negative electrode and copper is the positive electrode. Copper is not always the positive electrode. If copper is paired with silver in a chemical cell, the reaction:

$$Cu(s) \rightarrow Cu^{2+}(aq) + 2e^-$$

happens to a greater extent than the reaction:

$$Ag(s) \rightarrow Ag^+(aq) + e^-$$

The build-up of electrons on copper is greater than the build-up of electrons on silver. Copper is the negative electrode of the cell. Paired with a more reactive metal, e.g. zinc, copper is the positive electrode; paired with a less reactive metal, e.g. silver, copper is the negative electrode.

It is possible to construct a 'league table' of metals, arranging them in order of their tendency to lose electrons. The method is to choose one metal as a reference electrode, say copper, and measure the e.m.f. of a number of different metal–copper cells. The values of e.m.f. place the metals in order of their readiness to give electrons and form cations in aqueous solution. The order is called the **electrochemical series**. Table 20.1 overleaf shows a section of the electrochemical series including hydrogen. It also puts cations in the reverse order of the ease with which they accept electrons and are discharged in electrolysis (see Topic 9.5).

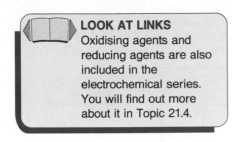

LOOK AT LINKS
Oxidising agents and
reducing agents are also
included in the
electrochemical series.
You will find out more
about it in Topic 21.4.

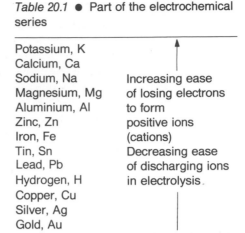

Table 20.1 ● Part of the electrochemical
series

Potassium, K	
Calcium, Ca	
Sodium, Na	Increasing ease
Magnesium, Mg	of losing electrons
Aluminium, Al	to form
Zinc, Zn	positive ions
Iron, Fe	(cations)
Tin, Sn	Decreasing ease
Lead, Pb	of discharging ions
Hydrogen, H	in electrolysis.
Copper, Cu	
Silver, Ag	
Gold, Au	

The reactivity series

The electrochemical series will remind you of the reactivity series of
metals (Topic 19.4). The reactivity series does not always show the true
ability of a metal to form ions. Some metals, e.g. aluminium, acquire a
coating of the metal oxide which prevents the metal from showing its
true reactivity. Other metals, e.g. calcium, form insoluble salts and this
slows down their reactions with solutions, e.g. with water and acids.
The electrochemical series, on the other hand, shows the true ability of a
metal to form ions.

CHECKPOINT

❶ Name the particles that conduct electricity through (a) a chemical cell and
(b) an electrical circuit outside the cell.

❷ The following pairs of metals are joined to form chemical cells. Place the
pairs in list (a) and the pairs in list (b) in order of the e.m.f. of the cells:
(a) zinc–copper, magnesium–copper, iron–copper
(b) magnesium–zinc, iron–magnesium, magnesium–tin

❸ Zinc and iron are connected to form a cell.
(a) Which of the two metals is the more able to form ions?
(b) Which metal will become the negative electrode of the cell and which
the positive?
(c) Sketch a zinc–iron cell showing the direction of flow of electrons in the
external circuit.

❹ Copper and lead are both low in the electrochemical series. When they are
paired up to form a chemical cell, which of the two metals will become
negative electrode?

The Daniell cell

Daniell invented the cell shown in Figure 20.1C. The overall cell reaction
is (as shown in Figure 20.1B):

Zinc atoms + Copper ions $\rightarrow$ Zinc ions + Copper atoms
$Zn(s)$ + $Cu^{2+}(aq)$ $\rightarrow$ $Zn^{2+}(aq)$ + $Cu(s)$

The Daniell cell has an e.m.f. of 1.1 volts (1.1 V).

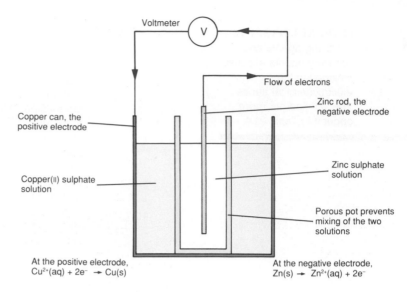

Figure 20.1C ● The Daniell cell

At the positive electrode,
$Cu^{2+}(aq) + 2e^- \rightarrow Cu(s)$

At the negative electrode,
$Zn(s) \rightarrow Zn^{2+}(aq) + 2e^-$

Many metals can be paired up in chemical cells. The further apart the metals are in the electrochemical series, the bigger is the e.m.f. produced by the cell. Cells can be combined in series to form a battery. Three cells each with an e.m.f. of 1.5 V combine to make a battery with an e.m.f. of 4.5 V. The word 'battery' is sometimes used incorrectly for a single cell. For example, what we call a torch 'battery' is often a single cell.

Figure 20.1D ● A car battery

SUMMARY

A vehicle battery consists of a number of chemical cells. Chemical reactions in the cells supply an electric current. When the chemicals have been used up they can be reformed by recharging the battery.
- During discharge, chemical energy is converted into electrical energy.
- During charge, electrical energy is converted into chemical energy.

The lead–acid accumulator

The type of battery commonly used in motor vehicles is the lead–acid accumulator. The battery delivers 6 V or 12 V, depending on the number of 2 V cells which it contains (see Figure 20.1A). In each cell, one pole is a lead plate and the other is a lead grid packed with lead(IV) oxide, PbO_2. The electrolyte is sulphuric acid. When the battery supplies a current, the reactions which occur convert lead and lead(IV) oxide into lead(II) ions. When the reactants have been used up, the battery can no longer supply a current; it is 'flat'. It can, however, be recharged. When the vehicle engine is running, the dynamo (the alternator) rotates and generates electricity which recharges the battery. If the dynamo is faulty, the battery will become flat. It can be recharged by connecting it to a battery charger, which is a transformer connected to the mains. This reverses the sign of each electrode and reverses the chemical reactions that have occurred at the electrodes. The battery is again able to supply a current.

CHECKPOINT

❶ What is the advantage of the lead–acid accumulator over the Daniell cell?

❷ What does the alternator in a vehicle engine do?

❸ (a) Why do car batteries sometimes become flat?
   (b) How can they be given new life?

20.2 Dry Cells

Figure 20.2A ● A pocket calculator uses a dry cell

Dry cells are chemical cells which are used in everyday life. For use in batteries for torches, radios etc., a Daniell cell (see Figure 20.1C) is not suitable because the electrolyte might leak out. A dry cell is used, one which has a damp paste instead of a liquid electrolyte.

The zinc–carbon dry cell

You will already be familiar with the zinc–carbon type of dry cell (see Figure 20.2B).

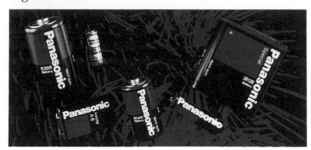

Figure 20.2B ● Dry cells

The zinc–carbon cell has an e.m.f. of 1.5 V. Figure 20.2C shows how it works.

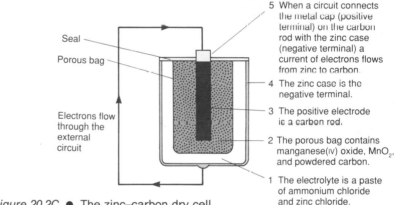

Seal
Porous bag
Electrons flow through the external circuit

5 When a circuit connects the metal cap (positive terminal) on the carbon rod with the zinc case (negative terminal) a current of electrons flows from zinc to carbon.

4 The zinc case is the negative terminal.

3 The positive electrode is a carbon rod.

2 The porous bag contains manganese(IV) oxide, MnO_2, and powdered carbon.

1 The electrolyte is a paste of ammonium chloride and zinc chloride.

Figure 20.2C ● The zinc–carbon dry cell

The chemical reactions that take place are complex. A summary is:
Anode: Zinc atoms are oxidised to zinc ions.
Cathode: Manganese(IV) oxide is reduced to manganese(III) oxide.
When the chemicals have been used up, the cell can operate no longer: it is flat.

Cells of this type are 1–2 cm in diameter and 3–10 cm in length. They are used in flashlights, portable radios, toys etc. A disadvantage is that with heavy use the cell quickly becomes flat. Zinc–carbon dry cells cannot be recharged and have a short shelf life.

The alkaline manganese cell

The alkaline manganese cell also uses zinc as anode and manganese(IV) oxide as cathode. The electrolyte contains potassium hydroxide. The zinc anode is slightly porous, giving it a larger surface area. The alkaline manganese cell can therefore deliver more current than the zinc–carbon cell; it has an e.m.f. of 1.5 V. The alkaline manganese cells do not become flat as quickly in use and have a longer shelf life than the zinc–carbon cells.

The silver oxide cell

Another type of dry cell is the silver oxide cell (Figure 20.2D).

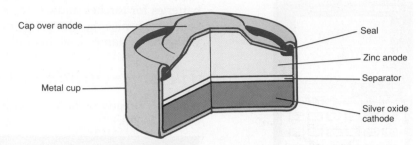

Cap over anode

Seal

Zinc anode

Separator

Metal cup

Silver oxide cathode

Figure 20.2D ● A silver oxide cell

As the cell operates, zinc is oxidised to zinc ions, and silver ions are reduced to silver. Silver oxide cells measure 0.5–1 cm in diameter and 0.25–0.5 cm in height. These tiny 'batteries' are used in electronic wrist watches, in cameras and in electronic calculators. They have a long life because little current is drawn from them. They are rather expensive.

The nickel–cadmium cell

The dry cells described above have a limited life. Once the chemicals have reacted, there is no further source of energy. The nickel–cadmium cell (see Figure 20.2E) need not be thrown away when the chemical reaction that gives rise to the e.m.f. is finished. The cell can be recharged. This is done by connecting the cell to a source of direct current. The chemical reaction is reversed, and the cell has a new source of e.m.f. As the cell operates, cadmium is oxidised to cadmium(II) ions, and nickel(IV) oxide is reduced to nickel(II) hydroxide. During recharging, these reactions are reversed.

The nickel–cadmium cell has an e.m.f. of 1.4 V. It has a longer life than the lead storage battery. It can be packaged in a sealed unit ready for use in rechargeable calculators and photographic flash units etc.

Figure 20.2E ● Nickel–cadmium cells and charger

Table 20.2 ● Dry cells

Type	Electrodes	Electrolyte	Voltage	Use	Characteristics
Zinc–carbon cell	Carbon (+) Zinc (−)	Ammonium chloride	1.5 V	Radios, cassette players, electrical toys, torches	Cheap. Relatively short life. Not rechargeable. Leakage of electrolyte occurs in the unsealed types of cell. Operates on low current; current drops gradually during discharge.
Alkaline manganese cell	Manganese(IV) oxide (+) Zinc (−)	Potassium hydroxide	1.5 V	Cassette players, electrical toys, appliances which are in use for long periods	Costs about twice as much as the zinc–carbon cell and lasts twice as long. Not rechargeable. Long shelf-life. Gives a large current; voltage remains steady during discharge.
Silver oxide cell	Silver oxide (+) Zinc (−)	Potassium hydroxide	1.5 V	Watches, calculators, cameras, hearing aids	High cost. Not rechargeable. Small in size; light-weight. Voltage remains constant in operation.
Nickel–cadmium cell	Nickel oxide (+) Cadmium (−)	Potassium hydroxide	1.5 V	Cassette players, electrical toys, radios	Expensive. Rechargeable. Gives a large current. There is a danger of exploding during charging if the current is too high or the rate of charging is too high.

Space craft use **fuel cells**. A fuel cell is another means of converting chemical energy into electrical energy. The fuel cell shown in Figure 20.2F uses the reaction:

Hydrogen + Oxygen ➔ Water

The big advantage of a fuel cell over other cells is that it operates continuously with no need for recharging. As long as reactants are fed in, the cell can supply energy.

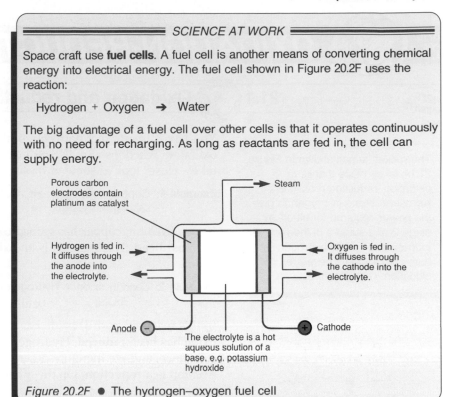

Porous carbon electrodes contain platinum as catalyst

Steam

Hydrogen is fed in. It diffuses through the anode into the electrolyte.

Oxygen is fed in. It diffuses through the cathode into the electrolyte.

Anode ⊖

Cathode ⊕

The electrolyte is a hot aqueous solution of a base, e.g. potassium hydroxide

Figure 20.2F ● The hydrogen–oxygen fuel cell

SUMMARY

Chemical cells transform the energy of chemical reactions into electrical energy. Examples of dry cells are: the carbon–zinc dry cell, the alkaline manganese cell, the silver oxide dry cell, the nickel–cadmium dry cell.

Fuel cells operate continuously. One kind uses the reaction between hydrogen and oxygen to make water.

CHECKPOINT

❶ Why is the silver oxide 'battery' suitable for use in a camera? Why is it expensive?

❷ Both the lead–acid accumulator and the nickel–cadmium cell are rechargeable.
(a) What does 'rechargeable' mean?
(b) What is the advantage of rechargeable cells over others?
(c) For what uses is the nickel–cadmium cell more suited than the lead–acid accumulator?
(d) What is the most widespread use of the lead–acid accumulator?

❸ Why is the alkaline manganese cell preferred to the silver oxide cell for use in radios?

TOPIC 21 OXIDATION-REDUCTION REACTIONS

FIRST THOUGHTS

21.1 Oxidation and reduction

The rocket launch shown in Figure 21.1A takes place thanks to oxidation–reduction reactions. Kerosene burns in oxygen to give the power required for lift-off. In stage 2 and stage 3 of the rocket's journey into space, hydrogen burns in oxygen. These are oxidation–reduction reactions.

Figure 21.1A ● Rocket launch

LOOK AT LINKS
for **oxidation** and **reduction**
See Topic 15.6.

SUMMARY

Oxidation is the gain of oxygen or loss of hydrogen by a substance. Reduction is the loss of oxygen or gain of hydrogen by a substance. An oxidising agent gives oxygen to or takes hydrogen from another substance. A reducing agent takes oxygen from or gives hydrogen to another substance. Oxidation and reduction occur together in oxidation–reduction reactions or redox reactions.

You have already met both chemical reactions that are described as oxidation reactions and also reduction reactions. In this topic, we shall take a closer look at some of these reactions.

Example 1: Copper + Oxygen → Copper(II) oxide
$$2Cu(s) + O_2(g) → CuO(s)$$

In this reaction, copper has gained oxygen, and we say that copper has been **oxidised**. This reaction is an **oxidation**, and oxygen is an **oxidising agent**.

Example 2: Lead(II) oxide + Hydrogen → Lead + Water
$$PbO(s) + H_2(g) → Pb(s) + H_2O(l)$$

In this reaction, lead(II) oxide has lost oxygen, and we say that lead(II) oxide has been **reduced**. Hydrogen has taken oxygen from another substance, and we therefore describe hydrogen as a **reducing agent**. The reaction is a **reduction**. On the other hand, notice that hydrogen has gained oxygen: hydrogen has been oxidised. You can also describe this reaction as an oxidation. Since lead(II) oxide has given oxygen to another substance, in this reaction lead(II) oxide is an oxidising agent. Oxidation and reduction are occurring together. One reactant is the oxidising agent and the other reactant is the reducing agent.

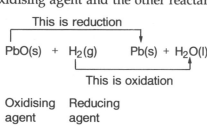

This is reduction

$$PbO(s) + H_2(g) \quad Pb(s) + H_2O(l)$$

This is oxidation

Oxidising   Reducing
agent   agent

Example 3: Zinc oxide + Carbon → Zinc + Carbon monoxide
$$ZnO(s) + C(s) → Zn(s) + CO(g)$$

Again, you see in this example that oxidation and reduction are occurring together. This is always true, however many examples you consider. Since oxidation never occurs without reduction, it is better to call these reactions **oxidation–reduction reactions** or **redox reactions**.

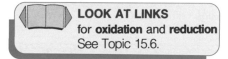

CHECKPOINT

❶ State which is the oxidising agent and which is the reducing agent in each of these reactions:
   (a) the thermit reaction:
      Aluminium + Iron oxide → Iron + Aluminium oxide
   (b) roasting the ore tin sulphide in air:
      Tin sulphide + Oxygen → Tin oxide + Sulphur dioxide
   (c) 'smelting' tin oxide with coke:
      Tin oxide + Carbon → Tin + Carbon monoxide
   (d) roasting the ore copper sulphide in air:
      $2CuS(s) + 3O_2(g) → 2CuO(s) + 2SO_2(g)$

21.2 Gain or loss of electrons

It is not obvious in Example 1 on p. 250 that reduction accompanies oxidation.

Copper + Oxygen → copper(II) oxide
$$2Cu(s) + O_2(g) → 2CuO(s)$$

Atoms of copper, Cu, have been converted into copper(II) ions, Cu^{2+}. Molecules of oxygen, O_2, have been converted into oxide ions, O^{2-}. This means that copper atoms have lost electrons:

$$Cu(s) → Cu^{2+}(aq) + 2e^-$$

and oxygen molecules have gained electrons:

$$O_2(g) + 4e^- → 2O^{2-}(s)$$

This picture holds for all oxidation–reduction reactions.
- The substance which is oxidised loses electrons.
- The substance which is reduced gains electrons.
- An oxidising agent accepts electrons.
- A reducing agent gives electrons.

> **HOW TO REMEMBER**
> Which is which: oxidation or reduction – loss or gain of electrons?
> OIL Oxidation is loss.
> RIG Reduction is gain.

> Oxidation is the gain of oxygen or the loss of hydrogen or the loss of electrons.
> Reduction is the loss of oxygen or the gain of hydrogen or the gain of electrons.

Metals are reducing agents. Metals react (see Topics 12.1, 19.3) by losing electrons to form positive ions, M^+, M^{2+} and M^{3+}. When a metal reacts with an acid, the reaction can be considered in two parts, and a half-equation can be written for each part.

Part (a)   Metal atom → Metal ion + electrons
Half-equation (a) $M(s)$ → $M^{2+}(aq) + 2e^-$

Part (b)   Hydrogen ions + electrons → Hydrogen molecule
Half-equation (b) $2H^+(aq) + 2e^- → H_2(g)$

The metal atoms have supplied electrons, and hydrogen ions have accepted electrons. The metal atoms have been oxidised, and hydrogen ions have been reduced. Adding the two half-equations (a) and (b),

$$M(s) + 2H^+(aq) + 2e^- → M^{2+}(aq) + 2e^- + H_2(g)$$

The electrons on the left-hand side (LHS) and the right-hand side (RHS) cancel so the equation can be written:

$$M(s) + 2H^+(aq) → M^{2+}(aq) + H_2(g)$$

Some metals form more than one type of ion. When iron(II) compounds are converted into iron(III) compounds, Fe^{2+} ions lose electrons: they are oxidised. The half-equation is:

(c) $Fe^{2+}(aq) → Fe^{3+}(aq) + e^-$

The oxidation is brought about by a substance which can accept electrons – an oxidising agent, e.g. chlorine. Chlorine is reduced to chloride ions. The half-equation is:

(d) $Cl_2(aq) + 2e^- → 2Cl^-(aq)$

To obtain the equation for the complete reaction, the two half-equations (c) and (d) must be added. So that the electrons on the LHS and the RHS will cancel, first multiply half-equation (c) by 2:

(e) $2Fe^{2+}(aq) \rightarrow 2Fe^{3+}(aq) + 2e^-$

Adding the half-equations (d) + (e),

$2Fe^{2+}(aq) + Cl_2(g) + 2e^- \rightarrow 2Fe^{3+}(aq) + 2e^- + 2Cl^-(aq)$

and cancelling the electrons on the LHS and RHS,

$2Fe^{2+}(aq) + Cl_2(g) \rightarrow 2Fe^{3+}(aq) + 2Cl^-(aq)$

21.3  Redox reactions and cells

LOOK AT LINKS
for **electrolysis**
See Topics 9.3 and 9.4.

Electrolysis cells

Redox reactions involve the gain and the loss of electrons. The reactions that happen at the electrodes in electrolysis involve either the gain or the loss of electrons. Let us look at the connection between electrode reactions and redox reactions.

In the electrolysis of copper(II) chloride, the electrode reactions are as follows.

Cathode (negative electrode): $Cu^{2+}(aq) + 2e^- \rightarrow Cu(s)$

Here copper ions, Cu^{2+}, gain electrons: they are reduced to copper atoms, Cu.

Anode (positive electrode): $Cl^-(aq) \rightarrow Cl(g) + e^-$

Here chloride ions, Cl^- give up electrons: they are oxidised to chlorine atoms, Cl. These immediately combine to form chlorine molecules, Cl_2. The cathode reaction is reduction; the anode reaction is oxidation.

Half-equations can be written for the electrode reactions and added to give the equation for the overall cell reaction.

Example 1: The electrolysis of copper(II) chloride:

Cathode: $Cu^{2+}(aq) + 2e^- \rightarrow Cu(s)$
Anode: $2Cl^-(aq) \rightarrow Cl_2(g) + 2e^-$

Adding these half-equations:

$Cu^{2+}(aq) + 2e^- + 2Cl^-(aq) \rightarrow Cu(s) + Cl_2(g) + 2e^-$

The electrons on the LHS and the RHS cancel to give:

$Cu^{2+}(aq) + 2Cl^-(aq) \rightarrow Cu(s) + Cl_2(g)$

> In an electrolysis cell, the anode reaction is oxidation, and the cathode reaction is reduction.

Chemical cells

A simple chemical cell can be made by immersing two different metals in an electrolyte and connecting them with a conducting wire. A current flows through the wire (see Figure 21.3A).

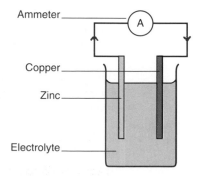

Ammeter

Copper
Zinc
Electrolyte

Figure 21.3A ● A chemical cell

LOOK AT LINKS
for **chemical cells**
See Topic 20.1, Figure 20.1B.

Example 1: The zinc–copper chemical cell

The half-equations for the two electrode reactions are:

Zinc electrode: $Zn(s) \rightarrow Zn^{2+}(aq) + 2e^-$

Oxidation (loss of electrons) occurs at this electrode; therefore this is the anode.

Copper electrode: $Cu^{2+}(aq) + 2e^- \rightarrow Cu(s)$

Reduction (gain of electrons) occurs at this electrode; therefore this is the cathode. Notice that, in the chemical cell, the anode is negative and the cathode is positive. Adding the two half-equations,

$Zn(s) + Cu^{2+}(aq) + 2e^- \rightarrow Zn^{2+}(aq) + 2e^- + Cu(s)$

The electrons on the LHS cancel those on the RHS, giving

$Zn(s) + Cu^{2+}(aq) \rightarrow Zn^{2+}(aq) + Cu(s)$

This is the equation for the whole cell reaction.

SUMMARY

- The reactions which take place in chemical cells are redox reactions.
- Oxidation takes place at the anode.
- Reduction takes place at the cathode.
- Half-equations are written for the electrode reactions.
- The half-equations can be added to give the overall cell reaction.

CHECKPOINT

❶ (a) Write half-equations for the electrode reactions in the zinc–tin chemical cell.
  (b) Add the half-equations to obtain the equation for the overall cell reaction.
  (c) Explain why the reaction which occurs at the zinc electrode is classified as an oxidation reaction.
  (d) Does this mean that zinc is an oxidising agent? Explain your answer.

❷ (a) Write half-equations for the electrode reactions in the aluminium–copper chemical cell.
  (b) Add the half-equations to obtain the equation for the overall cell reaction.
  (c) Which electrode is the anode? Explain your answer.

21.4 Tests for oxidising agents and reducing agents

Test for reducing agents

Acidified potassium manganate(VII) is a powerful oxidising agent. A reducing agent will change the colour from the purple of $MnO_4^-(aq)$ to the pale pink of $Mn^{2+}(aq)$. Since potassium manganate(VII) is a strong oxidising agent, it can oxidise – and therefore test for – many reducing agents.

For example, (see Figure 21.7A),

Soak a piece of filter paper in acidified potassium manganate(VII) solution, and hold it in a stream of sulphur dioxide.
The paper turns from purple to a very pale pink.

Test for oxidising agents

Many oxidising agents will oxidise the iodide ion, I^-, to iodine, I_2. The presence of iodine can be detected because it forms a dark blue

> **SUMMARY**
>
> Reducing agents decolourise acidified potassium manganate(VII) solution. Oxidising agents give a dark blue colour with a solution of potassium iodide and starch.

compound with starch. Potassium iodide solution and starch can therefore be used to test for oxidising agents.

$$2I^-(aq) \rightarrow I_2(aq) + 2e^-$$
colourless   brown, forms a dark blue compound with starch

For example,

Wet a filter paper with a solution of potassium iodide and starch, and hold it in a stream of chlorine.
The colour changes from white to dark blue.

CHECKPOINT

❶ (a) Say which of the following will decolourise acidified potassium manganate(VII):

$FeSO_4(aq)$, $ZnSO_4(aq)$, $SO_2(g)$

(b) Say which of the following will turn starch–iodide paper blue:

$Br_2(aq)$, $Fe_2(SO_4)_3$, Cl_2, I_2, $HCl(aq)$

21.5  Chlorine as an oxidising agent

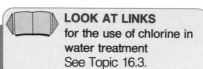
LOOK AT LINKS
for the **halogens**
See Topic 7.5.

LOOK AT LINKS
for the use of chlorine in water treatment
See Topic 16.3.

The reactions of chlorine are dominated by its readiness to act as an oxidising agent: to gain electrons and form chloride ions.

Chlorine molecules + Electrons → Chloride ions
$$Cl_2(aq) + 2e^- \rightarrow 2Cl^-(aq)$$

● Water treatment
Of major importance is the role played by chlorine in making our tap water safe to drink. The ability of chlorine to kill bacteria is due to its oxidising power.

● Metals
Sodium and chlorine react to form sodium chloride. Sodium atoms are oxidised (give electrons), becoming sodium ions, Na^+. Chlorine molecules are reduced (gain electrons) to become chloride ions, Cl^-.

$$Na(s) \rightarrow Na^+(s) + e^-$$

Adding the half-equations for sodium and chlorine,

$$2Na(s) + Cl_2(g) \rightarrow 2Na^+(s) + 2Cl^-(s)$$

● Ions of metals of variable valency
The oxidation of iron(II) salts to iron(III) salts by chlorine has been mentioned (Topic 21.2).

● Halogens
The reactions of chlorine with bromides and iodides are oxidation–reduction reactions. Chlorine displaces bromine from bromides. This happens because chlorine is a stronger oxidising agent than bromine is. Chlorine takes electrons away from bromide ions: bromide ions are oxidised to bromine molecules. Chlorine molecules are reduced to chloride ions.

$$\text{Cl}_2(aq) + 2\text{Br}^-(aq) \rightarrow 2\text{Cl}^-(aq) + \text{Br}_2(aq)$$

with REDUCTION and OXIDATION labelled.

The bromine formed can be detected more easily if a small volume of an organic solvent, e.g. trichloroethene, is added. Bromine dissolves in the organic solvent to form an orange solution.

Similarly, chlorine displaces iodine from iodides because chlorine is a stronger oxidising agent than iodine. Again, iodine can be detected readily by adding a small volume of an organic solvent. Iodine dissolves to give a purple solution.

Bromine displaces iodine from iodides because bromine is a stronger oxidising agent than iodine.

Fluorine is the most powerful oxidising agent of the halogens. It is dangerously reactive and is not used in schools and colleges. In order of oxidising power, the halogens rank:

$$\text{F}_2 > \text{Cl}_2 > \text{Br}_2 > \text{I}_2$$

● Oxidation–reduction

Oxidation is
- the gain of oxygen or
- the loss of hydrogen or
- the loss of electrons by a substance.

Reduction is
- the loss of oxygen or
- the gain of hydrogen or
- the gain of electrons by a substance.

An oxidising agent
- gives oxygen to or
- takes hydrogen from or
- takes electrons from a substance.

A reducing agent
- takes oxygen from or
- gives hydrogen to or
- gives electrons to a substance.

SUMMARY

Chlorine is an oxidising agent. It is used in water treatment to kill bacteria. It oxidises metals, metal ions, bromides and iodides. In order of oxidising power, the halogens rank

$$\text{F}_2 > \text{Cl}_2 > \text{Br}_2 > \text{I}_2$$

21.6 Chlorine as a bleach

LOOK AT LINKS
for the bleaching action of sulphur dioxide
See Topic 21.7.

If the oxidised form, DO, of a dye, D, is colourless, then chlorine can bleach that dye. Chlorine reacts with water in a reversible reaction to form chloric(I) acid, HClO, and hydrochloric acid.

Chlorine + Water ⇌ Chloric(I) acid + Hydrochloric acid
$$\text{Cl}_2(aq) + \text{H}_2\text{O}(l) \rightleftharpoons \text{HOCl}(aq) + \text{HCl}(aq)$$

It is the chlorate(I) ion, ClO^- that acts as the oxidising and bleaching agent.

Chlorate(I) + dye D → Chloride + Oxidised form of dye DO
$$\text{ClO}^-(aq) + \text{D} \rightarrow \text{Cl}^-(aq) + \text{DO}$$

If D is coloured and DO is colourless, the dye has been decolourised and therefore bleached. The chlorate(I) ion has acted as an oxidising agent and a bleaching agent.

Chlorine bleaches paper, grass, flowers, ink (except printers' ink which contains carbon) and other substances. It is used as a laundry bleach.

Manufacture of bleach

The electrolysis of sodium chloride solution (brine) is an important industrial process (see Topic 9.6, including Figures 9.6C and D). During electrolysis, chlorine and hydrogen are evolved, and the electrolyte changes from sodium chloride into sodium hydroxide. In some plants, the electrodes are rotated to allow the chlorine formed at the anode to react with the sodium hydroxide that has formed.

Chlorine + Sodium → Sodium chlorate(I) + Sodium + Water
hydroxide chloride
$Cl_2(g) + 2NaOH(aq) → NaClO(aq) + NaCl(aq) + H_2O(l)$

From the solution which is formed, sodium chlorate(I) is obtained. Being an oxidising agent, it kills bacteria, and it is sufficiently mild to be safe to use as an antiseptic on the skin. It is sold as Milton® which is widely used for sterilising babies' bottles. The chlorine bleaches sold for domestic use are solutions of sodium chlorate(I) or calcium chlorate(I).

Figure 21.6A ● Domestic bleaches and sterilising products

In other plants, the electrolysis of brine with rotating electrodes is carried out at a higher temperature. Then the reaction which takes place is:

Chlorine + Sodium → Sodium chlorate(V) + Sodium + Water
hydroxide chloride
$3Cl_2(g) + 6NaOH(aq) → NaClO_3(aq) + 5NaCl(aq) + 3H_2O(l)$

Sodium chlorate(V) is a powerful oxidising agent. It is used as a weedkiller, e.g. Tandol®.

Safety matters
Bleaches which contain chlorine are powerful oxidising agents. For this reason, you should not use a chlorine bleach together with other chemical cleaning fluids. Detergents which contain ammonia and bleaches which contain hydrogen peroxide will react violently with chlorine.

SUMMARY

Chlorine is a bleach. Sodium chlorate(I), NaClO, is used as a bleach and as an antiseptic. Sodium chlorate(V), NaClO$_3$, is used as a weedkiller.

CHECKPOINT

❶ Material which has been bleached by chlorine must be thoroughly rinsed. It is important to remove two substances which are present. What are they?

❷ (a) Write a word equation for the oxidation of sodium iodide by chlorine.
(b) Write a balanced chemical equation for the reaction in (a).
(c) Write a word equation for the oxidation of potassium iodide by bromine.
(d) Write a balanced chemical equation for the reaction in (c).

❸ The lab. technician has taken delivery of three bottles containing crystalline white solids. Unfortunately, their labels have come off.

| **Potassium chloride** | **Potassium bromide** | **Potassium iodide** |

The technician has to decide which label to stick on each of the three bottles. All she has to work with is a bottle of chlorine water and an organic solvent. How can she solve the problem?

❹ With which one of the following pairs of reagents would a displacement reaction take place?
(a) aqueous bromine and aqueous potassium chloride
(b) aqueous bromine and aqueous sodium chloride
(c) aqueous chlorine and aqueous potassium iodide
(d) aqueous iodine and aqueous potassium bromide.

21.7 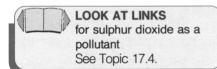 Sulphur dioxide as a reducing agent

> **LOOK AT LINKS**
> for sulphur dioxide as a pollutant
> See Topic 17.4.

Sulphur dioxide is a gas with a very unpleasant, penetrating smell. It is poisonous, causing congestion followed by choking, and at sufficiently high concentrations it will kill.

Sulphurous acid

Sulphur dioxide is extremely soluble in water. It 'fumes' in moist air as it reacts with water vapour to form a mist. It reacts with water to form sulphurous acid, H_2SO_3.

Sulphur dioxide + Water → Sulphurous acid
$$SO_2(g) \quad + H_2O(l) \quad \rightarrow \quad H_2SO_3(aq)$$

Sulphurous acid is a weak acid which forms salts called sulphites, e.g. sodium sulphite Na_2SO_3. Sulphurous acid is slowly oxidised by oxygen in the air to sulphuric acid, H_2SO_4. Sulphuric acid is formed whenever sulphurous acid accepts oxygen from an oxidising agent.

Sulphurous acid + Oxygen from the air or from → Sulphuric acid
an oxidising agent
$$H_2SO_3(aq) \quad + \quad (O) \quad \rightarrow \quad H_2SO_4(aq)$$

For example,

Sulphurous acid + Bromine water → Sulphuric acid + Hydrogen bromide
$$H_2SO_3(aq) + Br_2(aq) + H_2O(l) \quad \rightarrow \quad H_2SO_4(aq) \quad + \quad 2HBr(aq)$$

Tests for sulphur dioxide

The reducing action of sulphur dioxide is the basis of tests for the gas.
1 It reduces bromine water, $Br_2(aq)$, from brown bromine, Br_2 to colourless bromide ions, Br^-.
2 It reduces potassium dichromate(VI) from orange dichromate ions, $Cr_2O_7{}^{2-}$ to blue chromium(III) ions, Cr^{3+}, passing through the intermediate colour of green, which results from a mixture of orange ions and blue ions.

3 It reduces potassium manganate(VII) from purple manganate(VII) ions, $MnO_4^-(aq)$ to pale pink, almost colourless manganese(II) ions, Mn^{2+}

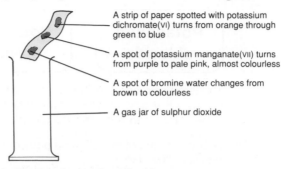

A strip of paper spotted with potassium dichromate(VI) turns from orange through green to blue

A spot of potassium manganate(VII) turns from purple to pale pink, almost colourless

A spot of bromine water changes from brown to colourless

A gas jar of sulphur dioxide

Figure 21.7A ● Testing for sulphur dioxide

Figure 21.7B ● Straw hats

Sulphur dioxide as a bleach

Sulphur dioxide is a useful bleach. It is mild in its action, and can be used on wool and silk, on flour and cheese, as well as on paper and straw. The substances which colour these materials are dyes which contain oxygen. Sulphur dioxide acts as a reducing agent. It bleaches dyes if the oxidised form of the dye is coloured and the reduced form of the dye is colourless.

Coloured form + Sulphurous acid → Colourless form + Sulphuric acid
of dye   of dye
(Dye + oxygen) + H_2SO_3(aq)   →   (Dye)   + H_2SO_4(aq)

The colourless substance formed is gradually oxidised by air to form the coloured dye again. This is why straw, silk and paper turn yellow with age.

Food preservation

To keep food wholesome, the growth of micro-organisms must be prevented. One way is to deprive micro-organisms of the oxygen they need and delay their growth. Another way is to provide an acidic environment which inhibits the growth of bacteria and moulds. Sulphur dioxides and sulphites are used as food preservatives. They are given E numbers to show that they are food additives permitted by the European Community, e.g. E220 is sulphur dioxide, E221 is sodium sulphite, E222 is sodium hydrogensulphite. In solution, sulphites yield sulphurous acid. Sulphur dioxide and sulphites can therefore act in both these ways: as reducing agents and as weak acids. They also prevent the oxidation of fats and oils to sour-smelling products. Sulphur dioxide, as a mild bleach, prevents the formation of a brown colour in fruits and vegetables.

Sulphur dioxide is used as a sterilising agent to make sure that food containers are free of micro-organisms.

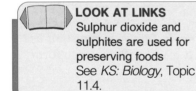

LOOK AT LINKS
Sulphur dioxide and sulphites are used for preserving foods
See *KS: Biology*, Topic 11.4.

SUMMARY

Sulphur dioxide is an acidic, poisonous gas with a pungent smell. It dissolves in water to form sulphurous acid, which is a reducing agent. Sulphur dioxide is used as a bleach. Sulphur dioxide and sulphites are used as food preservatives and for sterilising containers.

CHECKPOINT

❶ In what way does the bleaching action of sulphur dioxide differ from that of chlorine?

❷ Explain why old straw hats are yellowish-white in colour.

❸ Makers of home-brewed beer and wine rinse their bottles with a solution of sodium sulphite before filling them. Explain why they do this.

TOPIC 22 ANALYTICAL CHEMISTRY

FIRST THOUGHTS

Does a sample of drinking water contain nitrates? Does a brand of paint contain lead compounds? Which metals are present in an alloy steel? To find out the answers to such questions, we use analytical chemistry.

LOOK AT LINKS
for **volumetric analysis**
See Topic 24.11;
for an example of
gravimetric analysis
See Topic 24.7.

To **analyse** means to separate something into its component parts in order to learn more about the nature of these components. The branch of chemistry that deals with the analysis of chemical compounds and mixtures is called **analytical chemistry**. Sometimes we need to know only which substances are present in a sample; not the quantities of those substances. Which elements are present in this sample of soil? Which pigments are present in this food dye? Does this sample of zinc contain a trace of zinc sulphide? To find answers to such questions, we use **qualitative analysis**. If we need to know the precise quantity of one or more components of a sample, we use **quantitative analysis**. This might entail finding the percentage of nickel in a nickel ore or the number of parts per million of mercury in a fish. Quantitative analysis can be done by means of **volumetric analysis**, in which reactions take place in solution. An alternative is **gravimetric analysis**, in which the masses of reacting substances are found by the use of an accurate balance. In this topic, we shall be dealing with qualitative analysis.

22.1 Flame colour

Table 22.1 ● Some flame colours

Metal	Colour of flame
Barium	Apple-green
Calcium	Brick-red
Copper	Green with blue streaks
Lithium	Crimson
Potassium	Lilac
Sodium	Yellow

Some metals and their compounds give characteristic colours when heated in a Bunsen flame. The colour can be used to identify the metal ion present in a compound.

22.2 Tests for ions in solution

The following tests can be done on a solution. A soluble solid can be analysed by dissolving it in distilled water and then applying the tests.

Table 22.2 ● Tests for anions in solution

Anion	Test and observation
Chloride, Cl^- (aq)	Add a few drops of dilute nitric acid followed by a few drops of silver nitrate solution. A white precipitate of silver chloride is formed. The precipitate is soluble in ammonia solution.
Bromide, Br^- (aq)	Add a few drops of dilute nitric acid followed by a few drops of silver nitrate solution. A pale yellow precipitate of silver bromide is formed. The precipitate is slightly soluble in ammonia solution.

(continued on following page)

Table 22.2 ● Tests for anions in solution (continued)

Anion	Test and observation
Iodide, I^-(aq)	Add a few drops of dilute nitric acid followed by a few drops of silver nitrate solution. A yellow precipitate of silver iodide is formed. It is insoluble in ammonia solution.
Sulphate, SO_4^{2-}(aq)	Add a few drops of barium chloride solution followed by a few drops of dilute hydrochloric acid. A white precipitate of barium sulphate is formed.
Sulphite, SO_3^{2-}(aq)	Test as for sulphate. A white precipitate of barium sulphite appears and then reacts with acid to give sulphur dioxide.
Carbonate, CO_3^{2-}(aq)	Add dilute hydrochloric acid to the solution (or add it to the solid). Bubbles of carbon dioxide are given off.
Nitrate, NO_3^-(aq)	Make the solution strongly alkaline. Add Devarda's alloy and warm. Ammonia is given off. (If the solution contains ammonium ions, warm with alkali to drive off ammonia before testing for nitrate.)

Table 22.3 ● Tests for cations in solution

Cation	Add sodium hydroxide solution	Add ammonia solution
Ammonium, NH_4^+(aq)	Warm. Ammonia is given off	—
Copper, Cu^{2+}(aq)	Blue jelly-like precipitate, $Cu(OH)_2$(s)	Blue jelly-like ppt., dissolves in excess to form a deep blue solution
Iron(II), Fe^{2+}(aq)	Green gelatinous ppt., $Fe(OH)_2$(s)	Green gelatinous ppt.
Iron(III), Fe^{3+}(aq)	Rust-brown gelatinous ppt., $Fe(OH)_3$(s)	Rust-brown gelatinous ppt.
Lead(II), Pb^{2+}(aq)	White ppt., $Pb(OH)_2$(s) dissolves in excess NaOH(aq)	White ppt., $Pb(OH)_2$
Zinc, Zn^{2+}(aq)	White ppt., $Zn(OH)_2$(s) dissolves in excess NaOH(aq)	White ppt., $Zn(OH)_2$(s) dissolves in excess NH_3(aq)
Aluminium, Al^{3+}(aq)	Colourless ppt., $Al(OH)_3$(s) dissolves in excess NaOH(aq)	Colourless ppt., $Al(OH)_3$(s)

Note: ppt. = precipitate

22.3 Identifying some common gases

If you are testing the smell of a gas, you should do so cautiously (see Figure 22.3A). Tests for common gases are given in Table 22.3.

Figure 22.3A ● How to smell a gas

Table 22.4 ● Testing gases

Gas	Colour and smell	Test and observation
Ammonia[†]	Colourless Pungent smell	Hold damp red litmus paper or universal indicator paper in the gas. The indicator turns blue.
Bromine	Reddish-brown Choking smell	Hold damp blue litmus paper in the gas. Litmus turns red and is then bleached.
Carbon dioxide	Colourless Odourless	Bubble the gas through limewater (calcium hydroxide solution). A white solid precipitate appears, making the solution appear cloudy.
Chlorine[†]	Poisonous Green gas Choking smell	Test a very small quantity in a fume cupboard. Hold damp blue litmus paper in the gas. Litmus turns red and is quickly bleached. Chlorine turns damp starch-iodide paper blue-black.
Hydrogen	Colourless Odourless when pure	Introduce a lighted splint. Hydrogen burns with a squeaky 'pop'.
Hydrogen chloride	Colourless Pungent smell	Hold damp blue litmus paper in the gas. Litmus turns red. With ammonia, a white smoke of NH_4Cl forms.
Iodine	Black solid or purple vapour	Dissolve in trichloroethane (or other organic solvent). Gives a violet solution.
Nitrogen dioxide[†]	Reddish-brown Pungent smell	Hold damp blue litmus paper in the gas. Litmus turns red but is not bleached.
Oxygen	Colourless Odourless	Hold a glowing wooden splint in the gas. The splint bursts into flame.
Sulphur dioxide[†]	Colourless Choking smell	Dip a filter paper in potassium dichromate solution, and hold it in the gas. The solution turns from orange through green to blue. Potassium manganate(VII) solution turns very pale pink.

[†]These gases are poisonous. Test with care.

22.4 Solubility of some compounds

The solubilities of some compounds are tabulated in Table 22.4.

Table 22.5 ● Soluble and insoluble compounds

Soluble	Insoluble
All sodium, potassium and ammonium salts	
All nitrates	
Most chlorides, bromides and iodides	Chlorides, bromides and iodides of silver and lead
Most sulphates	Sulphates of lead, barium and calcium
Sodium, potassium and ammonium carbonates	Most other carbonates
Sodium, potassium and calcium oxides	Most other oxides
Sodium, potassium and calcium hydroxides	Most other hydroxides

CHECKPOINT

❶ A body is found at the bottom of a clay pit. The dead man is known to have quarrelled violently with a neighbour, and the neighbour has clay on his shoes. The clay in the clay pit contains a high percentage of iron(II). How can you test the clay on the neighbour's shoes for Fe^{2+}?

❷ After a visit to Somerset, some students bring home a sample of rock from an underground cavern. Luke says that the rock is calcium carbonate. Natalie believes that it may be magnesium carbonate. Rosalie suggests that it may be barium carbonate. How can the students investigate whether the rock is (a) a carbonate (b) a calcium compound (c) a barium compound (d) a magnesium compound?

❸ A packet is labelled 'bicarb of soda'. How can you test to see whether it contains (a) a sodium compound (b) either a carbonate or a hydrogencarbonate?

❹ A solution contains the metal ions, Q^{2+}(aq) and R^{2+}(aq). When a solution of sodium hydroxide is added, a precipitate is obtained. Addition of excess of sodium hydroxide to this precipitate gives a green precipitate containing Q and a colourless solution containing R. When dilute hydrochloric acid is added to this solution, a white precipitate is obtained. Deduce what Q and R may be, explaining your reasoning.

❺ Old Sir Joshua Vellof often falls asleep after meals. He is a suspicious old gentleman, and he wonders whether his no-good nephew Jake is putting some of the tranquilliser potassium bromide into the salt cellar. Sir Joshua asks you to analyse the contents of the salt cellar to see whether it contains sodium chloride or potassium bromide. What do you do?

❻ A factory orders calcium oxide, magnesium oxide and barium oxide. A lorry driver deposits three sacks at the factory and drives off without saying which is which. The works chemist has to sort it out. Suggest two tests which he could use to tell which sack is which.

❼ A solution contains the sulphates of three metals, A, B and C. Explain the reactions shown in the flow chart and identify A, B and C.

Solution of sulphates of A, B and C

| NaOH(aq)

Blue-green precipitate

| Excess NaOH(aq)

Blue-green precipitate + Colourless solution

| NH$_3$(aq)　　　　| HCl(aq)

Blue solution　　　　White precipitate
+
Green precipitate　　　　| NH$_3$(aq)

Colourless solution

22.5　Methods of collecting gases

Gases can be bought in cylinders from industrial manufacturers or made in the laboratory by a chemical reaction. You may need to obtain gas jars full of gas. The following methods can be used to collect gases from either source.

● *Collecting over water*

A gas which is insoluble in water can be collected over water (see Figure 22.5A). The gas displaces the water in the gas jar. When the gas jar is full, it must be replaced by another. This method can be used for hydrogen, oxygen, chlorine, carbon dioxide and other gases. A little of the gas will dissolve in the water.

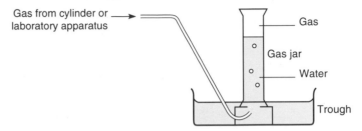

Figure 22.5A ● Collecting over water (for an insoluble gas)

● *Collecting by downward displacement*

A gas which is denser than air can be collected by downward displacement of air (see Figure 22.5B). The gas displaces the air in the gas jar. This method can be used for chlorine, carbon dioxide, sulphur dioxide and other dense gases.

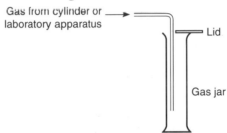

Figure 22.5B ● Collecting by downward displacement (for a dense gas)

● *Collecting by upward displacement*

A gas which is less dense than air can be collected by upward displacement of air (see Figure 22.5C). This method can be used for ammonia.

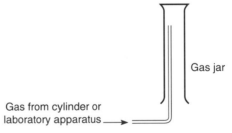

Figure 22.5C ● Collecting by upward displacement (for a gas of low density)

Safety matters
When collecting hydrogen, take care that there are no flames about. Hydrogen forms an explosive mixture with air.

● *Collecting in a gas syringe*

A gas supply can be connected to an empty gas syringe. As gas enters, it drives the plunger down the barrel (see Figure 22.5D). When the syringe is full, it must be replaced by another.

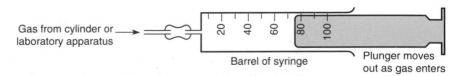

Figure 22.5D ● Using a gas syringe

22.6 Methods of drying gases

Methods of drying gases are shown in Figure 22.6A. They can be used when gases are collected by upward or downward displacement of air or in a gas syringe.

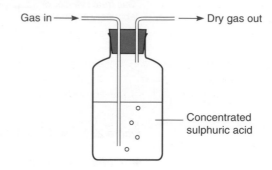

(a) Bubble trough concentrated sulphuric acid (used for all gases except ammonia)

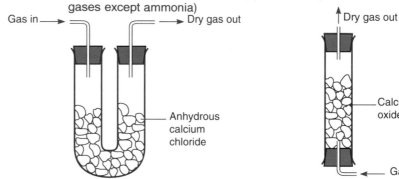

(b) Pass over anhydrous calcium chloride (used for all gases except ammonia, which forms a complex with calcium chloride)

(c) Pass over calcium oxide (used for ammonia and for neutral gases)

Figure 22.6A ● Methods of drying gases

CHECKPOINT

❶ (a) Why is the method shown in Figure 22.5A not used for collecting ammonia?
   (b) Why is the method shown in Figure 22.5C not used for collecting sulphur dioxide?

❷ (a) Why is the method shown in Figure 22.6A(a) not used for drying ammonia?
   (b) Why is the method shown in Figure 22.6A(c) not used for drying sulphur dioxide?

❸ Janet wants to fill some gas jars with oxygen from a cylinder of oxygen. She decides to bubble the gas through concentrated sulphuric acid, as in Figure 22.6A(a), and then collect it as shown in Figure 22.5A.
   (a) What does Janet hope to achieve by bubbling the gas through concentrated sulphuric acid?
   (b) Can you see anything wrong with her plan?
   (c) If you see a mistake, what change would you advise?

❹ Hydrogen can be collected by either of the methods shown in Figures 22.5A and C.
   (a) State an advantage of the method shown in Figure 22.5A over that in Figure 22.5C.
   (b) Sketch the apparatus which you would use to collect dry hydrogen.

? THEME QUESTIONS

1 Iron is extracted from its ores in a blast furnace. Most iron is converted into steel in a basic oxygen furnace.
(a) Name the two substances, in addition to iron ore, which are fed into the top of the blast furnace.
(b) Explain how the ore haematite, Fe_2O_3, is reduced to iron.
(c) Molten iron and slag form at the base of the blast furnace. Explain how slag is formed.
(d) What impurities are removed from iron in the basic oxygen furnace?
(e) Explain the chemical reaction that removes them.

2 Aluminium is mined as its oxide. It is extracted from its ore in a plant called an aluminium smelter.
(a) Name this aluminium oxide ore.
(b) State what process is used to extract aluminium from its oxide.
(c) Why can a blast furnace not be used for the extraction of aluminium?
(d) An aluminium ore called cryolite is used in the extraction of aluminium from aluminium oxide. What part does it play in the process?
(e) What environmental damage can be caused by an aluminium smelter?
(f) What two economies are made when aluminium is recycled?

3 The table shows part of the reactivity series of metals.

| Aluminium |
| Zinc |
| Iron |
| Tin |
| Lead |
| Copper |

Use the table to explain the following.
(a) Iron food cans are coated with tin.
(b) Zinc bars are attached to the hulls of ships below the waterline.
(c) When zinc powder is dropped into a solution of copper(II) sulphate, the colour of the solution fades.
(d) Galvanised (zinc-coated) steel does not rust.
(e) A mixture of aluminium and iron oxide is used to mend gaps in railway lines.

4 Explain why:
(a) electrical wiring is made of copper
(b) saucepans are made from aluminium
(c) aeroplanes are made from aluminium alloys
(d) bells are made from bronze
(e) trumpets are made from brass
(f) bridges are made from steel
(g) baking foil is made from aluminium
(h) solder is made from brass and tin
(i) lead was for many years used for water pipes
(j) lead is no longer used for water pipes
(k) dental amalgams are made from mercury
(l) teeth can be fitted with gold caps.

5 A geologist finds a green compound in a sample of rock. He does a number of experiments on the sample. The results are shown in the diagram.

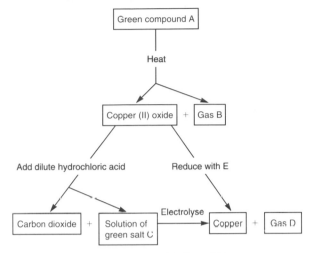

(a) Name the substances A, B, C and D.
(b) Name one substance which could be E. Describe (with a diagram if you wish) how this reaction could be carried out. Say, with an explanation, whether this reaction is an oxidation or a reduction.
(c) Is the formation of copper from C by electrolysis oxidation or reduction? Explain your answer.

6 (a) The table shows the prices and the lifetimes of two kinds of AA-size dry cell.

	Alkaline manganese dry cell	Zinc-carbon dry cell
Price of cell	£0.45	£0.20
Life when used in		
(i) electronic flash	180 flashes	20 flashes
(ii) walkman	14 hours	7 hours

Which of the two dry cells is the better buy when used in (i) an electronic flash (ii) a walkman? Explain your choices.

7 Read the following passage and answer the questions on it.

Mercury

In ancient times, the material used for filling cavities in teeth was a cement made of metal oxides. In medieval times, the method of clearing the cavity of decayed matter and then filling it with gold leaf was employed. Early in the nineteenth century the first dental amalgam was used. (An amalgam is an alloy of mercury with other metals.) The original dental amalgam of bismuth, lead, tin and mercury melted at about 100 °C and was poured into the cavity at this temperature. Later the composition was altered to make an amalgam which melted at 66 °C.

In the late nineteenth century, dentistry was re-volutionised by G. V. Black. He measured the force required to chew different foods and the pressure which fillings could stand without cracking. Black's formula, 65% silver, 27% tin, 6% copper and 2% zinc is still used today, although many manufacturers use more copper and less tin. Dentists mix this powder with mercury, put it into the cavity, 'carve' it into shape, and allow it to set for 5–10 minutes. After 48 hours, the amalgam has hardened.

(a) What is the name for a mixture of metals?

(b) What is the name for a mixture of mercury with other metals?

(c) Why is gold a good metal to use for filling teeth?

(d) Why does a cavity have to be cleaned before it is filled?

(e) What is the advantage of gold over the metal oxides used to fill cavities?

(f) What is the drawback to gold?

(g) How did scientific instruments help dentistry?

(h) What was the disadvantage of the original dental amalgam?

(i) Give three reasons why mercury is a good basis for a dental amalgam.

(j) Your little brother finds out that his fillings contain mercury. He becomes very alarmed because he has heard of mercury poisoning. How would you explain to him why the mercury in his teeth does not poison him?

8 The following metals are listed in order of decreasing reactivity. X and Y are two unknown metals.

K   X   Ca   Mg   Al   Zn   Y   Fe   Cu

(a) Will X react with cold water?

(b) Will Y react with cold water?

(c) Will Y react with dilute hydrochloric acid?

Explain how you arrive at your answers.

(d) What reaction would you expect between zinc sulphate solution and (i) X (ii) Y?

(e) Which is more easily decomposed, XCO_3 or YCO_3?

9 Look at the following pairs of chemicals. If a reaction happens, copy the word equation, and complete the right hand side.

(a) Copper + Oxygen →

(b) Calcium + Hydrochloric acid →

(c) Copper + Sulphuric acid →

(d) Carbon + Lead(II) oxide →

(e) Hydrogen + Calcium oxide →

(f) Aluminium + Tin(II) oxide →

(g) Gold + Oxygen →

(h) Zinc + Copper(II) sulphate solution →

(i) Magnesium + Sulphuric acid →

(j) Hydrogen + Silver oxide →

(k) Carbon + Magnesium oxide →

(l) Lead + Copper(II) sulphate solution →

(m) Hydrogen + Potassium oxide →

10 Explain these statements about aluminium.

(a) Although aluminium is a reactive metal, it is used to make doorframes and windowframes.

(b) Although aluminium conducts heat, it is used to make blankets which are good thermal insulators.

(c) Although aluminium oxide is a common mineral, people did not succeed in extracting aluminium from it until seven thousand years after the discovery of copper.

(d) Recycling aluminium is easier than recycling scrap iron.

11 (a) Explain how steel is made from cast iron.

(b) Explain what advantages steel has over cast iron.

(c) Explain how the following methods protect iron against rusting: painting, galvanising, tin-plating, sacrificial protection.

12 Imagine that you live in a beautiful part of Northern Ireland. A firm called Alumco wants to build a new aluminium plant in your area so that they can use a river as a source of hydroelectric power.

(a) Write a letter from a local farmer to the Secretary of State for Northern Ireland. Say what you fear may happen as a result of pollution from the plant.

(b) Write a letter from a group of environmentalists to the Secretary of State, opposing the plan and giving your reasons.

(c) Write a letter from an unemployed couple to the Secretary of State saying that you welcome the coming of new industry to the area.

(d) Write a letter from the local Council to the Secretary of State. Tell him or her that there is very little unemployment in the area. Say that the new plant would have to bring in workers from outside the region. Explain that there is not enough housing in the area for newcomers.

(e) Write a letter from Alumco to the Secretary of State for Northern Ireland. Tell the Secretary of State of the importance of aluminium. Point out the many uses of aluminium. Explain that to keep up with increasing demand you have to build another plant to supply aluminium.

(If five letters are too many for you, divide the work among the class. Then get together to read out the letters. Have a discussion to decide what the Secretary of State ought to do.)

VG INSTRUM

How fast will a chemical reaction take place?
Can we speed it up or slow it down?
Will the reaction go to completion or reach an equilibrium?
How much product will be formed?
Can we increase the yield of product?
Which fuel shall we use to heat the plant?
These are questions to which the manager of an industrial plant must find answers.
In this theme, we show how the answers can be found.

CHEMICAL REACTION SPEEDS

FIRST THOUGHTS

23.1 Why reaction speeds are important

Who is interested in the speeds of chemical reactions? If you were a cheese manufacturer, you would be interested in speeding up the chemical reactions which produce cheese. The more tonnes of cheese you could produce in a month, the more profit you would make. If you were a butter manufacturer, you would want to slow down the chemical reactions which make your product turn rancid.

In a chemical reaction, the starting materials are called the **reactants**, and the finishing materials are called the **products**. It takes time for a chemical reaction to happen. If the reactants take only a short time to change into the products, that reaction is a **fast reaction**. The **speed** or **rate** of that reaction is high. If a reaction takes a long time to change the reactants into the products, it is a **slow reaction**. The speed or rate of that reaction is low.

Many people are interested in knowing how to alter the speeds of chemical reactions. The factors which can be changed are:
- the size of the particles of a solid reactant,
- the concentrations of reactants in solution,
- the temperature,
- the presence of light,
- the addition of a catalyst.

A **catalyst** is a substance which can alter the rate of a chemical reaction without being used up in the reaction.

23.2 Particle size and reaction speed

Carbon dioxide can be prepared in the laboratory by the reaction

Calcium carbonate	+	Hydrochloric acid	→	Carbon dioxide	+	Calcium chloride	+	Water
$CaCO_3(s)$	+	$2HCl(aq)$	→	$CO_2(g)$	+	$CaCl_2(aq)$	+	$H_2O(l)$

One of the reactants, calcium carbonate (marble) is a solid. You can use this reaction to find out whether large lumps of a solid react at the same speed as small lumps of the same solid. Figure 23.2A shows a method for finding the rate of the reaction. It can be used when one of the products is a gas. As carbon dioxide escapes from the flask, the mass of the flask and contents decreases.

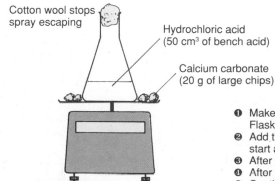

Cotton wool stops spray escaping

Hydrochloric acid (50 cm³ of bench acid)

Calcium carbonate (20 g of large chips)

❶ Make a note of the mass of Flask + Acid + Marble chips
❷ Add the chips to the acid and start a stopwatch
❸ After 10 seconds, note the mass
❹ After 30 seconds, note the mass
❺ Continue for 5–10 minutes, noting the mass every 30 seconds

Top-loading balance

Figure 23.2A ● Apparatus for following the loss in mass when a gas is evolved

The reaction starts when the marble chips are dropped into the acid. The mass of the flask and contents is noted at various times after the start of the reaction. The mass can be plotted against time. Figure 23.2B shows typical results.

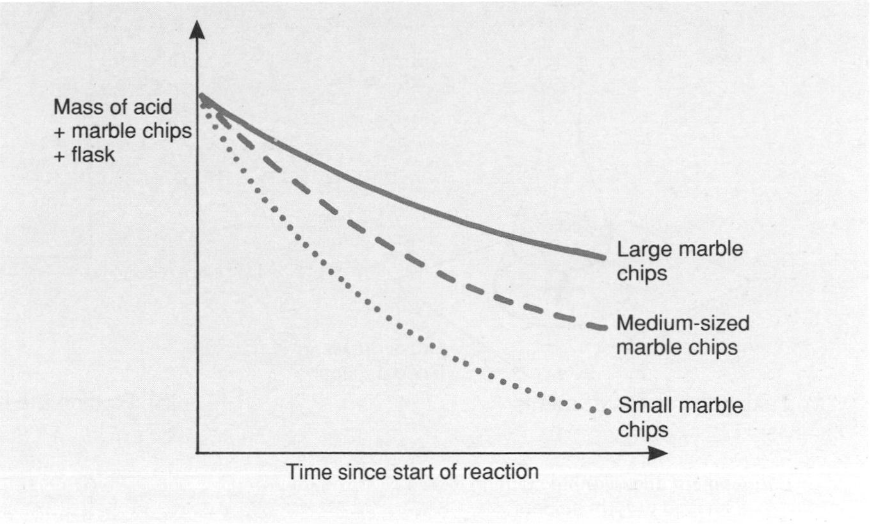

Figure 23.2B ● Results obtained with different sizes of marble chips

The results show that the smaller the size of the particles of calcium carbonate, the faster the reaction takes place. The difference is due to a difference in surface area. There is a larger surface area in 20 g of small chips than in 20 g of large chips. The acid attacks the surface of the marble. It can therefore react faster with small chips than with large chips.

SUMMARY

Reactions in which one reactant is a solid take place faster when the solid is divided into small pieces. The reason is that a certain mass of small particles has a larger surface area than the same mass of large particles.

CHECKPOINT

1. When potatoes are cooked, a chemical reaction occurs. What can you do to increase the speed at which potatoes cook?
2. There is a danger in coal mines that coal dust may catch fire and start an explosion. Explain why coal dust is more dangerous than coal.
3. 'Alko' indigestion tablets and 'Neutro' indigestion powder are both alkalis. Which do you think will act faster to cure acid indigestion? Describe how you could test the two remedies in the laboratory with a bench acid to see whether you are right.

23.3 Concentration and reaction speed

RESOURCE ACTIVITY PACK

Many chemical reactions take place in solution. One such reaction is

Sodium thiosulphate + Hydrochloric acid → Sulphur + Sodium chloride + Sulphur dioxide + Water

$$Na_2S_2O_3(aq) + 2HCl(aq) \rightarrow S(s) + 2NaCl(aq) + SO_2(g) + H_2O$$

Sulphur appears in the form of very small particles of solid. The particles do not settle: they remain in suspension. Figure 23.3A overleaf shows how you can follow the speed at which sulphur is formed.

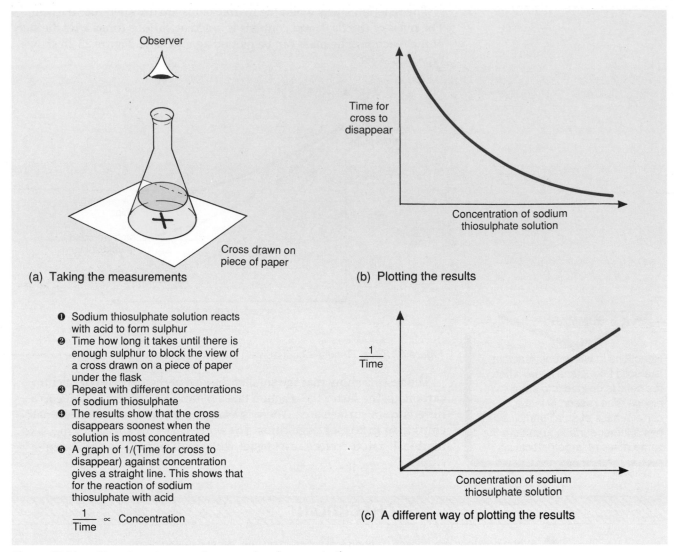

(a) Taking the measurements

Observer

Cross drawn on piece of paper

(b) Plotting the results

Time for cross to disappear

Concentration of sodium thiosulphate solution

❶ Sodium thiosulphate solution reacts with acid to form sulphur
❷ Time how long it takes until there is enough sulphur to block the view of a cross drawn on a piece of paper under the flask
❸ Repeat with different concentrations of sodium thiosulphate
❹ The results show that the cross disappears soonest when the solution is most concentrated
❺ A graph of 1/(Time for cross to disappear) against concentration gives a straight line. This shows that for the reaction of sodium thiosulphate with acid

$$\frac{1}{Time} \propto Concentration$$

(c) A different way of plotting the results

$\frac{1}{Time}$

Concentration of sodium thiosulphate solution

Figure 23.3A ● Experiment on reaction speed and concentration

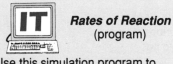

Rates of Reaction
(program)

Use this simulation program to study how a chemical reaction is affected by temperature, concentration and particle size.

SUMMARY

For the reactions in solution mentioned here, the speed of the reaction is proportional to the concentration of the reactant (or reactants). That is, the speed doubles when the concentration is doubled.

The faster a reaction takes place, the shorter is the time needed for the reaction to finish. To be more precise, the speed of the reaction is **inversely proportional** to the time taken for the reaction to finish

$$Speed\ of\ reaction \propto 1/Time$$

You can see from Figure 23.3A(c) above that

$$1/Time \propto Concentration$$

Therefore

$$Speed\ of\ reaction \propto Concentration$$

In this experiment, only one concentration was altered. A variation is to keep the concentration of sodium thiosulphate constant and alter the concentration of acid. Then the speed of the reaction is found to be proportional to the concentration of the acid. If the acid concentration is doubled, the speed doubles. The reason for this is that the ions are closer together in a concentrated solution. The closer together they are, the more often the ions collide. The more often they collide, the more chance they have of reacting.

❶ Molly is asked to investigate the marble chips–acid reaction. She must find out what effect changing the concentration of the acid has on the speed of the reaction. Explain how Molly could adapt the experiment shown in Figure 23.2A to carry out her investigation.

23.4 Pressure and reaction speed

> **LOOK AT LINKS**
> The effect of pressure on gas density is discussed in Topic 4.5.

Pressure has an effect on reactions between gases. The speed of the reaction increases when the pressure is increased. The reason is that increasing the pressure pushes the gas molecules closer together. The molecules therefore collide more often, and the gases react more rapidly.

23.5 Temperature and reaction speed

> **LOOK AT LINKS**
> The effect of temperature on the motion of molecules is discussed in Topic 4.4.

You met the reaction between sodium thiosulphate and acid in Topic 23.3. This reaction can also be used to study the effect of temperature on the speed of a chemical reaction. Warming the solutions makes sulphur form faster. There is a steep increase in the speed of the reaction as the temperature is increased.

> **IT**
> *Chemical Collisions* (program)
>
> *What factors affect the rate of reaction between two gases?* Use this program to investigate.

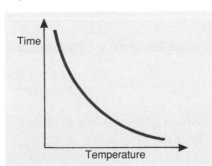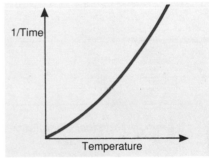

Figure 23.5A ● The effect of temperature on the speed of reaction

> **SUMMARY**
>
> The speed of a reaction between gases increases when the pressure is increased. The speed of a reaction increases when the temperature is raised. Some chemical reactions are speeded up by light.

This reaction goes approximately twice as fast at 30 °C as it does at 20 °C. It doubles in speed again between 30 °C and 40 °C and so on.

At the higher temperature, the ions have more kinetic energy. Moving through the solution more rapidly, they collide more often and more vigorously and so there is a greater chance that they will react.

23.6 Light and chemical reactions

> **LOOK AT LINKS**
> for **photographic film**
> See Topic 13.4.
> for **photosynthesis**
> See Topic 11.1 of *KS: Biology*.

Heat is not the only form of energy that can speed up reactions. Some chemical reactions take place faster when they absorb light. The formation of silver from silver salts takes place when a **photographic film** is exposed to light. In sunlight, green plants are able to carry on the process of **photosynthesis**.

❶ Polly's project is to find out what effect temperature has on the speed at which milk goes sour. Suggest what measurements she should make.

❷ Ismail's project is to find out whether iron rusts more quickly at higher temperatures. Suggest a set of experiments which he could do to find out.

❸ You are asked to study the reaction

Magnesium + Sulphuric acid ➔ Hydrogen + Magnesium sulphate

You are provided with magnesium ribbon, dilute sulphuric acid, a thermometer and any laboratory glassware you need. Describe how you would find out what effect a change in temperature has on the speed of the reaction. Say what apparatus you would use, what measurements you would make and what you would do with your results.

❹ Magnesium reacts with cold water slowly

Magnesium + Water ➔ Magnesium hydroxide + Hydrogen

If there is phenolphthalein in the water, it turns pink, showing that an alkali has been formed.
Describe experiments which you could do to find the effect of increasing the temperature on the speed of this reaction. Say what you would measure and what you would do with your results. With your teacher's approval, try out your ideas.

23.7 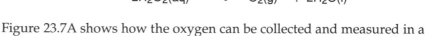 Catalysis

A reaction used to prepare oxygen is

Hydrogen peroxide ➔ Oxygen + Water
$$2H_2O_2(aq) \quad ➔ \quad O_2(g) \quad + 2H_2O(l)$$

Figure 23.7A shows how the oxygen can be collected and measured in a gas syringe.

Physical Chemistry Pack
(program)

Use the *Rates of Reaction* part of this program to study the effect of different reaction conditions on the decomposition of hydrogen peroxide.

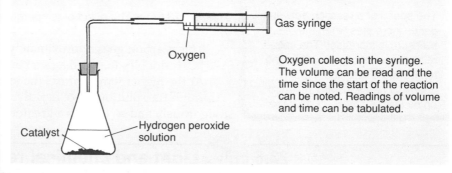

Gas syringe

Oxygen

Oxygen collects in the syringe. The volume can be read and the time since the start of the reaction can be noted. Readings of volume and time can be tabulated.

Catalyst — Hydrogen peroxide solution

Figure 23.7A ● Collecting and measuring a gas

The formation of oxygen is very slow at room temperature. The reaction can be speeded up by the addition of certain substances, for example manganese(IV) oxide. When manganese(IV) oxide is added to hydrogen peroxide, the evolution of oxygen takes place much more rapidly (see Figure 23.7B). Manganese(IV) oxide is not used up in the reaction. At the end of the reaction, the manganese(IV) oxide can be

LOOK AT LINKS
You will learn how the chemical reactions that take place in animals and plants are catalysed by **enzymes**.
See Topic 31.6.

filtered out of the solution and used again. A substance which increases the speed of a chemical reaction without being used up in the reaction is called a **catalyst**. Different reactions need different catalysts.

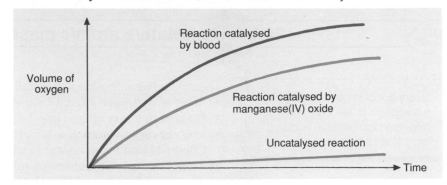

Figure 23.7B ● Catalysis of the decomposition of hydrogen peroxide

LOOK AT LINKS
You will see how valuable catalysts are in industry in Topic 29.

Any individual catalyst will only catalyse a certain reaction or group of reactions. Platinum catalyses a number of oxidation reactions. Nickel catalyses hydrogenation reactions. Industries make good use of catalysts. If manufacturers can produce their product more rapidly, they make bigger profits. If they can produce their product at a lower temperature with the aid of a catalyst, they save on fuel. The reactions which make plastics take place under high pressure. The construction of industrial plastics plants which are strong enough to withstand high pressure is expensive. A plastics manufacturer therefore tries to find a catalyst which will enable the reaction to give a good yield of plastics at a lower pressure. Then the plant will not have to withstand high pressures. Less costly materials can be used in its construction. Industrial chemists are constantly looking for new catalysts.

Some reactions only give good yields of product at high temperatures. Such a reaction costs a manufacturer high fuel bills. Industrial chemists look for a catalyst which will make the reaction take place more readily at a lower temperature. This will cut running costs and increase profits.

SUMMARY

Some chemical reactions can be speeded up by adding a substance which is not one of the reactants, which is not used up in the reaction. Such a substance is called a catalyst.

CHECKPOINT

❶ Catalysts A and B both catalyse the decomposition of hydrogen peroxide. The following figures were obtained at 20 °C for the volume of oxygen formed against the time since the start of the reaction.

Time (minutes)	0	5	10	15	20	25	30	35
Volume of oxygen with catalyst A (cm³)	0	4	8	12	16	17	18	18
Volume of oxygen with catalyst B (cm³)	0	5	10	15	16.5	18	18	18

(a) Plot a graph to show both sets of results.
(b) Say which is the better catalyst, A or B.
(c) Explain why both experiments were done at the same temperature.
(d) Explain why both sets of figures stop at 18 cm³ of oxygen.
(e) Add a line to your graph to show the shape of the graph you would obtain for the uncatalysed reaction.

❷ Someone tells you that nickel oxide will catalyse the decomposition of hydrogen peroxide to give oxygen. How could you find out whether this is true? Draw the apparatus you would use and state the measurements you would make.

TOPIC 24 — CHEMICAL CALCULATIONS

24.1 Relative atomic mass

> Every dot of ink on this page is big enough to have a million hydrogen atoms fitted across it from side to side.

The masses of atoms are very small. Some examples are:
- Mass of hydrogen atom, H = 1.4×10^{-24} g.
- Mass of mercury atom, Hg = 2.8×10^{-22} g.
- Mass of carbon atom, C = 1.7×10^{-23} g.

Chemists find it convenient to use **relative atomic masses**. The hydrogen atom is the lightest of atoms, and the masses of other atoms can be stated *relative to* that of the hydrogen atom. On the original version of the relative atomic mass scale:
- Relative atomic mass of hydrogen = 1.
- Relative atomic mass of mercury = 200 (a mercury atom is 200 times heavier than a hydrogen atom).
- Relative atomic mass of carbon = 12 (a carbon atom is 12 times heavier than a hydrogen atom).

Chemists now take the mass of one atom of carbon-12 as the reference point for the relative atomic mass scale. On the present scale

$$\text{Relative atomic mass of element} = \frac{\text{Mass of one atom of the element}}{(1/12) \text{ Mass of one atom of carbon-12}}$$

Since relative atomic mass (symbol A_r) is a ratio of two masses, the mass units cancel, and relative atomic mass is a number without a unit.

The relative atomic masses of some common elements are listed in Table 24.1. You will find a complete list in the Appendix.

Table 24.1 ● Some relative atomic masses

Element	A_r	Element	A_r
Aluminium	27	Magnesium	24
Barium	137	Mercury	200
Bromine	80	Nitrogen	14
Calcium	40	Oxygen	16
Chlorine	35.5	Phosphorus	31
Copper	63.5	Potassium	39
Hydrogen	1	Sodium	23
Iron	56	Sulphur	32
Lead	207	Zinc	65

> **SUMMARY**
>
> The relative atomic mass, A_r, of an element is the mass of one atom of the element compared with $1/12$ the mass of one atom of carbon-12.

CHECKPOINT

❶ Refer to Table 24.1.
  (a) How many times heavier is one atom of nitrogen than one atom of hydrogen?
  (b) What is the ratio

  Mass of one atom of mercury / Mass of one atom of bromine?

  (c) How many atoms of oxygen are needed to equal the mass of one atom of bromine?
  (d) How many atoms of sodium are needed to equal the mass of one atom of lead?

24.2 Relative molecular mass

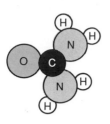

Figure 24.2A ● Atoms in a molecule of urea

The mass of a molecule is the sum of the masses of all the atoms in it. The relative molecular mass (symbol M_r) of a compound is the sum of the relative atomic masses of all the atoms in a molecule of the compound (see Figure 24.2A).

Worked example Find the relative molecular mass of urea.

Solution Formula of compound is CON_2H_4
1 atom of C (A_r 12) = 12
1 atom of O (A_r 16) = 16
2 atoms of N (A_r 14) = 28
4 atoms of H (A_r 1) = 4
Total = 60
Relative molecular mass, M_r, of urea = 60

Many compounds consist of ions, not molecules. The formula of an ionic compound represents a **formula unit** of the compound; for example, $CaSO_4$ represents a formula unit of calcium sulphate, not a molecule of calcium sulphate. The term relative molecular mass can be used for ionic compounds as well as molecular compounds.

> **SUMMARY**
>
> The relative molecular mass of a compound is equal to the sum of the relative atomic masses of all the atoms in one molecule of the compound or in one formula unit of the compound.
> M_r = Sum of A_r values

CHECKPOINT

❶ Work out the relative molecular masses of these compounds:

CO CO_2 SO_2 SO_3 $NaOH$ $NaCl$ CaO $Mg(OH)_2$ Na_2CO_3

$CuSO_4$ $CuSO_4.5H_2O$ $Ca(HCO_3)_2$

24.3 Percentage composition

From the formula of a compound you can find the percentage by mass of the elements in the compound.

Worked example 1 Find the percentage by mass of (a) calcium (b) chlorine in calcium chloride.

Solution First find the relative molecular mass of calcium chloride, formula $CaCl_2$.

$M_r = A_r(Ca) + 2A_r(Cl)$
   $= 40 + (2 \times 35.5) = 111$

Percentage of calcium $= \dfrac{40}{111} \times 100 = 36\%$

Percentage of chlorine $= \dfrac{71}{111} \times 100 = 64\%$

You can see that the two percentages add up to 100%.

Worked example 2 Find the percentage of water in crystals of magnesium sulphate-7-water.

Solution Find the relative molecular mass of $MgSO_4.7H_2O$.

1 atom of magnesium (A_r 24)   = 24
1 atom of sulphur (A_r 32)   = 32
4 atoms of oxygen (A_r 16)   = 64
7 molecules of water = 7 x [(2 x 1) + 16] = 126
Total = M_r = 246

SUMMARY

You can calculate the percentage by mass composition of a compound from its formula.

$$\text{Percentage of water} = \frac{\text{Mass of water in formula}}{\text{Relative molecular mass}} \times 100$$

$$= \frac{126}{246} \times 100 = 51.2\%$$

The percentage of water in magnesium sulphate crystals is 51%.

CHECKPOINT

You do not need calculators for these problems.

❶ Find the percentage by mass of;
    (a) calcium in calcium bromide, $CaBr_2$,
    (b) iron in iron(III) oxide, Fe_2O_3,
    (c) carbon and hydrogen in ethane, C_2H_6,
    (d) sulphur and oxygen in sulphur trioxide, SO_3,
    (e) hydrogen and fluorine in hydrogen fluoride, HF,
    (f) magnesium, sulphur and oxygen in magnesium sulphate, $MgSO_4$.

❷ Calculate the percentage by mass of water in:
    (a) copper(II) sulphate-5-water, $CuSO_4.5H_2O$ (take A_r (Cu) = 64),
    (b) sodium sulphide-9-water, $Na_2S.9H_2O$.

24.4 The mole

FIRST THOUGHTS

Chemical equations tell us which products are formed when substances react. Equations can also be used to tell us what mass of each product is formed. The key to success is the mole concept.

A reaction of industrial importance is

Calcium carbonate → Calcium oxide + Carbon dioxide
   $CaCO_3(s)$ → $CaO(s)$ + $CO_2(g)$

Cement manufacturers use this reaction to make calcium oxide (quicklime) from calcium carbonate (limestone). The mole concept makes it possible to calculate what mass of calcium oxide will be formed when a certain mass of calcium carbonate dissociates.

    The mole concept dates back to the nineteenth century Italian chemist called Avogadro. This is how he argued:

The relative atomic masses of magnesium and carbon are: A_r (Mg) = 24, A_r (C) = 12.
Therefore we can say:
Since one atom of magnesium is twice as heavy as one atom of carbon, then one hundred Mg atoms are twice as heavy as one hundred C atoms, and five million Mg atoms are twice as heavy as five million C atoms.

Imagine a piece of magnesium that has twice the mass of a piece of carbon. It follows that the two masses must contain equal numbers of atoms: two grams of magnesium and one gram of carbon contain the same number of atoms; ten tonnes of magnesium and five tonnes of carbon contain the same number of atoms.

The same argument applies to the other elements. Take the relative atomic mass in grams of any element:

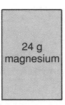    

| 12 g carbon | 24 g magnesium | 32 g sulphur | 40 g calcium | 56 g iron | 80 g bromine | 207 g lead |

All these masses contain the same number of atoms. The number is 6.022×10^{23}.

> The amount of an element that contains 6.022×10^{23} atoms (the same number of atoms as 12 g of carbon-12) is called one **mole** of that element.

The symbol for mole is **mol**. The ratio 6.022×10^{23}/mol is called the Avogadro constant. When you weigh out 12 g of carbon, you are counting out 6×10^{23} atoms of carbon. This amount of carbon is one mole (1 mol) of carbon atoms. Similarly, 46 g of sodium is two moles (2 mol) of sodium atoms. You can say that the **amount** of sodium is two moles (2 mol). One mole of the compound ethanol, C_2H_6O, contains 6×10^{23} molecules of C_2H_6O, that is, 46 g of C_2H_6O (the molar mass in grams). To write 'one mole of oxygen' will not do. You must state whether you mean one mole of oxygen atoms, O (with a mass of 16 grams) or one mole of oxygen molecules, O_2 (with a mass of 32 grams).

Molar mass

The mass of one mole of a substance is called the **molar mass**, symbol M. The molar mass of carbon is 12 g/mol. The molar mass of sodium is 23 g/mol. The molar mass of a compound is the relative molecular mass expressed in grams per mole. Urea, CON_2H_4, has a relative molecular mass of 60; its molar mass is 60 g/mol. Notice the units: relative molecular mass has no unit; molar mass has the unit g/mol.

> Amount (in moles) of substance = $\dfrac{\text{Mass of substance}}{\text{Molar mass of substance}}$
>
> Molar mass of element = Relative atomic mass in grams per mole
>
> Molar mass of compound = Relative molecular mass in grams per mole

Worked example 1   What is the amount (in moles) of sodium present in 4.6 g of sodium?

Solution A_r of sodium = 23. Molar mass of sodium = 23 g/mol

Amount of sodium = $\dfrac{\text{Mass of sodium}}{\text{Molar mass of sodium}}$ = $\dfrac{4.6 \text{ g}}{23 \text{ g/mol}}$ = 0.2 mol

The amount (moles) of sodium is 0.2 mol.

The Mole Concept (program)

Use this program to investigate the idea of a 'mole'. Listen to the tape and work through the program at the same time.

SUMMARY

The number of atoms in 12.000 g of carbon-12 is 6.022×10^{23}. The same number of atoms is present in a mass of any element equal to its relative atomic mass expressed in grams. This amount of any element is called **one mole** (1 mol) of the element. The ratio 6.022×10^{23}/mol is called the Avogadro constant. The number of moles of a substance is called the **amount** of that substance. The mass of one mole of an element or compound is the **molar mass**, M, of that substance.

M of an element = A_r expressed in g/mol

M of a compound = M_r expressed in g/mol

Worked example 2   If you need 2.5 mol of sodium hydroxide, what mass of sodium hydroxide do you have to weigh out?

Solution   Relative molecular mass of NaOH = 23 + 16 + 1 = 40

Molar mass of NaOH = 40 g/mol

$$\text{Amount of substance} = \frac{\text{Mass of substance}}{\text{Molar mass of substance}}$$

$$2.5 = \frac{\text{Mass}}{40}$$

Mass = 40 x 2.5 = 100 g

You need to weigh out 100 g of sodium hydroxide.

CHECKPOINT

❶ State the mass of:

    (a) 1.0 mol of aluminium atoms,
    (b) 3.0 mol of oxygen molecules, O_2,
    (c) 0.25 mol of mercury atoms,
    (d) 0.50 mol of nitrogen molecules, N_2,
    (e) 0.25 mol of sulphur atoms, S,
    (f) 0.25 mol of sulphur molecules, S_8.

❷ Find the amount (moles) of each element present in:
    (a) 100 g of calcium,
    (b) 9.0 g of aluminium,
    (c) 32 g of oxygen, O_2,
    (d) 14 g of iron.

❸ State the mass of:
    (a) 1.0 mol of sulphuric acid, H_2SO_4,
    (b) 0.5 mol of nitrogen dioxide molecules, NO_2,
    (c) 2.5 mol of magnesium oxide, MgO,
    (d) 0.10 mol of calcium carbonate, $CaCO_3$.

24.5 The masses of reactant and product

Have you grasped the mole concept? In this section you will find out how it is used to obtain information about chemical reactions.

As well as knowing what products are formed in a chemical reaction, chemists want to know **what mass** of product is formed from a given mass of starting material. For example, calcium oxide (quicklime) is made by heating calcium carbonate (limestone). Manufacturers need to know what mass of limestone to heat to yield the mass of quicklime they want.

Worked example 1   What mass of limestone (calcium carbonate) must be decomposed to yield 10 tonnes of calcium oxide (quicklime)?

Solution   First write the equation for the reaction

Calcium carbonate → Calcium oxide + Carbon dioxide
$$CaCO_3(s) \quad \rightarrow \quad CaO(s) \quad + \quad CO_2(g)$$

The equation tells us that

    1 mol of calcium carbonate forms 1 mol of calcium oxide.

Using the molar masses $M(CaCO_3) = 100$ g/mol, $M(CaO) = 56$ g/mol,

100 g of calcium carbonate forms 56 g of calcium oxide.

The mass of calcium carbonate needed to make 10 tonnes of calcium oxide is therefore given by

$$\text{Mass of CaCO}_3 = \frac{100}{56} \text{ x (Mass of CaO)} = \frac{100}{56} \text{ x 10} = 17.8 \text{ tonnes}$$

Worked example 2 What mass of aluminium can be obtained by the electrolysis of 60 tonnes of pure aluminium oxide, Al_2O_3?

Solution The equation comes first.

$$\text{Aluminium oxide} \rightarrow \text{Aluminium} + \text{Oxygen}$$
$$2Al_2O_3(s) \rightarrow 4Al(s) + 3O_2(g)$$

From the equation you can see that

1 mole of aluminium oxide forms 2 mol of aluminium.

Using the molar masses $M(Al) = 27$ g/mol, $M(Al_2O_3) = 102$ g/mol,

102 g of aluminium oxide form 54 g of aluminium.

The mass of aluminium obtained from 60 tonnes of aluminium oxide is therefore given by

$$\text{Mass of aluminium} = \frac{54}{102} \text{ x Mass of aluminium oxide}$$

$$= \frac{54}{102} \text{ x 60} = 31.8 \text{ tonnes}$$

SUMMARY

The equation for a chemical reaction shows how many moles of product are formed from one mole of reactant. Using the equation and the molar masses of the chemicals, you can find out what mass of product is formed from a certain mass of reactant. In chemical calculations, the balanced equation for the reaction is the key to success.

CHECKPOINT

You do not need calculators for these problems.

❶ What mass of magnesium oxide, MgO, is formed when 4.8 g of magnesium are completely oxidised?

❷ Hydrogen will reduce hot copper(II) oxide, CuO, to copper:

Hydrogen + Copper(II) oxide → Copper + Water

(a) Write the balanced chemical equation for the reaction.
(b) Calculate the mass of copper that can be obtained from 4.0 g of copper(II) oxide. Use A_r (Cu) = 64

❸ What mass of sodium bromide, NaBr, must be electrolysed to give 8 g of bromine, Br_2?

❹ Ammonium chloride can be made by neutralising hydrochloric acid with ammonia:

$HCl(aq) + NH_3(aq) \rightarrow NH_4Cl(aq)$

What mass of ammonium chloride is formed when 73 g of hydrochloric acid are completely neutralised by ammonia?

24.6 Percentage yield

Calculations based on chemical equations give the **theoretical yield** of product to be expected from a reaction. Often the actual yield is less than the calculated yield of product. The reason may be that some product has remained in solution or on a filter paper and has not been weighed with the final yield. The percentage yield of a product is given by:

$$\text{Percentage yield} = \frac{\text{Actual mass of product}}{\text{Calculated mass of product}} \times 100$$

Worked example   A student calculates that a certain reaction will yield 7.0 g of a salt. Her product weighs 6.3 g. What percentage yield has she obtained?

Solution

$$\text{Percentage yield} = \frac{\text{Actual mass of product}}{\text{Calculated mass of product}} \times 100$$

$$= \frac{6.3}{7.0} \times 100 = 90\%$$

CHECKPOINT

You will need a calculator to solve some of these problems.

❶ When 6.4 g of copper were heated in air, 7.6 g of copper(II) oxide, CuO, were obtained.

$$2Cu(s) + O_2(g) \rightarrow 2CuO(s)$$

(a) Calculate the mass of copper(II) oxide that would be formed if the copper reacted completely. (Use A_r (Cu) = 64)
(b) Calculate the percentage yield that was actually obtained.

❷ When 28 g of nitrogen and 6 g of hydrogen were mixed and allowed to react, 3.4 g of ammonia formed.

$$N_2(g) + 3H_2(g) \rightarrow 2NH_3(g)$$

(a) What is the maximum mass of ammonia that could be formed?
(b) What percentage of this yield was obtained?

❸ A student passed chlorine over heated iron until all the iron had reacted. He collected 16.0 g of iron(III) chloride, $FeCl_3$. What percentage yield had he obtained?

❹ A student neutralised 98 g of sulphuric acid, H_2SO_4, with ammonia, NH_3. On evaporating the solution until the salt crystallised, she obtained 120 g of ammonium sulphate, $(NH_4)_2SO_4$.
(a) Write a balanced chemical equation for the reaction.
(b) Calculate the theoretical yield of ammonium sulphate.
(c) Calculate the actual percentage yield.
(d) What do you think happened to the rest of the ammonium sulphate?

❺ Copper(II) sulphate can be made by neutralising sulphuric acid with copper(II) oxide:

$$CuO(aq) + H_2SO_4(aq) \rightarrow CuSO_4(aq) + H_2O(l)$$

The salt crystallises as copper(II) sulphate-5-water, $CuSO_4.5H_2O$.
(a) Calculate the mass of crystals that can be made from 8.0 g of copper(II) oxide and an excess (more than enough) of sulphuric acid. Use A_r (Cu) = 64.
(b) A student obtained 22 g of crystals from this preparation. What percentage yield was this?

24.7 Finding formulas

All those fascinating formulas in chemistry books – where do they come from? In this section you can find out.

The formula of a compound is worked out from the percentage composition by mass of the compound.

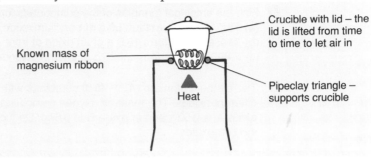

Crucible with lid – the lid is lifted from time to time to let air in

Known mass of magnesium ribbon

Heat

Pipeclay triangle – supports crucible

Figure 24.7A ● Heating magnesium

Worked example Finding the formula of magnesium oxide. First, an experiment must be done to find the mass of oxygen that combines with a weighed amount of magnesium. Figure 24.7A shows a weighed quantity of magnesium being heated until it has been converted completely into magnesium oxide. Then the magnesium oxide must be weighed. The mass of oxygen that has combined with the magnesium is found by subtraction. A typical set of results is given below.

[1] Mass of crucible = 19.24 g
[2] Mass of crucible + magnesium = 20.68 g
[3] **Mass of magnesium** = [2] − [1] = 1.44 g
[4] Mass of crucible + magnesium oxide = 21.64 g
[5] **Mass of oxygen combined** = [4] − [2] = 0.96 g

Solution The results are used in this way:

Element	Magnesium	Oxygen
Mass	1.44 g	0.96 g
A_r	24	16
Amount in moles	1.44/24	0.96/16
	= 0.060	= 0.060
Divide through by 0.060	1 mole Mg	1 mole O
Formula is		MgO

The formula MgO is the simplest formula which fits the results. Other formulas, such as Mg_2O_2, Mg_3O_3, etc. also fit the results. MgO is the **empirical formula** for magnesium oxide.

> The empirical formula of a compound is the simplest formula which represents the composition by mass of the compound.

SUMMARY

The empirical formula of a compound shows the symbols of the elements present and the ratio of the number of atoms of each element present in the compound.

CHECKPOINT

You do not need calculators for these problems.

❶ Find the empirical formulas of the following compounds:
   (a) a compound of 3.5 g of nitrogen and 4.0 g of oxygen,
   (b) a compound of 14.4 g of magnesium and 5.6 g of nitrogen,
   (c) a compound of 5.4 g of aluminium and 9.6 g of sulphur.

❷ Calculate the empirical formulas of the compounds which have the following percentage compositions by mass:
   (a) 40% sulphur, 60% oxygen,
   (b) 50% sulphur, 50% oxygen,
   (c) 20% calcium, 80% bromine,
   (d) 39% potassium, 1% hydrogen, 12% carbon, 48% oxygen.

❸ Find the empirical formulas of the compounds formed when
   (a) 18 g of beryllium form 50 g of beryllium oxide,
   (b) 11.2 g of iron form 16.0 g of an oxide of iron,
   (c) 2.800 g of iron form 5.325 g of an iron chloride.

❹ The element titanium (A_r = 48.0) combines with oxygen to form two different oxides. The mass of oxygen combining with 1.00 g of titanium in oxide A is 0.50 g and in oxide B is 0.67 g. Deduce the empirical formulas of the two oxides.

Finding the molecular formula from the empirical formula

The empirical formula is the simplest formula that represents the composition of a compound. It may not be the **molecular formula**, that is the formula that shows the number of atoms present in a molecule of the compound. The molecular formula of ethanoic acid is CH_3CO_2H, which can be written as $C_2H_4O_2$. The empirical formula is CH_2O. To work out the relative molecular mass for ethanoic acid, you have to use the molecular formula $C_2H_4O_2$, and then you obtain a value of 60.

Worked example 1   What is the molecular formula of a compound which has an empirical formula of CH_2 and a relative molecular mass of 70?

Method   Relative molecular mass = 70
Empirical formula mass = 12 + 2 = 14
The relative molecular mass is 5 x the relative empirical formula mass
The molecular formula is 5 x the empirical formula

Answer   The molecular formula is C_5H_{10}.

Worked example 2   What is the molecular formula of a compound which has an empirical formula of C_2H_6O and a relative molecular mass of 46?

Method   Relative molecular mass = 46
Relative empirical formula mass = 24 + 6 + 16 = 46

The empirical formula gives the correct relative molecular mass therefore the molecular formula is the same as the empirical formula.

Answer   The molecular formula is C_2H_6O.

CHECKPOINT

❶ Find the molecular formula of each of the following compounds from the empirical formula and the relative molecular mass.

Compound	Empirical formula	M_r	Compound	Empirical formula	M_r
A	CH_4O	32	B	CH_2	42
C	C_2H_4O	88	D	CH_3O	62
E	CH_3	30	F	CH_2Cl	99
G	CH	78	H	C_2HNO_2	213

❷ Calculate (a) the empirical formula and (b) the molecular formula of:
(i) a hydrocarbon which contains 80% by mass of carbon and has a relative molecular mass of 30
(ii) a hydrocarbon which contains 85.7% of carbon and has a relative molecular mass of 28.

❸ What is (a) the empirical formula and (b) the molecular formula of a compound which contains 4.04% H, 24.24% C, 71.72% Cl and has a relative molecular mass of 99?

24.8 Reacting volumes of gases

LOOK AT LINKS
for the **mole concept**
See Topic 24.4;
for the effect of
temperature and pressure
on gas volume
See Topic 4.5.

The volume of a certain mass of gas depends on its temperature and on its pressure. We therefore state the temperature and the pressure at which a volume was measured. It is usual to state gas volumes either at standard temperature and pressure (s.t.p., 0 °C and 1 atm) or at room temperature and pressure (r.t.p., 20 °C and 1 atm).

One mole of a gas is the amount of the gas that contains 6×10^{23} molecules of that gas. To measure one mole of gas, you take the molar mass expressed in grams, e.g. 2 g of hydrogen, H_2. Measurements show that one mole of gas occupies $24.0 \, dm^3$ at r.t.p. For all gases, the volume is the same.

- 2 g of hydrogen
- 28 g of nitrogen
- 64 g of sulphur dioxide

All these are one mole of gas (the molar mass expressed in grams) and all occupy $24 \, dm^3$ at r.t.p.

The volume of one mole of gas, $24.0 \, dm^3$ at r.t.p., is called the **gas molar volume**.

Calculations on the reacting volumes of gases start with the equation for the reaction. After that, it's as easy as one, two, three. Take the reaction:

$$A(g) + 3B(g) \rightarrow 2C(g)$$

The equation tells you that

1 mole of A reacts with 3 moles of B to form 2 moles of C

therefore, at r.t.p.,

$24 \, dm^3$ of A react with $3 \times 24 \, dm^3$ of B to form $2 \times 24 \, dm^3$ of C

and in general,

1 volume of A reacts with 3 volumes of B to form 2 volumes of C.

Worked example 1   Nitrogen and hydrogen combine to form ammonia:

$$N_2(g) + 3H_2(g) \rightleftharpoons 2NH_3(g)$$

If hydrogen is fed into the plant at $12\,m^3/second$, at what rate should nitrogen be fed in?

Method   From the equation, 1 mole of nitrogen combines with 3 moles of hydrogen, therefore 1 volume of nitrogen combines with 3 volumes of hydrogen, therefore the rate of flow of nitrogen should be one third that of hydrogen, that is $\frac{1}{3} \times 12\,m^3/s = 4\,m^3/s$.

Answer   Nitrogen should be fed in at a rate of $4\,m^3/s$.

Worked example 2   What volume of carbon dioxide (at r.t.p.) is formed by the complete combustion of 3 g of carbon?

Method   From the equation,

$$C(s) + O_2(g) \rightarrow CO_2(g)$$

we can see that 1 mole of carbon forms 1 mole of carbon dioxide that is 12 g carbon form $24\,dm^3$ of carbon dioxide therefore 3 g carbon form $\frac{3}{12} \times 24\,dm^3 = 6\,dm^3$ of carbon dioxide.

Answer   The combustion of 3 g of carbon produces $6\,dm^3$ of carbon dioxide at r.t.p.

SUMMARY

Calculations on reacting volumes of gases always start with the equation for the chemical reaction. One mole of any gas occupies $24.0\,dm^3$ at r.t.p. This volume is called the gas molar volume.

CHECKPOINT

❶   Calculate the volume at r.t.p. of oxygen needed for the complete combustion of 8 g of sulphur.

❷   Calculate the volume at r.t.p. of oxygen needed for the complete combustion of $250\,cm^3$ of methane, CH_4. What volume of carbon dioxide is formed?

❸   A power station burns coal which contains sulphur. The sulphur burns to form sulphur dioxide, SO_2. If the power station burns 28 tonnes of sulphur in its coal every day, what volume of sulphur dioxide does it send into the air? (1 tonne = 1000 kg)

❹   What volume of hydrogen at r.t.p. is formed when 7.0 g of iron react with an excess of sulphuric acid? The equation is

$$Fe(s) + H_2SO_4(aq) \rightarrow H_2(g) + FeSO_4(aq)$$

❺   A cook puts 3 g of sodium hydrogencarbonate into a cake mixture. Calculate the volume at r.t.p. of carbon dioxide that will be produced in the reaction:

$$2NaHCO_3(s) \rightarrow Na_2CO_3(s) + CO_2(g) + H_2O(g)$$

❻   The equation for the reaction between marble and hydrochloric acid is

$$CaCO_3(s) + 2HCl(aq) \rightarrow CaCl_2(aq) + CO_2(g) + H_2O(l)$$

What mass of marble is needed to give $6.0\,dm^3$ of carbon dioxide?

24.9  Calculations on electrolysis

LOOK AT LINKS
for **electrolysis**
See Topic 9.

When a current passes through a solution of a salt of a metal which is low in the electrochemical series, metal ions are discharged and metal atoms are deposited on the cathode (see Figure 24.9A).

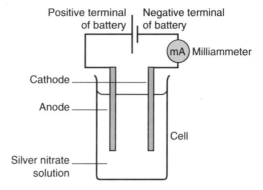

Positive terminal of battery
Negative terminal of battery
mA) Milliammeter
Cathode
Anode
Cell
Silver nitrate solution

Figure 24.9A ● The electrolysis of silver nitrate solution

The cathode process is

$$Ag^+(aq) + e^- \rightarrow Ag(s)$$

This equation tells us that 1 silver ion accepts 1 electron to form 1 atom of silver. Therefore 1 mole of silver ions accept 1 mole of electrons to form 1 mole of silver atoms.

When 1 mole of silver has been deposited, the cathode has increased in mass by 108 g (the mass of 1 mole of silver). This process must have needed the passage of 1 mole of electrons through the cell. We can measure the quantity of electric charge that passes through the cell when 108 g of silver are deposited. This quantity is the charge on 1 mole of electrons. Electric charge is measured in coulombs (C). One coulomb is the electric charge that passes when 1 ampere flows for 1 second.

Charge in coulombs = Current in amperes x Time in seconds
$$Q = I \times t$$

Using a cell such as that in Figure 24.9A, one can pass a current through a milliammeter in the circuit for a known time. By weighing the cathode before and after the passage of a current, the mass of silver deposited can be found. Experiments of this kind show that to deposit 108 g of silver requires the passage of 96 500 coulombs. This quantity of electricity must be the charge on one mole of electrons (6×10^{23} electrons). The value 96 500 C/mol is called the Faraday constant after Michael Faraday who did much of the early work on electrolysis.

Now in the deposition of copper in electrolysis, the cathode process is

$$Cu^{2+}(aq) + 2e^- \rightarrow Cu(s)$$

1 copper ion needs 2 electrons to become 1 atom of copper;
1 mole of copper ions need 2 moles of electrons for discharge;
1 mole of copper ions need 2 x 96 500 coulombs for discharge.

When gold is deposited during the electrolysis of gold(III) salts, the cathode process is:

$$Au^{3+}(aq) + 3e^- \rightarrow Au(s)$$

1 gold ion needs 3 electrons to form 1 mole of gold atoms;
1 mole of gold ions need 3 moles of electrons for discharge;
3 x 96 500 coulombs are needed to deposit 1 mole of gold.

It's as easy as one-two-three. First work out whether
1 mole of electrons discharge 1 mole of the element, e.g. silver, or
$\frac{1}{2}$ mole of the element, e.g. copper, or
$\frac{1}{3}$ mole of the element, e.g. gold.

$$\text{No. of moles of element discharged} = \frac{\text{No. of moles of electrons}}{\text{No. of charges on one ion of the element}}$$

$$= \frac{\text{No. of coulombs/96 500}}{\text{No. of charges on one ion of the element}}$$

Worked example 1   What masses of the following elements are deposited by the passage of one mole of electrons through solutions of their salts? (a) silver (b) copper (c) gold

Method   Start with the equations.

$$Ag^+(aq) + e^- \;\rightarrow\; Ag(s)$$
$$Cu^{2+}(aq) + 2e^- \;\rightarrow\; Cu(s)$$
$$Au^{3+}(aq) + 3e^- \;\rightarrow\; Au(s)$$

As already argued,
1 mol electrons deposit 1 mol silver = 108 g silver
1 mole electrons deposit $\frac{1}{2}$ mol copper = $\frac{1}{2}$ x 63.5 g = 31.8 g copper
1 mole electrons deposit $\frac{1}{3}$ mol gold = $\frac{1}{3}$ x 197 g = 65.7 g gold

Answer   (a) 108 g silver, (b) 31.8 g copper, (c) 65.7 g gold

Worked example 2   A current of 10.0 milliamps (mA) passes for 4.00 hours through a solution of silver nitrate, a solution of copper(II) sulphate and a solution of gold(III) nitrate connected in series. What mass of metal is deposited in each?

Method   Charge in coulombs = Current in amperes x Time in seconds
Charge = 0.010 x 4.00 x 60 x 60 = 144 C
$$\text{Charge} = \frac{144}{96\,500} \text{ moles of electrons}$$

Equations are:

$$Ag^+(aq) + e^- \;\rightarrow\; Ag(s)$$
$$Cu^{2+}(aq) + 2e^- \;\rightarrow\; Cu(s)$$
$$Au^{3+}(aq) + 3e^- \;\rightarrow\; Au(s)$$

1 mol electrons deposit 1 mol silver; therefore $\frac{144}{96\,500}$ mol electrons

deposit $\frac{144}{96\,500}$ mol silver = 144 x $\frac{108}{96\,500}$ g silver = 0.161 g silver

1 mol electrons deposit $\frac{1}{2}$ mol copper; therefore $\frac{144}{96\,500}$ mol electrons

deposit $\frac{1}{2}$ x $\frac{144}{96\,500}$ mol copper = $\frac{1}{2}$ x 144 x $\frac{63.5}{96\,500}$ g copper = 0.0474 g copper

1 mol electrons deposit $\frac{1}{3}$ mol gold; therefore $\frac{144}{96\,500}$ mol electrons

deposit $\frac{1}{3}$ x $\frac{144}{96\,500}$ mol gold = $\frac{1}{3}$ x 144 x $\frac{197}{96\,500}$ g gold = 0.0980 g gold

Answer   0.161 g silver, 0.0474 g copper, 0.0980 g gold

Worked example 3   Aluminium is extracted by the electrolysis of molten aluminium oxide. How many coulombs of electricity are needed to produce 1 tonne of aluminium (1 tonne = 1000 kg)?

Method   The equation

$$Al^{3+}(l) + 3e^- \rightarrow Al(s)$$

shows that 3 mol electrons are needed to give 1 mol aluminium

$1.00 \times 10^6 \times \dfrac{3}{27}$ mol electrons are needed to give 1 tonne of aluminium

$1.00 \times 10^6 \times \dfrac{3}{27}$ mol of electrons $= 1.00 \times 10^6 \times \dfrac{3}{27} \times 96\,500\,C$

$$= 1.07 \times 10^{10}\,C$$

Answer $1.07 \times 10^{10}\,C$ will deposit 1 tonne of aluminium.

Worked example 4   How long will it take for a current of 0.156 A to deposit 0.100 g of lead from a solution of lead(II) nitrate?

Method   First the equation

$$Pb^{2+}(aq) + 2e^- \rightarrow Pb(s)$$

2 mol electrons are needed to deposit 1 mol lead, 207 g lead

therefore $\dfrac{2}{207}$ mol electrons deposit 1.00 g lead, and

$0.100 \times \dfrac{2}{207}$ mol electrons deposit 0.100 g lead, and

$0.100 \times \dfrac{2}{207} \times 96\,500\,C$ are needed to deposit 0.100 g lead

Charge (C) = Current (A) x Time (s)

$0.100 \times \dfrac{2}{207} \times 96\,500 = 0.156 \times$ Time (s)

Time = 6000 s = 100 minutes

Answer   The current must pass for 100 minutes (1 hour 40 minutes).

Worked example 5   A metal of relative atomic mass 27 is deposited by electrolysis. If 0.201 g of the metal is deposited when 0.200 A flow for 3.00 hours, what is the charge on the ions of this element?

Method
Charge = 0.200 x 3.00 x 60 x 60 C = 2160 C
If 2160 C deposit 0.201 g of the metal,

then 96 500 C deposit $96\,500 \times \dfrac{0.201}{2160}\,g = 8.98\,g$

Since 8.98 g metal are deposited by 1 mol electrons, 27.0 g of metal are

deposited by $\dfrac{27}{8.98}$ mol electrons = 3 mol electrons

Answer   The charge on the metal ions is +3.

Calculate the volume of gas evolved during electrolysis

● Hydrogen

When a current is passed through a solution of a salt of a metal which is high in the electrochemical series, hydrogen ions are discharged at the cathode.

$$H^+(aq) + e^- \rightarrow H(g)$$
followed by $2H(g) \rightarrow H_2(g)$

Each hydrogen molecule needs 2 electrons for its evolution, and 1 mole of hydrogen needs 2 moles of electrons for its evolution. Thus 2 moles of electrons (2 x 96 500 C) will result in the evolution of 24 dm^3 at r.t.p. (the gas molar volume) of hydrogen.

● Chlorine

When chlorine is evolved at the anode,

$$Cl^-(aq) \rightarrow Cl(g) + e^-$$
followed by $2Cl(g) \rightarrow Cl_2(g)$

Each chlorine molecule gives 2 electrons when it is evolved, and 1 mole of chlorine gives 2 moles of electrons when it is evolved. Thus 2 moles of electrons (2 x 96 500 C) accompany the evolution of 24 dm^3 at r.t.p. (the gas molar volume) of chlorine.

● Oxygen

When oxygen is evolved at the anode,

$$OH^-(aq) \rightarrow OH(g) + e^-$$
followed by $4OH(g) \rightarrow O_2(g) + 2H_2O(l)$

Thus 4 moles of electrons must pass with the evolution of 1 mole of oxygen (24 dm^3 at r.t.p.).

Worked example 6 Name the gases formed at each electrode when 15.0 mA of current passes for 6.00 hours through a solution of sulphuric acid and calculate their volumes (at r.t.p.).

Method
At the cathode hydrogen is evolved

$$H^+(aq) + e^- \rightarrow H(g)$$
followed by $2H(g) \rightarrow H_2(g)$

so that 2 mol electrons discharge 1 mol hydrogen gas.
At the anode oxygen is evolved

$$OH^-(aq) \rightarrow OH(g) + e^-$$
followed by $4OH(g) \rightarrow O_2(g) + 2H_2O(l)$

so that 4 mol electron discharge 1 mol oxygen gas.

Charge (C) = Current (A) x Time (s)
= $15.0 \times 10^{-3} \times 6.00 \times 60 \times 60 = 324$ C

No. of mol electrons = $\dfrac{324 \text{ C}}{96\,500 \text{ C/mol}} = 3.36 \times 10^{-3}$ mol

Amount of hydrogen discharged = $\frac{1}{2} \times 3.36 \times 10^{-3}$ mol
Volume of hydrogen = $1.68 \times 10^{-3} \times 24.0$ dm^3 = 40.3 cm^3 at r.t.p.
Amount of oxygen = $\frac{1}{4} \times 3.36 \times 10^{-3}$ mol = 0.84×10^{-3} mol
Volume of oxygen = $0.84 \times 10^{-3} \times 24.0$ dm^3 = 20.2 cm^3 at r.t.p.

Answer At the cathode 40.4 cm^3 (at r.t.p.) of hydrogen are evolved. At the anode 20.2 cm^3 (at r.t.p.) of oxygen are evolved.

CHECKPOINT

(See relative atomic masses given in the margin.)

❶ Calculate the mass of each element discharged when 0.250 mol of electrons passes through each of the solutions listed.
(a) copper from copper(II) sulphate solution
(b) nickel from nickel chloride solution
(c) lead from lead(II) nitrate solution
(d) bromine from potassium bromide solution
(e) tin from tin(II) nitrate solution

❷ Calculate the volume (at r.t.p.) of each gas evolved when 48 250 C pass through a solution of (a) dilute hydrochloric acid (b) dilute nitric acid.

❸ A current passes through two cells in series. The cells contain solutions of silver nitrate and lead(II) nitrate. In the first, 0.540 g of silver is deposited. What mass of lead is deposited in the second?

❹ When a current passes through a solution of copper(II) sulphate, 0.635 g of copper is deposited on the cathode. What volume of oxygen is evolved at the anode?

❺ A current passes through two cells in series. In the first, 0.2160 g of silver is deposited. In the second, 0.1125 g of cadmium is deposited. Use this information to calculate the charge on the cadmium ion.

❻ A current of 2.00 A passes for 96.5 hours through molten aluminium oxide. What mass of aluminium is deposited?

❼ A current of 2.01 mA passed for 8.00 minutes through aqueous nickel sulphate. The mass of the cathode increased by 0.295 g.
(a) How many coulombs of electricity passed during the experiment?
(b) How many moles of nickel were deposited?
(c) How many moles of electrons are needed to discharge 1 mole of nickel ions?
(d) Write an equation for the cathode reaction.

Some relative atomic masses

Ag	108
Al	27
Br	80
Cd	112
Cl	35.5
Cu	63.5
Fe	56
H	1
Ni	59
O	16
Pb	207
Sn	119

❽ Choose the correct answers.
A student electrolysed a solution of silver nitrate in series with a solution of sulphuric acid. A steady current passed for 30 minutes, and 24 cm³ of hydrogen collected at the cathode of the sulphuric acid cell.
(a) The volume of oxygen evolved at the anode is
(i) 6 cm³ (ii) 12 cm³ (iii) 24 cm³ (iv) 32 cm³ (v) 48 cm³
(b) If the student doubled the current and passed it for 15 minutes, the volume of hydrogen evolved would be
(i) 6 cm³ (ii) 12 cm³ (iii) 24 cm³ (iv) 48 cm³ (v) 96 cm³
(c) The mass of silver deposited on the cathode of the silver nitrate cell is
(i) 0.108 g (ii) 0.216 g (iii) 0.0270 g (iv) 0.0540 g (v) 0.0135 g

❾ Which one of the following requires the largest quantity of electricity for discharge at an electrode?
(a) 1 mol Ni^{2+} ions   (b) 2 mol Fe^{3+} ions   (c) 3 mol Ag^+ ions
(d) 4 mol Cl^- ions   (e) 5 mol OH^- ions

❿ A current of 0.010 A passed for 5.00 hours through a solution of a salt of the metal M, which has a relative atomic mass of 52. The mass of M deposited on the cathode was 0.0323 g. What is the charge on the ions of M?

FIRST THOUGHTS

24.10 Concentration of solution

Remember:
1 litre = 1000 cm³ = 1 dm³

One way of stating the concentration of a solution is to state the mass of solute in one litre of solution, for example in grams per litre, g/l. Chemists find it more useful to state the amount in moles of a solute present in one litre of solution (see Figure 24.10A).

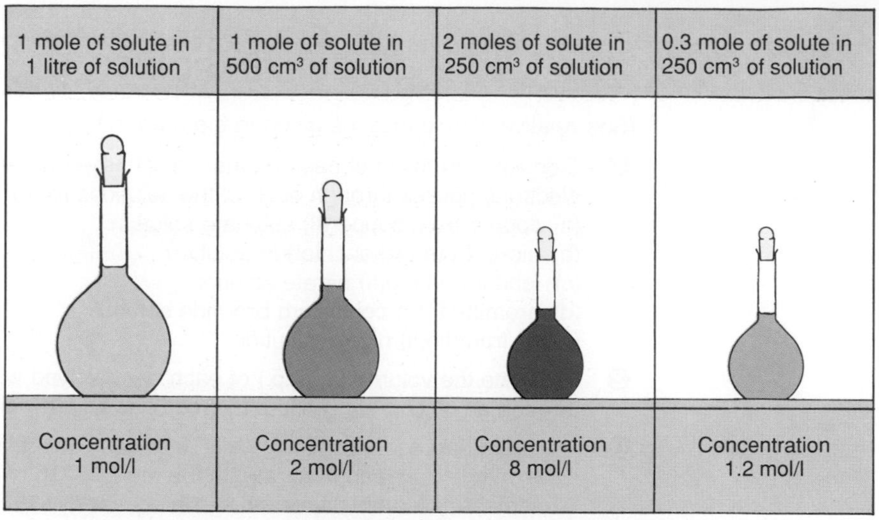

1 mole of solute in 1 litre of solution	1 mole of solute in 500 cm³ of solution	2 moles of solute in 250 cm³ of solution	0.3 mole of solute in 250 cm³ of solution
Concentration 1 mol/l	Concentration 2 mol/l	Concentration 8 mol/l	Concentration 1.2 mol/l

Figure 24.10A ● Concentrations of solutions in moles per litre (mol/l)

Concentration in moles per litre = $\dfrac{\text{Amount of solute in moles}}{\text{Volume of solution in litres}}$

Rearranging,

Amount of solute (mol) = Volume of solution (l) x Concentration (mol/l)

Worked example 1 Calculate the concentration of a solution that was made by dissolving 100 g of sodium hydroxide and making the solution up to 2.0 litres.

Solution Molar mass of sodium hydroxide, NaOH = 23 + 16 + 1 = 40 g/mol

Amount in moles $= \dfrac{\text{Mass}}{\text{Molar mass}} = \dfrac{100\ g}{40\ g/mol} = 2.5$ mol

Volume of solution = 2.0 litre

Concentration $= \dfrac{\text{Amount in moles}}{\text{Volume in litres}} = \dfrac{2.5\ mol}{2.0\ l} = 1.25$ mol/l

Worked example 2 Calculate the amount in moles of solute present in 75 cm³ of a solution of hydrochloric acid which has a concentration of 2.0 mol/l.

Solution Amount (mol) = Volume (l) x Concentration (mol/l)

Amount of solute, HCl $= (75 \times 10^{-3}) \times 2.0$
$= 0.15$ mol

Note that the volume in cm³ has been changed into litres to make the units correct:

Amount (**moles**) = Volume (**litres**) x Concentration (**moles per litre**)

Worked example 3 What mass of sodium carbonate must be dissolved in 1 l of solution to give a solution of concentration 1.5 mol/l (a 1.5 M solution)?

Solution Amount (mol) = Volume (l) x Concentration (mol/l)
$= 1.00\ l \times 1.5$ mol/l
$= 1.5$ mol

Molar mass of sodium carbonate, $Na_2CO_3 = (2 \times 24) + 12 + (3 \times 16)$
$= 106$ g/mol

Mass of sodium carbonate = 1.5 mol x 106 g/mol = 159 g

CHECKPOINT

❶ Calculate the concentrations of the following solutions.
   (a) 30 g of ethanoic acid, $C_2H_4O_2$, in 500 cm³ of solution.
   (b) 8.0 g of sodium hydroxide in 2.0 l of solution.
   (c) 8.5 g of ammonia in 250 cm³ of solution.
   (d) 12.0 g of magnesium sulphate in 250 cm³ of solution.

❷ Find the amount of solute in moles present in the following solutions.
   (a) 1.00 l of a hydrochloric acid solution of concentration 0.020 mol/l.
   (b) 500 cm³ of a solution of potassium hydroxide of concentration 2.0 mol/l.
   (c) 500 cm³ of sulphuric acid of concentration 0.12 mol/l.
   (d) 100 cm³ of a 0.25 mol/l solution of sodium hydroxide.

❸ Solutions which are injected into a vein must be isotonic with the blood. They contain 8.43 g/l of sodium chloride.
   (a) What is the concentration of sodium chloride in mol/l?
   (b) What does 'isotonic' mean? Why must the solutions be isotonic with blood?

❹ A woman mixes a drink containing 9.2 g of ethanol, C_2H_6O, in 100 cm³ of solution. What is the concentration of ethanol in the solution (in mol/l)?

❺ Calculate the concentrations of the following solutions.
   (a) 4.0 g of sodium hydroxide in 500 cm³ of solution
   (b) 7.4 g of calcium hydroxide in 5.0 l of solution
   (c) 49.0 g of sulphuric acid in 2.5 l of solution
   (d) 73 g of hydrogen chloride in 250 cm³ of solution

❻ Find the amount of solute in moles present in the following solutions.
   (a) 1.00 l of a solution of sodium hydroxide of concentration 0.25 mol/l
   (b) 500 cm³ of hydrochloric acid of concentration 0.020 mol/l
   (c) 250 cm³ of 0.20 mol/l sulphuric acid
   (d) 10 cm³ of a 0.25 mol/l solution of potassium hydroxide

Standard solutions

A solution of known concentration is called a **standard solution**. A standard solution can only be made from a solid which can be obtained 100% (almost) pure. Anhydrous sodium carbonate and sodium hydrogencarbonate can be used to make standard solutions. Ethanedioic acid (oxalic acid) can be used to make a standard acid solution. These substances are called **primary standards**. For other substances, a solution of approximately known concentration is made and then the concentration is found by analysis (see Topic 24.11). For example, you could not make a standard solution of sodium hydroxide. As you were weighing it out, it would absorb water vapour from the air and react with carbon dioxide in the air.

To make a standard solution, the required mass of solid is weighed in a weighing bottle. With a wash bottle, it is transferred carefully into a beaker containing distilled water. When all the solid has dissolved, the solution is poured into a volumetric flask and the volume is made up to the mark with distilled water.

A standard solution can be prepared by diluting a more concentrated solution. The required volume of the more concentrated solution is run from a burette into a volumetric flask, and distilled water is added to make it up to the mark. If you want to know the concentration only approximately, the concentrated solution can be measured in a measuring cylinder and diluted in a graduated beaker. This method is used for diluting solutions, e.g. concentrated sulphuric acid and glacial ethanoic acid, which cannot be used in a burette.

SUMMARY

A standard solution is one of known concentration. You can prepare a standard solution by dissolving a calculated mass of pure solid in distilled water and making the volume of the solution up to a certain volume. Alternatively, a measured volume of a more concentrated standard solution can be diluted to a certain volume. See *Key Science Resource Bank*.

CHECKPOINT

❶ (a) Jim's teacher gives him some 1.00 mol/l acid and instructs him to make a solution which is exactly 0.100 mol/l. Say what apparatus Jim should use and describe what he should do.

   (b) Jan is given the same 1.00 mol/l acid. She is asked to prepare quickly a solution which is between 0.09 mol/l and 0.11 mol/l. Say what apparatus she should use and what she should do.

❷ Explain what dangers you would risk by using a very concentrated acid in a burette.

❸ The concentrated hydrochloric acid in the store has a concentration of 12 mol/l. The college technician has to fill all the reagent bottles in the lab. with approximately 2 mol/l hydrochloric acid. There are 24 bottles, each of which holds 250 cm³.

   (a) What volume of dilute hydrochloric acid does the technician need?
   (b) What volume of the concentrated acid should she measure out?
   (c) What apparatus should she use for measuring?
   (d) Describe how she should prepare the solution of dilute hydrochloric acid.

❹ Citric acid, $C_6H_8O_7$, is a solid. Say how you would prepare 2.00 l of a 2.00 mol/l solution.

❺ Ammonia is bought as '880 ammonia' (a solution of density 0.88 g/cm³), which contains 245 g ammonia per litre of solution. What volume of the concentrated solution would you need to prepare 1.0 l of 2.0 mol/l ammonia solution?

❻ Concentrated sulphuric acid is bought as a solution that contains 98% by mass of sulphuric acid. 100 cm³ of this solution contain 180 g of sulphuric acid. The lab. technician wants to prepare 10 l of bench sulphuric acid, which is roughly 2 mol/l.

   (a) What volume of concentrated sulphuric acid is needed?
   (b) What apparatus should the technician use to measure this volume?
   (c) Describe the precautions which he or she should take in diluting the concentrated acid.

❼ Sodium hydrogencarbonate is used to prepare standard solutions because it can be obtained almost 100% pure and because it is very stable and does not absorb water vapour from the air.

   (a) What mass of sodium hydrogencarbonate would you weigh out to prepare 500 cm³ of a 0.0100 M solution?
   (b) Describe how you would make up the solution as accurately as possible.

❽ You have a large stock bottle of ethanoic acid of concentration 4.00 mol/l. How can you make up 1.00 l of a 0.250 mol/l solution of the acid? Say what quantities you would measure and what apparatus you would use.

❾ 'Glacial' ethanoic acid is the name given to a concentrated solution of ethanoic acid which freezes at 10 °C. It has a concentration of about 17 mol/l. Explain how you would prepare 2 l of 2 mol/l ethanoic acid.

24.11 Volumetric analysis

Volumetric analysis is a means of finding out the concentration of a solution. The method is to add a solution of, say an acid, to a solution of, say a base in a careful way until there is just enough of the acid to react with the base. This method is called **titration**. The concentration of one of the two solutions must be known, and the volumes of both must be measured. You can use a standard solution of a base to find out the concentration of a solution of an acid. You have to find out what volume of the acid solution of unknown concentration is needed to neutralise a known volume, usually $25\,cm^3$, of the standard solution of a base. The steps in volumetric analysis are as follows.

1 Pipette $25.0\,cm^3$ of the alkali solution into a clean conical flask. Add a few drops of indicator.
2 Read the burette, $V_1\,cm^3$. Run the acid solution from the burette drop by drop into the alkali until the indicator just changes colour. This is the 'end-point' of the titration.
3 Read the burette again, $V_2\,cm^3$. The volume of acid used, $(V_2 - V_1)\,cm^3$, is the volume of acid needed to neutralise $25.0\,cm^3$ of the alkali.

Worked example 1   A solution of hydrochloric acid is titrated against a standard sodium hydroxide solution. $15.0\,cm^3$ of hydrochloric acid neutralise $25.0\,cm^3$ of a $0.100\,mol/l$ solution of sodium hydroxide. What is the concentration of hydrochloric acid?

Solution

(1) First write the equation for the reaction:

Hydrochloric acid + Sodium hydroxide ➔ Sodium chloride + Water
$HCl(aq)$ + $NaOH(aq)$ ➔ $NaCl(aq)$ + $H_2O(l)$

The equation tells you that 1 mole of HCl neutralises 1 mole of NaOH.

(2) Now work out the amount in moles of base. You must start with the base because you know the concentration of base, and you do not know the concentration of acid.

Amount (mol) = Volume (l) x Concentration (mol/l)
Amount (mol) of NaOH = Volume ($25.0\,cm^3$) x Concentration ($0.100\,mol/l$)
= $25.0 \times 10^{-3}\,l \times 0.100\,mol/l = 2.50 \times 10^{-3}\,mol$

(3) Now work out the concentration of acid

Amount (mol) of HCl = Amount (mol) of NaOH = $2.50 \times 10^{-3}\,mol$
also Amount (mol) of HCl = Volume of HCl(aq) x Concentration of HCl(aq)
Therefore, if c is the concentration of HCl,
$25.0 \times 10^{-3}\,mol = 15.0 \times 10^{-3}\,l \times c$

$$c = \frac{2.50 \times 10^{-3}\,mol}{15.0 \times 10\,l}$$

$$= 0.167\,mol/l$$

The concentration of hydrochloric acid is $0.167\,mol/l$.

Worked example 2   In a titration, $25.0\,cm^3$ of sulphuric acid of concentration $0.150\,mol/l$ neutralised $30.0\,cm^3$ of potassium hydroxide solution. Find the concentration of the potassium hydroxide solution.

Solution

(1) First write the equation:

Sulphuric acid + Potassium hydroxide → Potassium sulphate + Water

$H_2SO_4(aq)$ + $2KOH(aq)$ → $K_2SO_4(aq)$ + $2H_2O(l)$

The equation tells you that 1 mole of H_2SO_4 neutralises 2 moles of KOH.

(2) Now work out the amount (moles) of acid. You start with the acid because you know the concentration of the acid.

Amount (mol) of acid = Volume ($25.0\,cm^3$) x Concentration (0.150 mol/l)

= 25.0×10^{-3} l x 0.150 mol/l = 3.75×10^{-3} mol

(3) Now work out the concentration of base.

Amount (mol) of KOH = 2 x amount (mol) of H_2SO_4

= 7.50×10^{-3} mol

also Amount (mol) of KOH = Volume of KOH(aq) x Concentration of KOH(aq)

Therefore, if c mol /l is the concentration of KOH

$$c \text{ mol/dm}^3 = \frac{7.50 \times 10^{-3} \text{ mol}}{30.0 \times 10^{-3} \text{ l}}$$

= 0.250 mol/l

The concentration of potassium hydroxide is 0.250 mol/l.

SUMMARY

A standard solution is a solution of known concentration. The concentration of a solution can be found by volumetric analysis. The concentration of a solution of an acid can be found by titrating it against a standard solution of a base. Similarly, the concentration of a solution of a base can be found by titrating it against a standard solution of an acid.

CHECKPOINT

❶ $25.0\,cm^3$ of sodium hydroxide solution are neutralised by $15.0\,cm^3$ of a solution of hydrochloric acid of concentration 0.25 mol/l. Find the concentration of the sodium hydroxide solution.

❷ A solution of sodium hydroxide contains 10 g/l.
(a) What is the concentration of the solution in mol/l?
(b) What volume of this solution would be needed to neutralise $25.0\,cm^3$ of 0.10 mol/l hydrochloric acid?

❸ $25.0\,cm^3$ of hydrochloric acid are neutralised by $20.0\,cm^3$ of a solution of 0.15 mol/l sodium carbonate solution.
(a) How many moles of sodium carbonate are neutralised by 1 mol HCl?
(b) What is the concentration of the hydrochloric acid?

❹ The class decide to test some antacid indigestion tablets. They dissolve tablets and titrate the alkali in them against a standard acid. The table shows their results.

Brand	Price of 100 tablets (£)	Volume of 0.01 mol/l acid required to neutralise 1 tablet (cm^3)
Paingo	0.60	2.8
Relievo	0.70	3.0
Alko	0.80	3.3
Clearo	0.90	3.6

(a) Which antacid tablets offer the best value for money?
(b) What other factors would you consider before choosing a brand?

TOPIC 25 FUELS

25.1 Methane

LOOK AT LINKS
for the **energy crisis**
See Topic 7.4 of *KS: Physics*.

FIRST THOUGHTS

Developing countries need more fuel. Industrial countries have difficulty in disposing of all their waste. One solution to both problems is biogas. You can find out about biogas in this section.

Father Conlon and Father Williams are monks at Bethlehem Abbey in Northern Ireland. In 1987 they won a 'pollution abatement technology award'. They feed manure from their farm into a **biomass digester**. Biomass is material of plant or animal origin. Bacteria feed on biomass and make it decay. Under anaerobic conditions (in the absence of air), a gas called **biogas** forms. Biogas is a fuel, and the monks burn it to provide the monastery with heating. The solid remains of the biomass form an odourless compost which they bag and sell as a fertiliser for use on gardens, potato farms and golf courses.

Many farms in the UK now have biomass digesters. There are millions of such digesters in India and China. India's huge cattle population supplies plenty of dung for digestion. The biogas produced is used as a fuel for cooking and heating. It burns with a hotter, cleaner flame than cattle dung. The residual sludge which gathers in the digester is a better fertiliser than raw dung. Both the gas and the sludge are clean and odourless.

SCIENCE AT WORK

Every year, 25 million tonnes of organic waste goes into landfill sites in the UK. Biogas forms as the rubbish decays. Most landfill operators burn biogas to get rid of it. Others sink pipes into the landfill and pump out biogas for sale. In Bedfordshire, biogas from a landfill is used to heat the kilns in a brickworks. On Merseyside, biogas is used to heat the ovens in a Cadburys' biscuit factory.

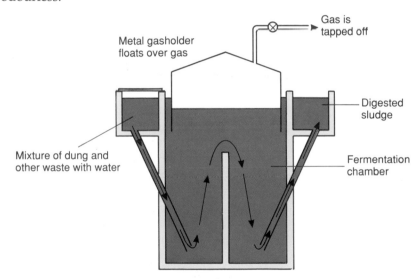

Figure 25.1A ● A biomass digester

SUMMARY

Biogas is formed when biomass (plant and animal matter) decays in the absence of air. The chief component of biogas is methane. Biogas can be collected from landfill rubbish sites, from sewage works and on farms. It can be burned as a fuel. Biogas generators are common in India and China. Methane accumulates in coal mines, where it can cause explosions.

Figure 25.1B ● A gas rig in the North Sea

The chief component of biogas is **methane**. Methane is the gas we burn in Bunsen burners and gas cookers. It is the valuable fuel we call **natural gas** or **North Sea gas**. Reserves of North Sea gas are usually found together with North Sea oil. Methane is also the gas called **marsh gas** which bubbles up through stagnant water. It collects in coalmines, and methane explosions have been the cause of many pit disasters.

❶ (a) What are the advantages of biogas generators for developing countries?
  (b) Why does India have plenty of cattle dung?
  (c) In what ways is biogas a more convenient fuel than dung?
  (d) What advantages do biogas generators have for industrial countries?

❷ (a) What is biomass? Briefly describe how biomass can be fermented to produce biogas. What other product is formed?
  (b) What are the problems in adapting the process to produce fuel gas for domestic use?
  (c) Can you point out any situations in which a biogas generator might both save money and also benefit the environment?

25.2 Alkanes

FIRST THOUGHTS

> Methane is one member of a family of compounds, the alkanes. Composed of hydrogen and carbon only, the alkanes are hydrocarbons. You will meet another family of hydrocarbons, the alkenes, in Topic 30.1.

Methane is a **hydrocarbon**, a compound of hydrogen and carbon. With formula CH_4, it is the simplest of the hydrocarbons. Figure 25.2A shows how, in a molecule of methane, four covalent bonds join hydrogen atoms to a carbon atom.

Hydrocarbons are **organic compounds**. Originally, the term 'organic compound' was applied to compounds which were found in plant and animal material, for example sugars, fats and proteins. All these compounds contain carbon, and the term 'organic compound' is now used for all carbon compounds, whether they have been obtained from plants and animals or made in a laboratory. However, simple compounds like carbon dioxide and carbonates are not usually described as organic compounds. Most organic compounds are covalent. The salts of organic acids contain ionic bonds.

Methane is one of a **series** of hydrocarbons called the **alkanes**. The next members of the series are ethane, C_2H_6, propane, C_3H_8 and butane, C_4H_{10} (see Figure 25.2B). Many hydrocarbons have much larger molecules.

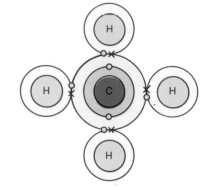

Figure 25.2A ● Bonding in methane

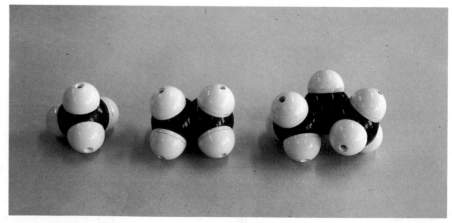

Figure 25.2B ● Models of methane, ethane and propane

As well as molecular formulas, CH_4, C_2H_6 and C_3H_8, the compounds have **structural formulas**. A structural formula shows the bonds between atoms.

$$
\begin{array}{ccc}
\text{Methane} & \text{Ethane} & \text{Propane}
\end{array}
$$

Each compound differs from the next in the series by the group

Table 48.1 ● The alkanes

Alkane	Formula C_nH_{2n+2}
Methane	CH_4
Ethane	C_2H_6
Propane	C_3H_8
Butane	C_4H_{10}
Pentane	C_5H_{12}
Hexane	C_6H_{14}
Heptane	C_7H_{16}
Octane	C_8H_{18}

A set of chemically similar compounds in which each member differs from the next in the series by a CH_2 group is called a **homologous series**. Table 25.1 lists the first members of the **alkane series**. Their formulas all fit the general formula C_nH_{2n+2}, for example for pentane $n = 5$, giving the formula C_5H_{12}. The alkanes are described as **saturated** hydrocarbons. This means that they contain only single bonds between carbon atoms. This is in contrast to the **alkenes**. The alkanes are unreactive towards acids, bases, metals and many other chemicals. Their important reaction is combustion.

Isomerism

Sometimes it is possible to write more than one structural formula for a molecular formula. For the molecular formula C_4H_{10}, there are two possible structures

(a) Butane

(b) Methylpropane

The difference is that (a) has an unbranched chain of carbon atoms and (b) has a branched chain. The formulas belong to different compounds, which differ in boiling point and other physical properties. The compound with formula (a) is called butane; the compound with formula (b) is called methylpropane. These compounds are **isomers**. Isomers are compounds with the same molecular formula and different structural formulas.

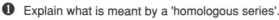

CHECKPOINT

❶ Explain what is meant by a 'homologous series'.

❷ (a) Explain what is meant by the term 'alkane'.
   (b) Why are alkanes very important in our way of life?
   (c) Where do we obtain alkanes?
   (d) The molecular formula for propane is C_3H_8. Write the structural formula for propane.
   (e) What information does the structural formula of a compound give that the molecular formula does not give?
   (f) Explain what is meant by the term 'isomerism'. Illustrate your answer by referring to pentane, C_5H_{12}.

LOOK AT LINKS
for **anaesthetics**
See Topic 31.8.

Substitution reactions

Alkanes react with chlorine in sunlight. The products are a **chloroalkane** and hydrogen chloride. For example,

Methane + Chlorine → Chloromethane + Hydrogen chloride
$$CH_4(g) + Cl_2(g) \rightarrow CH_3Cl(g) + HCl(g)$$

$$\begin{array}{c} H \\ | \\ H-C-H \\ | \\ H \end{array} + Cl-Cl \rightarrow \begin{array}{c} H \\ | \\ H-C-Cl \\ | \\ H \end{array} + H-Cl$$

The reaction is called a **substitution reaction** because one atom in the molecule of alkane is replaced by another atom. It is called **chlorination** because it is a chlorine atom that is substituted into the alkane molecule. It is described as a **photochemical reaction** because it takes place only in sunlight. The reaction can continue further.

Chloromethane + Chlorine → Dichloromethane + Hydrogen chloride
$$CH_3Cl(g) + Cl_2(g) \rightarrow CH_2Cl_2(g) + HCl(g)$$

A mixture of products, including trichloromethane, $CHCl_3$ (chloroform) and tetrachloromethane, CCl_4 (also called carbon tetrachloride) is formed. Trichloromethane, $CHCl_3$ is chloroform, which was used as an anaesthetic. The mixture of products formed in the chlorination of an alkane may make a useful solvent without the need to isolate the individual compounds. A much used solvent is 1,1,1-trichloroethane. The 1,1,1- in the name tells you that the three chlorine atoms are all on the same carbon atom, and the structure is

$$\begin{array}{ccc} Cl & & H \\ | & & | \\ Cl-C-&C-&H \\ | & & | \\ Cl & & H \end{array}$$

Bromine also reacts with alkanes in sunlight. The substitution product which is formed is called a **bromoalkane**. Hydrogen bromide is the other product, e.g.

Hexane + Bromine → Bromohexane + Hydrogen bromide
$$C_6H_{14}(l) + Br_2(l) \rightarrow C_6H_{13}Br(l) + HBr(g)$$

Iodination is very slow and is a reversible reaction. Fluorination is a dangerously violent reaction. The substitution of a halogen in a compound is called **halogenation**. Chlorination and bromination are

LOOK AT LINKS
Another series of hydrocarbons, the alkenes, take part in addition reactions.
See Topic 30.1.

SUMMARY

Alkanes take part in substitution reactions, e.g. halogenation.

both halogenation reactions. Compounds in which halogen atoms have been substituted for one or more of the hydrogen atoms in an alkane are called **halogenoalkanes**. Chloromethane, tetrachloromethane and bromohexane are all halogenoalkanes.

CHECKPOINT

1 (a) Write the equation for the reaction between ethane and chlorine to form chloroethane.
  (b) Write the structural formulas for all the substances.
  (c) Construct molecular models of all the substances.

2 (a) Write the structural formulas for iodomethane, dibromomethane, fluoroethane, tetrachloromethane.
  (b) Make molecular models of the molecules.

3 Is the chlorination of methane exothermic or endothermic? Explain your answer.

25.3 Combustion

SUMMARY

The combustion of hydrocarbons is exothermic. The products of complete combustion are carbon dioxide and water. Incomplete combustion produces carbon and poisonous carbon monoxide.

Hydrocarbons burn in a plentiful supply of air to form carbon dioxide and water. The reaction is **exothermic**: energy is released.

Methane + Oxygen → Carbon dioxide + Water; Energy is released
$CH_4(g) + 2O_2(g)$ → $CO_2(g) + 2H_2O(l)$

Butane + Oxygen → Carbon dioxide + Water; Energy is released
(in Camping Gaz)

Octane + Oxygen → Carbon dioxide + Water; Energy is released
(in petrol)

CHECKPOINT

Revise what you learned about combustion in Topic 15.7.

1 (a) What harmless combustion products are formed when hydrocarbons burn in plenty of air?
  (b) When does incomplete combustion take place?
  (c) What are the products of incomplete combustion? What is dangerous about incomplete combustion? How can it be avoided?

25.4 Petroleum oil and natural gas

FIRST THOUGHTS

Prospecting for oil is an even more important business than prospecting for gold. *How do prospectors find oil? And how was oil formed in the first place?* You can find out in this section.

In some parts of the world, a black, treacle-like liquid seeps out of the ground. At one time, farmers in Texas, USA, used to burn this substance to get rid of it when it formed troublesome pools on their land. The black liquid was called **petroleum oil** or **crude oil**. It is now regarded as one of the most valuable resources a country can have. The petrochemicals industry obtains valuable fuels such as petrol, and thousands of useful materials such as plastics, from crude oil. Natural gas is found in the same deposits as crude oil.

Oil and gas are **fossil fuels**. They were formed millions of years ago when a large area of the Earth was covered by sea. Tiny sea creatures called plankton died and sank to the sea bed, where they became mixed with mud. Bacteria in the mud began to bring about the decay of the creatures' bodies. Decay took place slowly because there is little oxygen dissolved in the depths of the sea. As the covering layer of mud and silt grew thicker over the years, the pressure on the decaying matter increased. Bacterial decay at high pressure with little oxygen turned the organic matter into crude oil and natural gas.

The sediment on top of the decaying matter became compressed to form rock. Rocks formed in this way are called **sedimentary rocks**. Some of these rocks are porous: they contain tiny passages through which liquid and gas can pass. Others are impermeable: they do not let any substances through. Ground water carries crude oil and natural gas upward through porous rocks. They may reach the surface, but more often they become trapped by a **cap rock** of impermeable rock (see Figure 25.4A). There the crude oil and natural gas remain unless an oil prospector drills down to them.

LOOK AT LINKS
for **sedimentary rocks**
See Topic 2.1.

IT'S A FACT

Have you heard of the petrol tree? The gopher tree grows well in desert areas. Its sap contains about 30% hydrocarbons. Petrol could be made from the sap of this tree.

LOOK AT LINKS

When a tanker has an accident at sea, oil can escape to form a huge oil slick. The damage which the oil may cause to wildlife and coastlines is described in Topic 18.6.

SUMMARY

Crude oil and natural gas were formed by the slow bacterial decay of animal and plant remains in the absence of oxygen. These fuels are found in many parts of the world.

LOOK AT LINKS
for **fractional distillation**
See Topic 11.8.

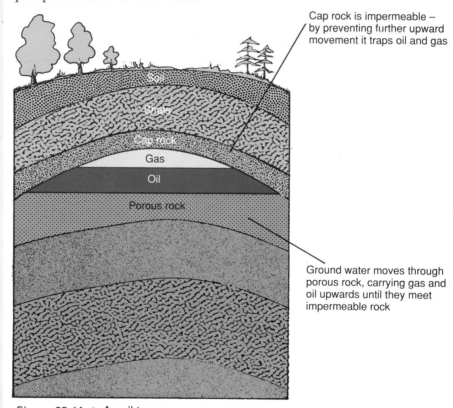

Cap rock is impermeable – by preventing further upward movement it traps oil and gas

Ground water moves through porous rock, carrying gas and oil upwards until they meet impermeable rock

Figure 25.4A ● An oil trap

Many countries have deposits of oil and gas, for example the USA, the USSR, Iran, Nigeria and the countries in the Arabian Gulf. The UK and Norway have oil and gas beneath the North Sea. The UK has piped ashore oil and gas since 1972. There are two ways of getting oil ashore from an oil well in the sea. One way is to lay a pipeline along the sea bed and pump the oil through it. The other method is to use 'shuttle tankers' which pick up the oil from the oilfield and transport it to a terminal on land. Natural gas is almost always brought ashore by pipeline.

Crude oil is transported from oil wells to refineries. Pipelines carry oil overland. Tankers carry oil overseas.

Crude oil does not burn very easily. **Fractional distillation** is used to separate crude oil into a number of important fuels. The fractions are separated on the basis of their boiling point range. They are not pure compounds: they are mixtures of alkanes with similar boiling points.

Table 25.2 ● Petroleum fractions and their uses

Fraction	Approximate boiling point range (°C)	Approximate number of carbon atoms per molecule	Use
Petroleum gases	below 25	1–4	Petroleum gases are liquefied and sold in cylinders as 'bottled gas' for use in gas cookers and camping stoves. They burn easily at low temperatures. Smelly sulphur compounds must be removed to make bottled gas pleasant to use and non-polluting.
Gasoline (petrol)	40–75	4–12	Petrol is liquid at room temperature, but vaporises easily at the temperature of vehicle engines.
Naphtha	75–150	7–14	Naphtha is used by the petrochemicals industry as the source of a huge number of useful chemicals, e.g. plastics, drugs, medicines, fabrics (see Figure 25.4C overleaf).
Kerosene	150–240	9–16	Kerosene is a liquid fuel which needs a higher temperature for combustion than petrol. The major use of kerosene is as aviation fuel. It is also used in 'paraffin' stoves.
Diesel oil	220–250	15–25	Diesel oil is more difficult to vaporise than petrol and kerosene. The diesel engine has a special fuel injection system which makes the fuel burn. It is used in buses, lorries and trains.
Lubricating oil	250–350	20–70	Lubricating oil is a viscous liquid. With its high boiling point range, it does not vaporise enough to be used as a fuel. Instead, it is used as a lubricant to reduce engine wear.
Fuel oil	250–350	Above 10	Fuel oil has a high ignition temperature. To help it to ignite, fuel oil is sprayed into the combustion chambers as a fine mist of small droplets. It is used in ships, heating plants, industrial machinery and power stations.
Bitumen	Above 350	Above 70	Bitumen has too high an ignition temperature to be used as a fuel. It is used to tar roads and to waterproof roofs and pipes.

Alkanes with small molecules boil at lower temperatures than those with large molecules. Alkanes with large molecules are more viscous than alkanes with small molecules. The fractions also differ in the ease with which they burn: in their **flash points** and **ignition temperatures**.
- When a fuel is heated, some of it vaporises. When the fuel reaches a temperature called the **flash point**, there is enough vapour to be set alight by a flame. Once the vapour has burned, the flame goes out.
- The **ignition temperature** of a fuel is higher than its flash point. It is the temperature at which a mixture of the fuel with air will ignite and continue to burn steadily.

The use that is made of each fraction depends on its boiling point range, flash point, ignition temperature and viscosity; see Table 25.2.

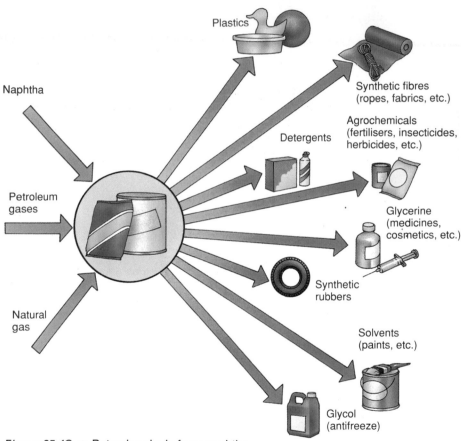

Figure 25.4C ● Petrochemicals from naphtha

Cracking

We use more petrol, naphtha and kerosene than heavy fuel oils. Fortunately, chemists have found a way of making petrol and kerosene from the higher-boiling fractions, of which we have more than enough. The technique used is called **cracking**. Large hydrocarbon molecules are cracked (split) into smaller hydrocarbon molecules. A heated catalyst (aluminium oxide or silicon(IV) oxide) is used.

Vapour of hydrocarbon with large molecules and high b.p.	CRACKING $\xrightarrow{\text{passed over a heated catalyst}}$	Mixture of hydrogen and hydrocarbons with smaller molecules and low b.p.

SUMMARY

Crude oil is separated into useful components by fractional distillation. The use that is made of each fraction depends on its boiling point range, ignition temperature and other properties. The fuels obtained from crude oil are listed in Table 25.2. Cracking is used to make petrol and kerosene from heavy fuel oils. The petrochemicals industry makes many useful chemicals from petroleum.

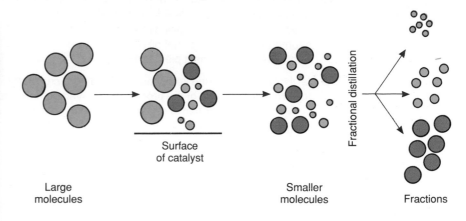

Figure 25.4D ● Cracking

CHECKPOINT

1 The use of crude oil fractions in the UK is as follows:
Road transport 37%
Heating and power for industry 21%
Heating and power for other consumers 19%
Making chemicals 11%
Power stations 8%
Heating houses 4%
Show this information in the form of a pie-chart or a bar graph.

2 The figure below shows the percentage of crude oil that distils at various temperatures.
(a) From the graph, read the percentage of crude oil that distils:
  (i)   below 70 °C,
  (ii) between 120 and 170 °C,
  (iii) between 170 and 220 °C,
  (iv) between 220 and 270 °C,
  (v)   between 270 and 320 °C,
  (vi) above 320 °C.
(b) Draw a pie-chart to show these figures.
(c) The percentages vary from one sample of oil to another. North Sea oil contains a larger percentage of lower boiling point compounds than Middle East oil does. Which oil should sell for a higher price? Explain your answer.

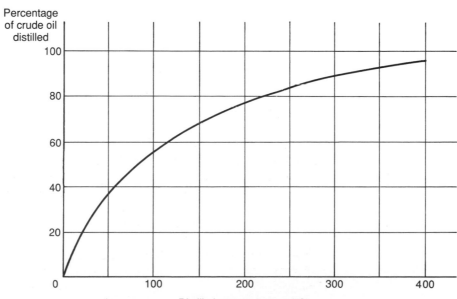

❸ A barrel of Middle East oil contains 160 litres. From it are obtained:
natural gas (7 litres)   kerosene (7 litres)
petrol (9 litres)   gas oil (41 litres)
naphtha (23 litres)   fuel oil (73 litres)
Show these figures as a pie-chart.

❹ (a) How are petrol and kerosene obtained from crude oil?
(b) What is the name of the process for converting heavy fuel oil into petrol?
What are the economic reasons for carrying out this reaction?

❺ (a) What is crude oil?
(b) Why is it described as a fossil fuel?
(c) What useful substances are obtained from crude oil?
(d) Why is an increase in the price of crude oil such a serious matter?

25.5 Coal

FIRST THOUGHTS

Coal is an important source of energy. It fuelled the Industrial Revolution and is still a major source of energy in the UK today.

Between 200 and 300 million years ago, the Earth was covered with dense forests. When plants and trees died, they started to decay. In the swampy conditions of that era, they formed peat. Gradually, the peat became covered by layers of mud and sand. The pressure of these layers and high temperatures turned the peat into coal. The mud became shale and the sand became sandstone. Coal is a fossil fuel: it was formed from the remains of living things.

Many countries have coal deposits. The largest coal-mining countries are the USSR, the USA, China, Poland and the UK. Coal is a mixture of carbon, hydrocarbons and other compounds. When it burns, the main products are carbon dioxide and water.

SUMMARY

Coal is a fossil fuel derived from plant remains. Most of the coal mined is burned in power stations. The destructive distillation of coal gives useful products.

Carbon (in coal) + Oxygen → Carbon dioxide
Hydrocarbons (in coal) + Oxygen → Carbon dioxide + Water

In the UK, three-quarters of the coal used is burned in power stations. The heat given out is used to raise steam, which drives the turbines that generate electricity.

If air is absent when coal is heated, coal does not burn. It decomposes to form coke and other useful products. The process is called **destructive distillation**.

CHECKPOINT

❶ (a) What is a fuel?
(b) Why are coal and oil called fossil fuels?
(c) State two properties that make a fuel a 'good' fuel.
(d) Where is most of the coal that we use burned?
(e) What use is made of coal apart from burning it?

TOPIC 26
CHEMICAL ENERGY CHANGES

26.1 Exothermic reactions

FIRST THOUGHTS

In many places in this book, we refer to the importance of energy. One means by which energy is converted from one form into another is by chemical reactions.

LOOK AT LINKS
for **carbohydrates**
See Topic 31.1.

LOOK AT LINKS
for **hydrates**
See Topic 13.3.

LOOK AT LINKS
for **acids** and **alkalis**
See Topic 12.

LOOK AT LINKS
for **aerobic respiration**
See Topic 12.1 of *KS: Biology*.

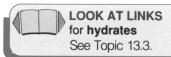

Our bodies obtain energy from the combustion of foods, and our vehicles obtain energy from the combustion of hydrocarbons. Reactions of this kind, which give out energy, are **exothermic reactions** (ex = out; therm = heat). You will meet many such reactions. Some are described below.

Combustion

Topic 25.3 described the combustion of the fuels:
• coal (carbon + hydrocarbons),
• natural gas (largely methane, CH_4),
• petrol (a mixture of hydrocarbons, e.g. octane, C_8H_{18})
We use the heat from these exothermic reactions to warm our homes and buildings and to drive our vehicles.

Figure 26.1A ● They all use exothermic reactions

• An oxidation reaction in which heat is given out is **combustion**. The oxidation of sugars in our bodies is combustion.
• Combustion accompanied by a flame is **burning**. When coal is burned in a fireplace, you see the flames.
• A substance which is oxidised with the release of energy is a fuel. Both sugars and coal are fuels.

Respiration

We use the combustion of foods to supply our bodies with energy. Important among 'energy foods' are **carbohydrates**. Glucose, a sugar, is oxidised to carbon dioxide and water with the release of energy. This exothermic reaction, which takes place inside plant cells and animal cells, is called **aerobic respiration**.

Neutralisation

When an acid neutralises an alkali, heat is given out.

Figure 26.1B ● Energy from respiration

| Hydrogen ion | + | Hydroxide ion | → | Water | Heat is given out |
| $H^+(aq)$ | + | $OH^-(aq)$ | → | $H_2O(l)$ | |

SUMMARY

In exothermic reactions, energy is released. Examples are:
* the combustion of hydrocarbons,
* respiration,
* neutralisation,
* hydration of anhydrous salts.

Hydration

Many anhydrous salts react with water to form hydrates. During hydration, heat is given out.

Anhydrous copper(II) sulphate	+	Water	→	Copper(II) sulphate-5-water	Heat is given out
$CuSO_4(s)$	+	$5H_2O(l)$	→	$CuSO_4.5H_2O(s)$	

26.2 **Endothermic reactions**

LOOK AT LINKS
for **photosynthesis**
See Topic 11.1 of *KS: Biology*.

LOOK AT LINKS
Calcium oxide is used in the manufacture of cement and concrete.
See Topic 28.1.
Calcium oxide is also used to make calcium hydroxide (slaked lime), which is used as a fertiliser.
See Topic 29.7.

RESOURCE
—ACTIVITY—
PACK

SUMMARY

In endothermic reactions, energy is taken in from the surroundings. Examples are:
* photosynthesis,
* thermal decomposition,
* the reaction between steam and coke.

Photosynthesis

Plants manufacture sugars in the process of **photosynthesis**. They convert the energy of sunlight into the energy of the chemical bonds in sugar molecules. Photosynthesis is an **endothermic reaction**: it takes in energy.

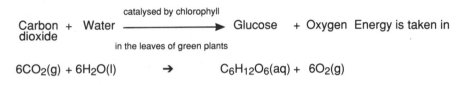

Carbon dioxide	+	Water	catalysed by chlorophyll → in the leaves of green plants	Glucose	+	Oxygen	Energy is taken in

$6CO_2(g) + 6H_2O(l)$ → $C_6H_{12}O_6(aq) + 6O_2(g)$

Thermal decomposition

Many substances decompose when they are heated. An example is calcium carbonate (limestone).

Calcium carbonate	→	Calcium oxide	+	Carbon dioxide	Heat is taken in
$CaCO_3(s)$	→	$CaO(s)$	+	$CO_2(g)$	

This is an important reaction because it yields **calcium oxide** (quick lime).

The reaction between steam and coke

The endothermic reaction between steam and hot coke produces carbon monoxide and hydrogen. This mixture of flammable gases is used as fuel.

Carbon (coke)	+	Steam	→	Carbon monoxide	+	Hydrogen	Heat is taken in
$C(s)$	+	$H_2O(g)$	→	$CO(g)$	+	$H_2(g)$	Heat is taken in

CHECKPOINT

❶ Give an example of a reaction which is of vital importance in everyday life and which is (a) exothermic and (b) endothermic. Explain why the reactions you mention are so important.

❷ The reaction shown below is exothermic.

Anhydrous copper(II) sulphate + Water → Copper(II) sulphate-5-water

Describe or illustrate an experiment by which you could find out whether this statement is true.

26.3 Heat of reaction

LOOK AT LINKS
for **chemical bonds**
See Topic 10.4.

Chemical bonds are forces of attraction between the atoms or ions or molecules in a substance. To break these bonds, energy must be supplied. When bonds are created, energy is given out. In a chemical reaction, bonds are broken, and new bonds are made.

Figure 26.3A shows the breaking and making of bonds when methane burns.

$$\text{Methane} + \text{Oxygen} \rightarrow \text{Carbon dioxide} + \text{Water}$$
$$CH_4(g) + 2O_2(g) \rightarrow CO_2(g) + 2H_2O(l)$$

The energy given out when the new bonds are made is greater than the energy taken in to break the old bonds. This reaction is therefore **exothermic**.

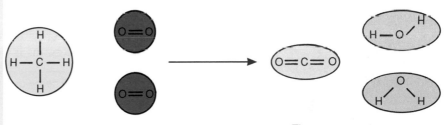

These bonds are broken.
Energy must be taken in

These new bonds are made.
Energy is given out

In this reaction, the energy given out is greater than the energy taken in — this reaction is exothermic

Figure 26.3A ● Bonds broken and made when methane burns

Figure 26.3B shows the breaking and making of bonds in the reaction

$$\text{Carbon (coke)} + \text{Water (steam)} \rightarrow \text{Carbon monoxide} + \text{Hydrogen}$$
$$C(s) + H_2O(g) \rightarrow CO(g) + H_2(g)$$

In this reaction, the energy taken in to break the old bonds is greater than the energy given out when the new bonds form. The reaction is therefore **endothermic**.

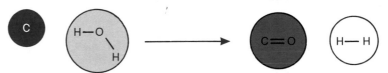

The bonds in H_2O must be broken.
Energy must be taken in

As the bonds in CO and H_2 are made, energy is given out

In this reaction, the energy taken in is greater than the energy given out. This reaction is endothermic

Figure 26.3B ● Bonds broken and made when steam reacts with hot coke

Energy diagrams

In a chemical reaction, the reactants and the products possess different chemical bonds. They therefore possess different amounts of energy. A diagram which shows the energy content of the reactants and the products is called an **energy diagram**.

Figure 26.3C is an energy diagram for an exothermic reaction. It shows the energy content of the reactants and the products.

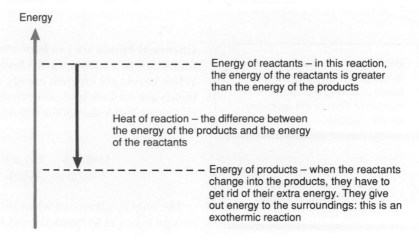

Energy

Energy of reactants – in this reaction, the energy of the reactants is greater than the energy of the products

Heat of reaction – the difference between the energy of the products and the energy of the reactants

Energy of products – when the reactants change into the products, they have to get rid of their extra energy. They give out energy to the surroundings: this is an exothermic reaction

Figure 26.3C ● An energy diagram for an exothermic reaction (e.g. the combustion of methane)

Figure 26.3D shows an energy diagram for an endothermic reaction. Marked on both energy diagrams is the heat of reaction

Heat of reaction = Energy of products – Energy of reactants

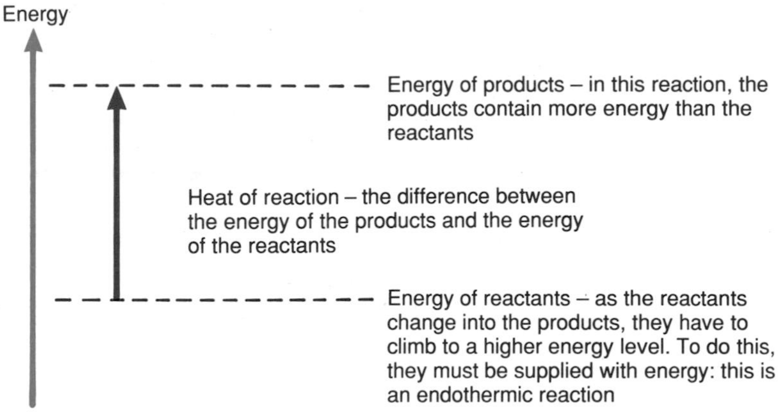

Energy

Energy of products – in this reaction, the products contain more energy than the reactants

Heat of reaction – the difference between the energy of the products and the energy of the reactants

Energy of reactants – as the reactants change into the products, they have to climb to a higher energy level. To do this, they must be supplied with energy: this is an endothermic reaction

Figure 26.3D ● An energy diagram for an endothermic reaction (e.g. the reaction between coke and steam)

SUMMARY

In a chemical reaction, energy must be supplied to break chemical bonds in the reactants. Energy is given out when new chemical bonds are made during the formation of the products. The difference between the energy content of the products and the energy content of the reactants is the heat of reaction.

CHECKPOINT

❶ Draw an energy diagram for the neutralisation of hydrochloric acid by sodium hydroxide solution. If you have forgotten whether it is exothermic or endothermic, see Topic 26.1. Mark the heat of reaction on your diagram.

❷ Draw an energy diagram for the combustion of petrol to form carbon dioxide and water. Mark the heat of reaction on your diagram.

❸ Is photosynthesis exothermic or endothermic? What are the reactants and what are the products? Illustrate your answer by an energy diagram.

❹ Describe the reaction that takes place when magnesium ribbon is heated in air. What do you see that shows the reaction is exothermic? What forms of energy are released? Draw an energy diagram for the reaction.

26.4 The value of the heat of reaction

The symbol H is used for the energy content of a substance. The symbol ΔH is used for a change in energy content. The energy diagrams in Figures 26.3C and D show the energy content of the reactants and the products in a chemical reaction. The difference (energy of products − energy of reactants) is called the **heat of reaction**. It is given the symbol ΔH. By definition,

Heat of reaction, ΔH = Energy of products − Energy of reactants

The products of the reaction shown in Figure 26.3C contain less energy than the reactants. The reactants must give out energy as they change into the products. This an exothermic reaction. Since by definition,

ΔH = Energy of products − Energy of reactants

in this reaction, ΔH is a negative quantity. For an exothermic reaction, ΔH is negative.

In the reaction shown in Figure 26.3D, the products contain more energy than the reactants. In order to change into the products, the reactants must take energy from the surroundings: they cool the surroundings. Since

ΔH = Energy of products − Energy of reactants

in this reaction ΔH is positive. For an endothermic reaction, ΔH is positive.

Defining the heat of reaction

How much heat is given out or taken in during a chemical reaction? Heat is measured in joules, J, and in kilojoules, kJ ($1\,kJ = 1000\,J$). The heat of reaction is measured in kilojoules per mole of reactant, kJ/mol.

Heat of neutralisation is the heat absorbed when 1 mole of hydrogen ions neutralises 1 mole of hydroxide ions.

$H^+(aq) + OH^-(aq) \;\rightarrow\; H_2O(l); \Delta H = -57\,kJ/mol$

Heat of combustion is the heat absorbed when 1 mole of reactant is completely burnt in oxygen. For the combustion of ethane,

$2C_2H_6(g) + 7O_2(g) \;\rightarrow\; 4CO_2 + 6H_2O(l); \Delta H = -1560\,kJ/(\text{mole ethane})$

The value of ΔH is the value per mole of ethane.

Heat of formation is the heat absorbed when 1 mole of reactant is formed from its elements. In the formation of sulphur dioxide,

$S(s) + O_2(g) \;\rightarrow\; SO_2(g); \Delta H = -297\,kJ/mol$

Heat of reaction is the heat absorbed when reaction occurs between the number of moles of reactants shown in the equation. For the reaction,

$4Al(s) + 2Fe_2O_3(s) \;\rightarrow\; 2Al_2O_3(s) + 4Fe(s); \Delta H = -1710\,kJ/mol$

The value of ΔH could be given per mole of aluminium or per mole of iron(III) oxide, but it is given per mole of the equation as written, that is, the reaction between 4 moles of aluminium and 2 moles of iron(III) oxide.

SUMMARY

The heat of reaction is shown by the symbol ΔH. For an exothermic reaction, ΔH is negative; for an endothermic reaction, ΔH is positive. Definitions are given for the heats of neutralisation, combustion, formation and reaction.

CHECKPOINT

❶ Draw an energy diagram for the cracking of heavy fuel oil to form kerosene. Mark the heat of reaction, and state whether it is positive or negative.

❷ The heat of combustion of propane = −2220 kJ/mol. Calculate the heat given out when 1000 g of propane burn completely. (Hint: propane, C_3H_8, has a molar mass of 44 g/mol; therefore combustion of 44 g of propane gives out 2220 kJ.)

❸ The heat of combustion of butane is −2880 kJ/mol. What mass of butane must be burnt to give out 1000 kJ? (Hint: what is the mass of 1 mole of butane?)

❹ When 0.290 g of propanone (C_3H_6O) is completely burnt, 8.95 kJ of heat is evolved. Calculate the heat of combustion of propanone. (Hint: answer in kJ/mol.)

❺ Solutions of sodium hydroxide and hydrochloric acid are added, and the temperature rise is measured. The volume of each solution is 100 cm³, and each concentration is 1.00 mol/dm³. The heat evolved is calculated as 5.7 kJ.
(a) Calculate the heat of neutralisation.
(b) Say whether the value for nitric acid would be greater, less or the same. (Hint: check the definition.)

26.5 ● Using bond energy values

Table 26.1 ● Bond energy values

Bond	Bond energy (kJ/mol)
H—H	436
O—O	496
C—C	348
C=C	612
C—H	412
C—O	360
C=O	743
H—O	463
Br—Br	193
C—Br	276

Chemists have drawn up tables which tell exactly how much energy it takes to break different chemical bonds. To break 1 mole of C—C bonds requires 348 kJ. We can say that the bond energy of the C—C bond is 348 kJ/mol. Table 26.1 lists the bond energies of some bonds in kJ/mol. To break a chemical bond, energy must be supplied so all the bond energies are positive. There are no bonds that fly apart by magic and release energy.

We can use bond energy values to calculate a value for the heat of reaction.

Worked example   Use bond energies to calculate ΔH for the reaction,

$$CH_4(g) + 2O_2(g) \rightarrow CO_2(g) + 2H_2O(l)$$

Solution   First show the bonds that are broken and the bonds that are made.

$$\begin{array}{c} H \\ | \\ H-C-H \\ | \\ H \end{array} + 2\,O=O \rightarrow O=C=O + 2\,H-O-H$$

Next list the bonds that are broken and the bonds that are made and their bond energies.

Bonds broken are 4(C—H) bonds; energy = 4 x 412 = 1648 kJ/mol
2(O=O) bonds; energy = 2 x 496 = 992 kJ/mol
Total energy required = +2640 kJ/mol

Bonds made are 2(C—O) bonds; energy = −2 x 743 = −1486 kJ/mol
2(H—O) bonds; energy = −4 x 463 = −1852 kJ/mol
Total energy given out = −3338 kJ/mol

Heat of reaction = Energy required to + Energy given out when
break old bonds   new bonds are made
= +2640 − 3338 = −698 kJ/mol

CHECKPOINT

❶ Find the heat of reaction for the reaction

$$H-\overset{\overset{\displaystyle H}{|}}{C}=\overset{\overset{\displaystyle H}{|}}{C}-H \;+\; H_2 \;\rightarrow\; H-\overset{\overset{\displaystyle H}{|}}{\underset{\underset{\displaystyle H}{|}}{C}}-\overset{\overset{\displaystyle H}{|}}{\underset{\underset{\displaystyle H}{|}}{C}}-H$$

(Hint: add the energies of the bonds that are broken (positive) and the energies of the bonds created (negative).)

❷ Find the heat of reaction for the reaction

$$2H_2(g) + O_2(g) \;\rightarrow\; 2H_2O(l)$$

❸ Calculate the heat of reaction for the reaction

$$CH_3-\overset{\overset{\displaystyle O}{\|}}{C}-CH_3(g) \;+\; H_2(g) \;\rightarrow\; CH_3-\overset{\overset{\displaystyle O-H}{|}}{\underset{\underset{\displaystyle H}{|}}{C}}-CH_3(g)$$

❹ Calculate the heat of reaction for the reaction

$$CH_2=CHCH_3(g) + Br_2(g) \;\rightarrow\; CH_2BrCHBrCH_3(g)$$

(Hint: draw structural formulas to show all the bonds.)

TOPIC 27 EQUILIBRIUM

Imagine that you are looking through the window of a popular swimming pool during a busy session. You can count 30 swimmers in the pool. You come back 30 minutes later, and you see that there are still 30 swimmers in the pool. However, you can see that people are diving into the pool and other people are leaving the pool. The same situation continues over the next 2 hours. The number of swimmers is constant at 30, while all the time people are entering and leaving the pool. There is a balance between the number leaving and the number entering. This state of balance can be described as a state of **equilibrium**. If the same 30 people occupied the pool all the time, one would say that the situation was **static** (unchanging). However, in the pool you are observing there is change as some swimmers arrive and others leave. The thing that remains constant is the balance between the number arriving and the number leaving. This is a **dynamic** (moving) equilibrium.

The pool you have been observing can be described as a system. The word **system** is used to describe a part of the universe which one wants to study in isolation from the rest of the universe. There are two kinds of systems:

- A system in a state of change: a change in the properties of the system is occurring.
- A system at equilibrium: no change in its properties is occurring. The equilibrium may be a **static equilibrium** or a **dynamic equilibrium**.

27.1 Physical equilibria

Vaporisation is a physical change. See what happens when a few cm³ of the brown liquid, bromine, are dropped into a gas jar and the lid is replaced (see Figure 27.1A).

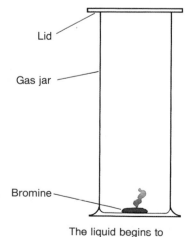

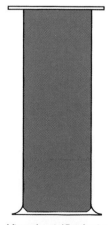

Lid

Gas jar

Bromine

The liquid begins to vaporise, and the mixture of gases inside the gas jar begins to look brown.

As minutes pass, the gas becomes darker and darker brown.

After about 15 minutes, the colour of the gas ceases to darken.

Figure 27.1A ● Vaporisation

The system has reached equilibrium. The equilibrium is a dynamic one. Molecules of bromine are still vaporising – leaving the liquid phase to enter the gas phase. But molecules of bromine are condensing – leaving the gas phase to enter the liquid phase. The rate of vaporisation is equal to the rate of condensation, so the properties of the system do not change: the system is at equilibrium.

Another physical change is dissolution (dissolving). Think about what happens when you stir copper(II) sulphate crystals into a beaker of water (see Figure 27.1B).

Copper(II) sulphate crystals and water.

The colour of the solution becomes deeper and deeper blue.

After a while, no more crystals will dissolve, and the colour of the solution ceases to change.

Figure 27.1B ● Dissolving

The saturated solution in contact with undissolved salt is at equilibrium. The equilibrium is dynamic. Although nothing seems to be happening, in fact copper(II) sulphate is still dissolving, but, as fast as crystals dissolve, dissolved copper(II) sulphate crystallises from solution. It is possible to prove that the equilibrium is dynamic. Some crystals of copper(II) sulphate containing radioactive sulphur can be added. After a time the radioactivity is found to be divided between the solution and the undissolved crystals. The reason is that some of the undissolved solid is constantly dissolving, while some solute crystallises from solution at the same rate.

27.2 Chemical equilibrium

Thermal dissociation

When calcium carbonate is heated strongly, it decomposes:

Calcium carbonate ➜ Calcium oxide + Carbon dioxide
$$CaCO_3(s) \quad ➜ \quad CaO(s) \quad + \quad CO_2(g)$$

However, the base calcium oxide reacts with the acidic gas carbon dioxide to form calcium carbonate.

$$CaO(s) + CO_2(g) \quad ➜ \quad CaCO_3(s)$$

We say that this reaction is reversible; it can go from right to left as well as from left to right. When a substance decomposes to form products which can recombine, we say that the substance **dissociates**. When the dissociation is brought about by heat, it is called **thermal dissociation**.

Think about what happens when calcium carbonate is heated inside a closed container. When heating starts, there is no carbon dioxide or calcium oxide, and the dissociation of calcium carbonate is the only reaction that takes place. Soon, the amounts of calcium oxide and carbon build up, and the combination of calcium oxide and carbon dioxide begins. As the concentration of carbon dioxide builds up inside the closed container, the rate of combination increases. Eventually a state is reached in which the rate of recombination is equal to the rate of dissociation of calcium carbonate. Then the amounts of calcium carbonate, calcium oxide and carbon dioxide remain constant. The system is at equilibrium. We use reverse arrows to denote an equilibrium:

$$CaCO_3(s) \rightleftharpoons CaO(s) + CO_2(g)$$

The reaction from left to right (as written) is called the forward reaction. The reaction from right to left is called the backward reaction.

The equilibrium is dynamic. Both the forward reaction and the backward reaction are taking place. The rate of the forward reaction is equal to the rate of the backward reaction. Therefore, once equilibrium is established, the amounts of the three substances remain constant.

A number of factors can disturb the equilibrium. If the container is opened, carbon dioxide escapes, and more calcium carbonate dissociates. When limestone is heated in a kiln (see Figure 28.1F on p. 322) to make calcium oxide (quicklime) and carbon dioxide, the kiln is open, and a draft of air is blown through the kiln to remove carbon dioxide and prevent equilibrium from becoming established. Then more and more calcium carbonate dissociates in an attempt to reach equilibrium, and more quicklime is produced.

LOOK AT LINKS
for the importance of the thermal dissociation of limestone in the manufacture of concrete See Topic 28.1.

Acids and bases

Another example of equilibrium is the dissociation of a weak acid, e.g. ethanoic acid. Pure ethanoic acid consists of CH_3CO_2H molecules. When ethanoic acid is added to water, some molecules ionise:

$$CH_3CO_2H(aq) \rightarrow H^+(aq) + CH_3CO_2{}^-(aq)$$

The concentrations of hydrogen ions and ethanoate ions increase rapidly, and collisions occur between hydrogen ions and ethanoate ions. Some of these collisions result in the formation of ethanoic acid molecules. Very soon (within a microsecond), the rate of ionisation of ethanoic acid is equal to the rate of recombination of the ions. A state of dynamic equilibrium has been reached.

$$\text{Ethanoic acid} \rightleftharpoons \text{Hydrogen ions} + \text{Ethanoate ions}$$
$$CH_3CO_2H(aq) \rightleftharpoons H^+(aq) + CH_3CO_2{}^-(aq)$$

LOOK AT LINKS
for **acids**
See Topic 12.1;
for **weak acids**
See Topic 12.5.

Molecules are constantly ionising, and ions are constantly combining, but the concentrations of all three species remain constant.

The position of equilibrium lies way over to the left-hand side. In a 1.00 mol/l solution of ethanoic acid, only 4 molecules in 1000 ionise. This is what is meant by describing ethanoic acid as a weak acid; it is only partially ionised in solution. In a solution of a strong acid, e.g. hydrochloric acid, there are no molecules; the acid is completely ionised. Chloroethanoic acid, $ClCH_2CO_2H$, is a weak acid, but it is stronger than ethanoic acid; that is, a larger fraction of the molecules are ionised.

A solution of ammonia is partially ionised.

$$\text{Ammonia} + \text{Water} \rightleftharpoons \text{Ammonium ions} + \text{Hydroxide ions}$$
$$NH_3(aq) + H_2O(l) \rightleftharpoons NH_4{}^+(aq) + OH^-(aq)$$

Equilibrium lies over to the left-hand side. Only a small fraction of the ammonia molecules ionise; ammonia is a weak base. The concentration of hydroxide ions is sufficient to allow aqueous ammonia to show the same reactions as sodium hydroxide solution, e.g. precipitating insoluble metal hydroxides from solutions of metal salts. As the concentration of hydroxide ions is used up, the equilibrium moves to the right-hand side, and more ammonia molecules ionise to give more hydroxide ions.

LOOK AT LINKS
for the **manufacture of ammonia**
See Topic 29.2.

LOOK AT LINKS
for **catalysis**
See Topic 23.7.

The Haber process

Nitrogen and hydrogen combine to form ammonia:

$$N_2(g) + 3H_2(g) \rightleftharpoons 2NH_3(g)$$

Equilibrium is reached as ammonia dissociates into nitrogen and hydrogen. Nitrogen is an unreactive gas, and the position of equilibrium is very far towards the left-hand side. Fritz Haber developed this reaction to manufacture ammonia. He had to try a number of ways to get the equilibrium to move towards the right-hand side and give a bigger yield of ammonia. He took note of **Le Chatelier's Principle**. This states that: **When conditions are changed, a system in a state of equilibrium adjusts itself in such a way as to minimise the effects of the change**. You can see that when the reaction goes from left to right there is a decrease in the number of moles of gas. If the reaction went to completion, the volume of ammonia would be only half that of the mixture of nitrogen and hydrogen used. So if the pressure on the mixture is increased, the system can absorb the increase in pressure by reducing its volume, that is, by reacting to form ammonia. The reaction is exothermic; that is, heat is given out in going from left to right. The reverse reaction, the dissociation of ammonia, is endothermic. If the temperature is raised, the system adjusts so as to absorb heat. It does this through the dissociation of ammonia because this reaction is endothermic. Running the process at high temperature therefore reduces the percentage conversion of the elements into ammonia. If the reactants are at a very low temperature, the system takes a long time to come to equilibrium. In practice, a high pressure (about 200 atm) and a moderate temperature (about 450 °C) are employed. A catalyst (iron or iron(III) oxide) is used to increase the speed of the reaction.

The Contact process

In the manufacture of sulphuric acid, an important reaction is:

Sulphur dioxide + Oxygen $\rightleftharpoons$ Sulphur trioxide
$$2SO_2(g) \quad + \quad O_2(g) \quad \rightleftharpoons \quad 2SO_3(g)$$

The formation of sulphur trioxide involves a decrease in volume as three moles of gas form two moles of gas. High pressure should favour the forward reaction. However, sulphur dioxide is easily liquefied, and industry uses a pressure of only 2 atm. The reaction from left to right is exothermic. If the temperature is raised, the reverse reaction, the dissociation of sulphur trioxide is favoured. Therefore the industry uses a moderate temperature (about 450 °C). The use of a catalyst, vanadium(V) oxide, helps to compensate for not using a very high temperature.

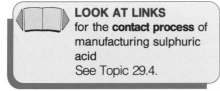

LOOK AT LINKS
for the **contact process** of manufacturing sulphuric acid
See Topic 29.4.

CHECKPOINT

❶ Reactions take place more rapidly as the temperature rises. In view of this, why is the manufacture of ammonia carried out at 450 °C, rather than at a very high temperature?

❷ A gas, A, reacts to form a mixture of two gases, B and C. The reaction is reversible, and the system reaches equilibrium.

$$A(g) \rightleftharpoons B(g) + 2C(g)$$

(a) Will an increase in pressure increase or decrease the yield of B and C?

(b) What information would you need before you could predict the effect of raising the temperature on the rate of equilibrium?

❸ Hydrocyanic acid dissociates partially in solution.

Hydrocyanic acid $\rightleftharpoons$ Hydrogen ions + Cyanide ions
$HCN(aq)$ $\rightleftharpoons$ $H^+(aq)$ + $CN^-(aq)$

If you add a little sodium hydroxide solution, what effect will this have on the concentration of cyanide ions? Explain your answer.

❹ An aqueous solution of bromine is called 'bromine water'. Some bromine molecules react with water molecules:

Bromine + Water $\rightleftharpoons$ Hydrobromic acid + Bromic(I) acid
$Br_2(aq)$ + $H_2O(l)$ $\rightleftharpoons$ $HBr(aq)$ + $HBrO(aq)$

The products are both strong acids. The reaction is reversible, and a solution of bromine in water reaches an equilibrium state in which the concentrations of all the species are constant.

Predict what change will happen if you add a small amount of sodium hydroxide. In which direction will the equilibrium be displaced, from left to right or from right to left? Predict what colour change you will see. How could you reverse the colour change?

Do an experiment to check your predictions.

? THEME QUESTIONS

1 The energy sources used in the UK are as follows:
Coal 31%
Oil 30%
Natural gas 21%
Nuclear power 17%
Hydroelectric power 1%
(a) Show these figures in the form of a pie-chart or a bar graph.
(b) Say which of these fuels are fossil fuels.
(c) Explain the meaning of the term 'fossil fuel', and say why such fuels are non-renewable.
(d) Name a product which is formed when all fossil fuels burn.
(e) Why is the contribution of hydroelectric power small in the UK?

2 You are provided with a solution of dilute hydrochloric acid of concentration 0.100 mol/l and a dilute solution of sodium hydroxide of unknown concentration.
(a) Briefly describe how you would find out what volume of the acid solution is needed to neutralise 25.0 cm³ of the sodium hydroxide solution.
(b) A student following your method finds that 35.0 cm³ of 0.100 mol/l hydrochloric acid neutralise 25.0 cm³ of sodium hydroxide solution. Calculate the concentration of the sodium hydroxide solution.

3 (i) What amount (in moles) of hydroxide ion is present in 25.0 cm³ of 0.050 mol/l sodium hydroxide solution?
(ii) Which of the following will neutralise exactly 25.0 cm³ of 0.050 mol/l sodium hydroxide solution? (More than one answer is correct.)
(a) 25.0 cm³ of 0.050 mol/l hydrochloric acid
(b) 25.0 cm³ of 0.050 mol/l nitric acid
(c) 25.0 cm³ of 0.050 mol/l sulphuric acid
(d) 50.0 cm³ of 0.025 mol/l hydrochloric acid
(e) 25.0 cm³ of 0.025 mol/l sulphuric acid

4 Aluminium is obtained by the electrolysis of molten aluminium oxide.
(a) Write the equation for the electrode process which yields aluminium.
(b) A current of 20 000 A is passed through a cell containing molten aluminium oxide for 24 hours. What mass of aluminium is produced? (Take the Faraday constant as 96 000 C/mol and $A_r(Al) = 27$

5 You are given a sample of copper(II) oxide mixed with carbon. You are asked to find the percentage purity of the copper oxide. You decide to add an excess of dilute sulphuric acid to 2.00 g of the mixture to react with all the copper(II) oxide. You filter, wash and dry the carbon that remains. The mass is 0.090 g. What is the percentage purity of the sample of copper(II) oxide?

6 The manufacture of ammonia uses the reaction:

$$N_2(g) + 3H_2(g) \rightleftharpoons 2NH_3(g)$$

The reaction is reversible, and the forward reaction is exothermic.

(a) What can be done to increase the rate of the reaction?
(b) What steps can be taken to increase the yield of ammonia?
(c) Is there any conflict between the requirements of (a) and (b)?
(d) What conditions does the industrial plant employ?

7 Zinc reacts with sulphuric acid to give hydrogen. Mention three ways in which you could speed up the reaction.

8 In a set of experiments, zinc was allowed to react with sulphuric acid. Each time, 0.65 g of zinc was used. The volume of acid was different each time. The volume of hydrogen formed each time is shown in the table.

Volume of sulphuric acid (cm³)	Volume of hydrogen (cm³)
5	45
15	135
20	180
25	215
30	235
35	240
40	240

Plot, on graph paper, the volume of hydrogen produced against the volume of acid used. From the graph, find out:
(a) where the reaction is most rapid (and explain why),
(b) what volume of sulphuric acid will produce 100 cm³ of gas,
(c) what volume of gas is produced if 10 cm³ of sulphuric acid are used,
(d) what volume of sulphuric acid is just sufficient to react with 0.65 g of zinc.

9 The diagram below shows part of the label from a packet containing tablets which can be used to treat headaches and upset stomachs.

$\bigcirc$ $\bigcirc$ $\bigcirc$ **TABLETONE** $\bigcirc$ $\bigcirc$ $\bigcirc$

Active Ingredients	Dose
Aspirin Citric acid Sodium hydrogencarbonate	2 tablets dissolved in a glass of water. The dose may be repeated every 4 hours up to 4 doses in 24 hours.

Warning: Do not exceed the stated dose.

(a) (i) When a tablet is added to water, carbon dioxide is produced. From which of the active ingredients does the carbon dioxide come?
(ii) Why is it dangerous to exceed the dose given on the label?

317

(b) Derek did an experiment in which he put a tablet in water at 20 °C. He measured the mass every twenty seconds. The balance he used was linked to a computer. His apparatus is shown in the diagram below.

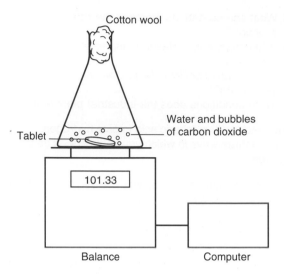

The computer started recording the mass when the tablet entered the water. The results for the experiments are given in this table.

Time(s)	0	20	40	60	80	100
Total mass(g)	101.33	101.12	101.00	100.93	100.90	100.90

(i) Why was the cotton wool placed in the top of the flask?

(ii) Suggest why Derek used a computer to record his weighings.

(iii) On a copy of the figure below plot a graph of total mass against time using the results given in the table above. Label the graph A.

(iv) Derek repeated the experiment using water at 30 °C. He used the same mass of water and of tablet as in the first experiment. Sketch the graph for this experiment on your copy of the figure below and label it B.

(v) In another experiment Derek crushed the tablet and then added it to the water at 30 °C. What effect would this have? Explain your answer.

(LEAG)

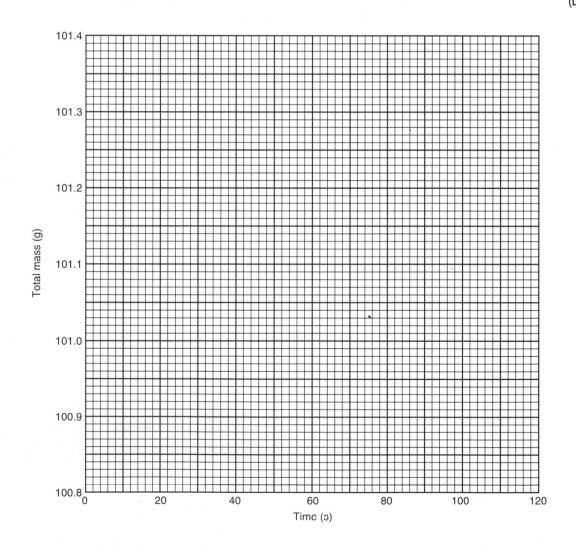

THEME H
The Chemical Industry

The chemical industry plays a central part in our lives. It provides the fertilisers which enable us to grow enough food, the pesticides which prevent crops from being eaten by insects or smothered by weeds and the treatments which make our water safe to drink. The chemical industry provides the building materials for our homes and the fabrics for the clothes we wear. When we are ill, we have a whole range of painkillers, antiseptics, antibiotics and anaesthetics to comfort us. When we want to enjoy a sport, the chemical industry offers us a choice of footballs, tennis racquets, golf balls, climbing ropes, canoes, sailing dinghies, all of them made from synthetic materials.

The chemical industry provides all these materials by using the Earth's resources and human ingenuity. Chemists study the rates of chemical reactions and the heat changes which take place; they calculate how much product a reaction should give and find out the best conditions in which to carry out a reaction. All these steps are involved in the design of the industrial plants which transform natural resources into the materials we need.

TOPIC 28 *BUILDING MATERIALS*

28.1 Concrete

The UK quarries 90 million tonnes of limestone and chalk each year. Most of it is used in the building industry.

Figure 28.1A ● Concrete and glass

All around you, you can see buildings made of concrete, steel and glass. Concrete is used in the construction of all kinds of homes, schools, factories, skyscrapers, power stations and reservoirs. All our forms of transport depend on the concrete which is used in roads and bridges, docks and airport runways. For extra strength, for example in bridges, reinforced concrete is strengthened by steel supports.

The starting point in the manufacture of concrete is limestone or chalk. Both these minerals are forms of calcium carbonate. First, limestone or chalk is used with shale or clay to make cement. A small percentage of calcium sulphate (gypsum) is also needed (see Figure 28.1B). Figure 28.1C shows how concrete is made from cement.

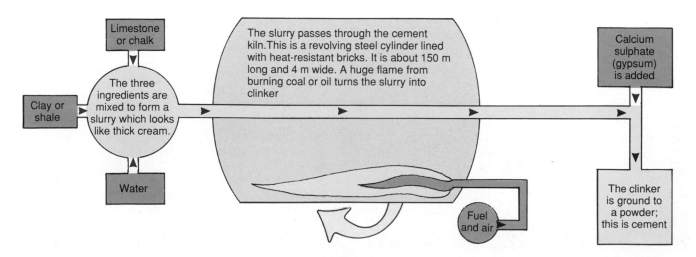

Figure 28.1B ● The manufacture of cement (after Blue Circle)

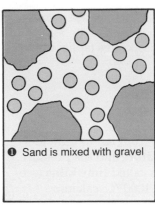
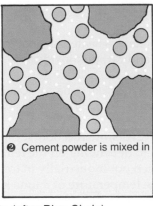
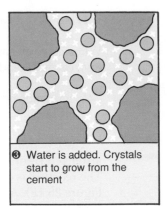
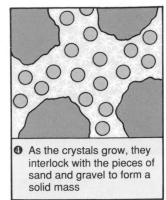

❶ Sand is mixed with gravel

❷ Cement powder is mixed in

❸ Water is added. Crystals start to grow from the cement

❹ As the crystals grow, they interlock with the pieces of sand and gravel to form a solid mass

Figure 28.1C ● Making concrete (after Blue Circle)

Chalk and limestone are calcium carbonate, $CaCO_3$. Clay and shale consist largely of silicon(IV) oxide, SiO_2, and aluminium oxide, Al_2O_3, with some other compounds. The chemical reactions which occur between them produce cement, which consists chiefly of calcium silicate, $CaSiO_3$, and calcium aluminate, $CaAl_2O_4$. A little calcium sulphate (gypsum) is added to slow down the rate at which concrete sets.

The UK has large deposits of both limestone and chalk. Limestone is quarried by blasting a hillside with an explosive. Chalk is dug out by mechanical excavators. Both methods devastate the landscape. It happens that these minerals often occur in regions of great natural beauty (Figure 28.1D). We have to balance the damage to our countryside against the useful materials which industry can obtain from limestone and chalk. Mining companies are required to restore the countryside after they exhaust a deposit of limestone or chalk (Figure 28.1E). Limestone is used in the manufacture of iron in blast furnaces, in the manufacture of glass and in the manufacture of lime.

SUMMARY

Concrete is a strong and versatile construction material. It is made from cement, sand, gravel and water. Cement is made from chalk or limestone and clay or shale. The quarrying of vast quantities of these raw materials creates environmental problems.

Figure 28.1D ● A limestone quarry

Figure 28.1E ● Restoring the scenery

Lime

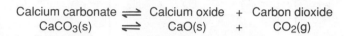

When calcium carbonate is heated strongly, it dissociates (splits up) to give calcium oxide and carbon dioxide.

Calcium carbonate $\rightleftharpoons$ Calcium oxide + Carbon dioxide
$$CaCO_3(s) \rightleftharpoons CaO(s) + CO_2(g)$$

The reaction is a **reversible** reaction: it can go from left to right and also from right to left, depending on the temperature and the pressure. The reaction is carried out industrially in towers called **lime kilns** (see Figure 28.1F). At the temperature of a lime kiln, 1000 °C, calcium carbonate dissociates. The through draft of air carries away carbon dioxide as it is formed and prevents it from recombining with calcium oxide. Otherwise there would be an acid–base reaction between the acid gas, carbon dioxide, and the base, calcium oxide.

LOOK AT LINKS
for the use of lime in agriculture.
See Topic 29.7.

LOOK AT LINKS
You can find out about the use of carbon dioxide in fire-extinguishers in Topic 15.8.

Figure 28.1G ● Enjoying a 'carbonated' drink

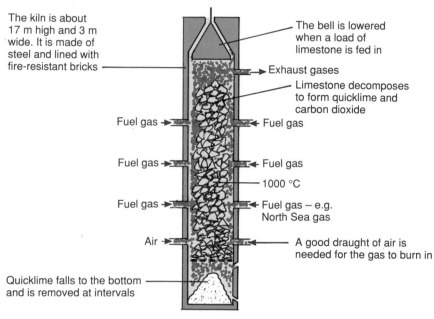

The kiln is about 17 m high and 3 m wide. It is made of steel and lined with fire-resistant bricks

The bell is lowered when a load of limestone is fed in

Exhaust gases

Limestone decomposes to form quicklime and carbon dioxide

Fuel gas → ← Fuel gas

Fuel gas → ← Fuel gas

1000 °C

Fuel gas → ← Fuel gas – e.g. North Sea gas

Air → ← A good draught of air is needed for the gas to burn in

Quicklime falls to the bottom and is removed at intervals

Figure 28.1F ● A lime kiln

Calcium oxide is called **lime** or **quicklime**. It reacts with water to form calcium hydroxide, which is called **slaked lime**.

Calcium oxide + Water → Calcium hydroxide
$$CaO(s) + H_2O(l) → Ca(OH)_2(s)$$

On building sites, calcium hydroxide, slaked lime, is mixed with sand to give **mortar**. Mortar is used to hold bricks together. As mortar is exposed to the air, it becomes gradually harder as it reacts with carbon dioxide in the air to form calcium carbonate.

Calcium oxide, lime, is used in agriculture. Farmers spread it on fields to neutralise excess acid in soils.

The carbon dioxide produced in lime kilns is also useful. Carbon dioxide dissolves to a slight extent in water to give a solution of the weak acid, carbonic acid, H_2CO_3. Under pressure, the solubility of carbon dioxide increases. The basis of the soft drinks industry is dissolving carbon dioxide in water under pressure, and adding flavourings. When the cap is taken off a bottle, the pressure is decreased, and carbon dioxide comes out of solution.

SUMMARY

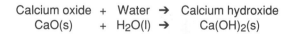

When calcium carbonate (limestone) is heated in lime kilns, calcium oxide (quicklime) and carbon dioxide are produced. On building sites, calcium oxide is used to make mortar. On farms, it is used to neutralise excessive acidity of soils. Carbon dioxide is used in the soft drinks industry and as a fire extinguisher.

28.2 🔬 Glass

The recipe for glass is 4500 years old. Egyptians discovered it when they melted sand with limestone and sodium carbonate. To their surprise, they obtained a transparent material, glass. Egypt is one of the few countries where sodium carbonate occurs naturally.

Glass is a mixture of calcium silicate, $CaSiO_3$, and sodium silicate, Na_2SiO_3. In structure, it resembles silicon(IV) oxide, SiO_2, (sand), which is a crystalline substance with a macromolecular structure (see Figure 28.2A). In glass, many of the Si—O bonds have been broken, and the structure is less regular. X-rays show that glass does not have the orderly packing of atoms found in other solids. Glass is neither a liquid nor a crystalline solid. It is a **supercooled liquid**: it appears to be solid, but has no sharp melting point.

● Some new uses for glass

Craftsmen worked with glass for thousands of years without knowing anything about its structure. When chemists began to study glass, they made many advances in glass technology. New types of glass were invented. Some of these are described in this section.

Pyrex® glass will stand sudden changes in temperature without cracking. It is made by adding boron oxide during the manufacture of glass.

Window glass consists of plates of glass which are the same thickness all over. Formerly, plates were made by grinding a sheet of glass to the required thickness and smoothness. In the process, 30% of the sheet was wasted. The Pilkington Glass Company invented the **float glass** process. Molten glass flows on to a bath of molten tin (Figure 28.2B). As the glass cools and solidifies, the top and bottom surfaces are both perfectly smooth and planar. While it is still soft, the glass is rolled to the required thickness and then cut into sections.

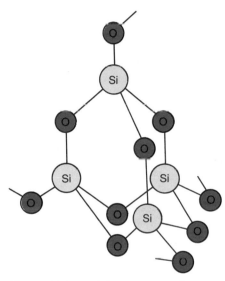

Figure 28.2A ● The structure of silicon(IV) oxide (silica)

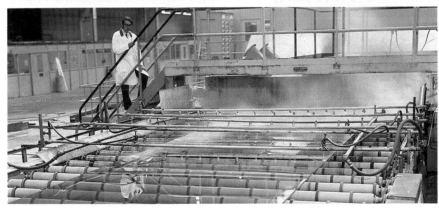

Figure 28.2B ● Float glass

Photosensitive glass is used in sunglasses which darken in bright light and lighten again when the light fades. The glass includes silver chloride.

Glass ceramic is almost unbreakable. It is made by heating photosensitive glass in a furnace. Glass ceramic is used in ovenware, electrical insulation, the nose cones of space rockets and the tiles of space shuttles.

Soluble glass has important uses. In many tropical countries, the disease, **bilharzia** (or schistosomiasis) is a blight. The snails which carry the disease can be killed by copper salts in quite low concentrations. The copper compounds can be incorporated into a soluble glass. Pellets of the copper-containing soluble glass can be put into the water. As they dissolve, they release copper compounds gradually into the snail-infested water. To make a soluble glass, phosphorus(V) oxide, P_2O_5, is used instead of silicon(IV) oxide.

SUMMARY

The glass industry makes:
* Pyrex® glass,
* plate glass,
* light-sensitive glass,
* glass ceramic,
* soluble glass.

❶ (a) Why must a bottle of a carbonated soft drink have a well-fitting cap?
   (b) Why can you not see bubbles inside the closed bottle?
   (c) Why can you see bubbles when the bottle is opened?

❷ 'Spreading calcium hydroxide on soil reduces the acidity of the soil.' Describe an experiment which you could do to check whether this statement is true.

❸ 'Mortar reacts with the air to form calcium carbonate.' Describe experiments which you could do on new mortar and old mortar to find out whether this statement is true.

28.3 Ceramics

Figure 28.3A ● Traditional ceramics

Ceramics are crystalline compounds which consist of an ordered arrangement of atoms. Many are compounds of metallic and non-metallic elements, e.g. silicon or aluminium combined with oxygen or nitrogen. Pottery is a ceramic. Brick is a ceramic. Traditional ceramics such as these are derived from the raw materials clay and silica. They are formed on, for example, a potter's wheel and hardened in a fire kiln. Examples of modern ceramics are silica glass, soda glass, aluminium oxide, Al_2O_3, silicon carbide, SiC, silicon nitride, SiN, zirconium oxide, ZrO_2, Titanium carbide, TiC, zirconium carbide, ZrC, tungsten carbide, WC. They are made by grinding the components to fine powders, mixing them and heating to high Temperatures under high pressure in electric kilns to cause 'densification'. New uses are being found for ceramics all the time. They are used in spacecraft, artificial bones, cutting tools, engine parts, turbine blades and electronic components (see next page).

The bonding in ceramics may be ionic or covalent. The arrangement of particles is more complicated than in metals. The bonds are directed in space and the structure is therefore more rigid than a metallic structure.

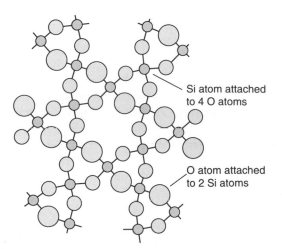

Figure 28.3B ● Bonding in crystalline silica, SiO_2

Si atom attached to 4 O atoms

O atom attached to 2 Si atoms

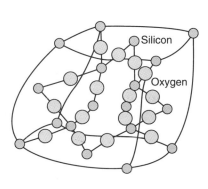

Figure 28.3C ● Bonding in silica glass (a distorted silica structure)

Silicon

Oxygen

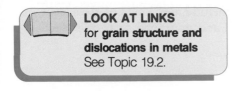
LOOK AT LINKS
for **grain structure and dislocations in metals**
See Topic 19.2.

LOOK AT LINKS
for the **structure of plastics**
See Topics 30.2 and 30.3.

LOOK AT LINKS
for a **comparison of metals, ceramics and plastics**
See Topic 30.3.

This structure makes ceramics
* crystalline
* harder than metals (able to cut steel and glass)
* of higher melting point than metals
* poor conductors of heat and electricity because they lack free electrons
* usually opaque because light is reflected by grain boundaries in the crystal structure
* not ductile or malleable or plastic (except at high temperature)
* rather brittle because the bonds are rigid, and dislocation movements are complicated so that it is difficult to change the shape without breaking

Ceramics can be made to behave like metals. Ceramic semiconductors and even superconductors can be made from suitable mixtures of metal oxides. Superplastic ceramics are more plastic than most metals. They are based on grains of silicon nitride and silicon carbide in a 'glue' of silicon(IV) oxide.

● *Applications of ceramics*

Vehicle components A ceramic coating, e.g. ZrO_2, on cast iron improves wear and tear and heat-resistance. It extends the life of diesel engine parts. Ceramic bearings can operate at high speeds without lubricants. Research is going on into the use of ceramics to allow diesel engines and jet engines to run at higher temperatures. The higher the temperature at which an engine operates, the more efficiently it runs and the less fuel it consumes.

Gas turbine blades made of ceramics, e.g. silicon nitride, can run at higher temperatures than those made of alloys.

Machine tools made of ceramics can rotate twice as fast as metal tools without deforming or wearing out. *Sialon* is a ceramic made of silicon, aluminium, oxygen and nitrogen. It is almost as hard as diamond, and is as strong as steel and as low in density as aluminium. It can be used up to 1300 °C and needs no lubricants.

Bioceramics Recently ceramics, e.g. aluminium oxide and silicon nitride, have been considered as implant materials. Alloys and polymers are used for replacement devices, e.g. hips, knees, teeth etc. Ceramics have a high strength to weight ratio, would wear better than alloys or polymers and would never corrode. For replacing damaged bone, a porous ceramic implant could be used, so as to allow bone to grow into and bond with the implant.

Space craft Ceramic tiles protect a space shuttle during re-entry into the Earth's atmosphere. The outside temperature can reach 1500 °C. Each tile consists of a cellular structure of very fine fibres coated with silica. The fibres are loosely packed: 95% of the tiles is air, which gives them their insulating property.

Figure 28.3D ● Space-age ceramics

SUMMARY

Ceramics are crystalline compounds of oxygen or nitrogen with other elements, e.g. silicon, aluminium and some transition elements. Their structure makes them hard-wearing, resistant to corrosion, and of high melting point. Their resistance to heat and corrosion finds them many uses.

Optical materials Recent developments include new kinds of glass (see Topic 28.2), glass fibres (see Topic 30.3) and improved solar cells (see Topic 7.3 of *KS: Physics*).

CHECKPOINT

❶ What is the advantage of a ceramic over (i) an alloy and (ii) a hard plastic for use as (a) a tooth implant (b) an elbow implant (c) a middle ear prosthesis?

❷ Explain why ceramic tiles are fitted to space shuttles.

❸ What is the chief difference between the methods used to fire traditional ceramics and modern ceramics?

❹ Why is glass transparent and china opaque?

❺ Say what structural features of ceramics make them (a) more brittle than metals (b) harder than metals (c) poor electrical conductors.

TOPIC 29 AGRICULTURAL CHEMICALS

29.1 Fertilisers

LOOK AT LINKS
for the **nitrogen cycle**
See Topic 14.5.

IT

Soil Analysis
(program)

Have you experimented with soil in a laboratory? You can try some computer experiments as well using this software.

LOOK AT LINKS
Plants need water
See *KS: Biology*, Topic 14.1.
and air
See *KS: Biology*, Topic 12.1.

As the world population has increased, the demand for food has increased. In 1965 it was 3.5 billion, and by 2000 it will have reached 6.5 billion. In many parts of the world, people are either hungry or undernourished. Farming has to be intensive, that is, able to grow large crops on the available land. Crops take nutrients from the soil, and these must be replaced before the next crop is sown.

Before the nineteenth century, farmers relied on the natural fertilisers, manure and compost, to replace nutrients in the soil. When the world population began to rise sharply in the nineteenth century, farmers needed more fertilisers. They asked the chemists to make some artificial fertilisers. Before the chemists could do this, they had to find out which chemicals in manure fertilised the soil. Then they would be able to make chemical fertilisers to do the same job.

The chemists' work showed that plants need these three groups of chemicals.

1. Large amounts of carbon, hydrogen and oxygen. There is no shortage of these elements. Carbon comes from carbon dioxide in the air, hydrogen and oxygen from water.

2. Small amounts of 'trace elements' (see Table 29.1). So little is needed that there is no need to add these elements to the soil.

3. Larger quantities of the other elements shown in Table 29.1.

Table 29.1 ● Elements obtained by plants from soil.

Major elements	Importance	Trace elements
Nitrogen	Needed for protein synthesis	Manganese
Phosphorus	Needed for development of roots, energy-transfer reactions, nucleic acids	Copper Iron Zinc
Potassium	Needed in photosynthesis	Chlorine
Magnesium	Present in chlorophyll	Boron
Sulphur	Needed for synthesis of some proteins	Molybdenum
Calcium	Needed for transport	

With 2–3 million tonnes of nitrogen (in the form of nitrates) going on to the soil of the UK each year, 600 000 tonnes of nitrate ion leach out every year.

LOOK AT LINKS
for the **eutrophication** (unintentional enrichment with nutrients) of lakes and rivers and the pollution of ground water by fertilisers. See Topic 18.4.

How much fertiliser?

Most fertilisers concentrate on supplying the necessary nitrogen (N), phosphorus (P) and potassium (K). **NPK fertilisers** are consumed in huge quantities: in 1990, the world consumption was over 20 million tonnes. In the UK, the consumption of NPK fertilisers is about seven million tonnes a year. The fertilisers cost about £60 a tonne. Farmers and market gardeners need to invest a lot of money in fertilisers. They want good crops, but they do not want to spend more than they need on fertilisers. They can obtain expert advice from the Ministry of Agriculture, Fisheries and Food. Agricultural chemists at the Ministry

Fertiliser
(program)

Investigate the chemistry of hydrogen, nitrogen and ammonium nitrate. You can also develop and operate a production system which is commercially viable.

IT'S A FACT

There are islands off the coast of South America which are inhabited by large flocks of sea birds. Mounds of their droppings, called guano, accumulate. Since sea birds eat a fish diet, which is high in protein, guano is rich in nitrogen compounds. For a century, European farmers imported guano from South America to use as a fertiliser.

will advise them on the type and quantity of fertiliser to apply and the best season of the year for applying it. Every farmer and grower has a different problem. The agricultural chemists must weigh up the type of crop and the type of soil before they can recommend the most suitable treatment.

Nitrogen

Nitrogen can be absorbed by plants when it is combined as nitrates. Ammonia and ammonium salts can also be used as fertilisers because ammonium salts are converted into nitrates by organisms which live in the soil. All nitrates and all ammonium salts are soluble so they are easily absorbed. Sometimes concentrated ammonia is used as a fertiliser. Solid fertilisers, ammonium nitrate, ammonium sulphate and urea, (CON_2H_4) are applied in pellet form (see Figure 29.1A). The manufacture of nitrogenous fertilisers is described below.

Figure 29.1A ● Applying pellets of ammonium salts to a field

29.2 Manufacture of ammonia

LOOK AT LINKS
for the **nitrogen cycle**
See Topic 14.4.

LOOK AT LINKS
for factors affecting equilibrium
See Topic 27.

LOOK AT LINKS
for soil
See Topic 2.4.

The first nitrogen-containing fertiliser which the farmers of Europe used was sodium nitrate. They had to import it across the Atlantic from Chile. There were disadvantages to this practice. European chemists tackled the problem of making nitrogenous fertilisers. Nitrogen was the obvious starting material because every country has plenty of it in the air. Making nitrogen combine with other elements proved to be a problem. The problem was solved by a German chemist called Fritz Haber. In 1908, he succeeded in combining nitrogen with hydrogen to form ammonia

Nitrogen + Hydrogen ⇌ Ammonia
$N_2(g)$ + $3H_2(g)$ ⇌ $2NH_3(g)$

This reaction is reversible: it takes place from right to left as well as from left to right. Some of the ammonia formed dissociates (splits up) into nitrogen and hydrogen. A mixture of nitrogen, hydrogen and ammonia is formed. To increase the percentage of ammonia formed in the reaction, a high pressure and a low temperature are used. A low temperature reduces the dissociation of ammonia, but it also makes the reaction very slow. Modern industrial plants use a compromise temperature and speed up the reaction by means of a catalyst. (The buildings and equipment in which a manufacturing process is carried out are called a **plant**.) The yield of ammonia is about 10%. Ammonia is condensed out of the mixture. With a boiling point of –33 °C, which is higher than that of most gases, ammonia is easily liquefied. The nitrogen and hydrogen which have not reacted are recycled through the plant (see Figure 29.2A).

The hydrogen for the Haber process is obtained from natural gas. The process takes place in a number of stages. The overall reaction is:

Natural gas (methane) + Steam → Hydrogen + Carbon monoxide
$$CH_4(g) + H_2O(g) \rightarrow 3H_2(g) + CO(g)$$

Carbon monoxide is oxidised to carbon dioxide and removed to leave hydrogen.

SUMMARY

The starting point in the manufacture of fertilisers is the manufacture of ammonia from nitrogen and hydrogen by the Haber process. Hydrogen is obtained from natural gas, and nitrogen is obtained from air.

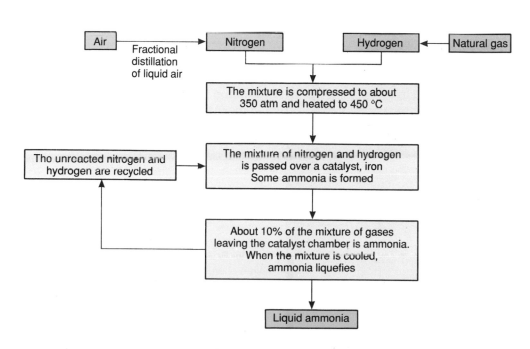

Figure 29.2A ● A flow diagram for the Haber process

● *Making fertilisers from ammonia*

A concentrated solution of ammonia can be used as a fertiliser. It is easier to store solid fertilisers, such as ammonium salts.
Being a base, ammonia is neutralised by acids to yield ammonium salts

Ammonia + Nitric acid → Ammonium nitrate
$$NH_3(aq) + HNO_3(aq) \rightarrow NH_4NO_3(aq)$$

Ammonia + Sulphuric acid → Ammonium sulphate
$$2NH_3(aq) + H_2SO_4(aq) \rightarrow (NH_4)_2SO_4(aq)$$

Ammonia + Phosphoric acid → Ammonium phosphate
$$3NH_3(aq) + H_3PO_4(aq) \rightarrow (NH_4)_3PO_4(aq)$$

SUMMARY

Ammonia is used as a fertiliser, but ammonium nitrate, sulphate and phosphate are preferred.

29.3 Manufacture of nitric acid

Nitric acid is made by the oxidation of ammonia (see Figure 29.3A).

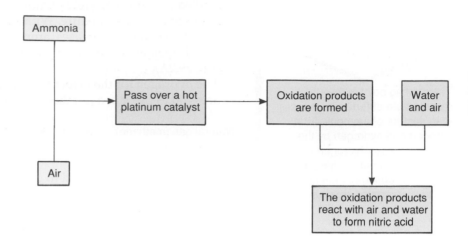

Figure 29.3A ● A flow diagram for the manufacture of nitric acid

> **SUMMARY**
>
> Nitric acid is made by the oxidation of ammonia.

In addition to its importance in the fertiliser industry, nitric acid is used in the manufacture of explosives, such as TNT (trinitrotoluene), and dyes.

Who's behind the science

In 1846, Professor Sobrero, an Italian chemist, had the idea of reacting concentrated nitric acid with glycerol. He obtained a liquid which explodes when ignited by a lighted fuse. He called it **nitro-glycerine.** The new compound was welcomed by miners who started to use it for blasting away rock. Unfortunately, there were accidents because nitro-glycerine can explode if it receives a sudden blow, and this sometimes happened in transport.

This drawback did not stop Immanuel Nobel from beginning to manufacture nitroglycerine in 1863 in Sweden. A year later, an explosion killed four people in the factory, including his son, Emil. The other son, Alfred, continued manufacturing, but he looked for a method of stabilising nitroglycerine to withstand mechanical shock. He found that a type of clay called kieselguhr would absorb nitro-glycerine to form a mixture which did not explode when struck and could only be detonated by a lighted fuse. He patented his invention as **dynamite**. Alfred Nobel made a fortune out of his new, safer explosive. As well as being used to make tunnels, roads and railways and in mining, dynamite was sometimes used in warfare. Alfred Nobel was distressed to see his invention being misused in this way, and he became interested in promoting peace between nations. He financed the Nobel Foundation in Stockholm. This awards prizes each year to the people who have done the most effective work for peace and the most outstanding work in medicine, physics, chemistry and literature.

29.4 Manufacture of sulphuric acid

Applied Chemistry
(program)

Use this program to study:
• The Haber Process
• The Contact Process
• The production of sodium hydroxide.

Sulphuric acid is made by the **contact process**. Sulphur dioxide and oxygen combine in contact with a catalyst, vanadium(V) oxide. Air is used as a source of oxygen. Sulphur dioxide is obtained by:
• Burning sulphur (deposits of sulphur occur in many countries).
• As a by-product of the extraction of metals from sulphide ores.
• As a by-product of the removal of unwanted, unpleasant-smelling sulphur compounds from petroleum oil and natural gas.
A flow diagram for the industrial process is shown in Figure 29.4A.

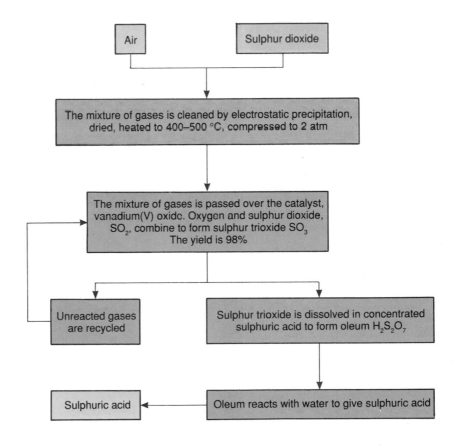

Figure 29.4A ● The contact process

Sulphur trioxide reacts with water to give sulphuric acid

Sulphur trioxide + Water → Sulphuric acid
$SO_3(s)$ + $H_2O(l)$ → $H_2SO_4(l)$

SUMMARY

Sulphuric acid is made by the contact process. Sulphur or a metal sulphide is used as a source of sulphur.

It is dangerous for industry to carry out the reaction in this way. As soon as sulphur trioxide meets water vapour, it reacts to form a mist of sulphuric acid. It is safer to absorb the sulphur trioxide in sulphuric acid to form oleum, $H_2S_2O_7$, and then convert the oleum into sulphuric acid.

Sulphuric acid is important in the fertiliser industry because it is needed for the manufacture of phosphoric acid. It has many other uses, such as the manufacture of paints, pesticides and plastics.

29.5 Manufacture of phosphoric acid

SUMMARY

Phosphoric acid is made from calcium phosphate.

Phosphoric acid is made from the widespread ore calcium phosphate by a reaction with sulphuric acid.

Calcium phosphate + Sulphuric acid → Phosphoric acid + Calcium sulphate

29.6 Making NPK fertilisers

LOOK AT LINKS
The importance of fertilisers is discussed in Topic 13.4. The most popular fertilisers are those that contain compounds of nitrogen, phosphorus and potassium: the NPK fertilisers.

Figure 29.6A shows a flow diagram for the manufacture of NPK fertilisers. It shows how ammonium nitrate and ammonium phosphate are manufactured. There is no need to make potassium chloride because there are plentiful deposits of it in the UK.

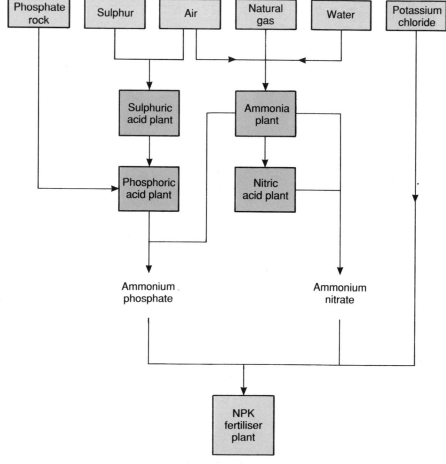

Figure 29.6A ● Manufacture of NPK fertilisers

SUMMARY

Ammonium nitrate and ammonium phosphate are manufactured. Potassium chloride is mined. The three salts are mixed to make NPK fertilisers.

CHECKPOINT

❶ State two advantages of the fertiliser ammonium phosphate over the insoluble salt calcium phosphate.

❷ What methods do farmers use for applying (a) ammonia and (b) ammonium sulphate?

❸ Why are farmers prepared to pay a lot for fertilisers? Why do all fertilisers contain nitrogen?

❹ What is the advantage of using nitrogen as a starting material in the manufacture of fertilisers? What difficulty had to be overcome before manufacture started? Who solved the difficulty and how?

❺ NPK fertilisers are popular. What do the letters NPK stand for? Explain why it is important that fertilisers contain (a) N, (b) P and (c) K.

❻ State three routes by which nitrogen from the air finds its way into the soil. Why is the nitrogen content of the air not used up?

❼ The figure below shows how the yields of winter wheat and grass increase as more nitrogenous fertiliser is used. After studying the graphs, say what mass of nitrogen you would apply to a 100 hectare field to avoid waste and to give a maximum yield of (a) winter wheat and (b) grass.

Grass response to nitrogen fertiliser

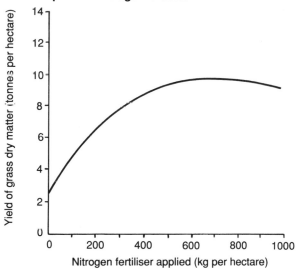

Winter wheat response to nitrogen fertiliser

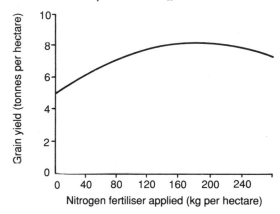

❽ A dairy farmer spreads 240 tonnes of nitrochalk on his pasture. It costs £50 per tonne. If his milk cheque is £6000 a month, how long will it take him to recoup the cost of the fertiliser?

❾ (a) The table below shows the mass of nitrogen, phosphorus and potassium (in kg/hectare) removed from the soil when different crops are grown. Name the crops which take out a large quantity of potassium. How do these crops differ from the rest?

Crop	Mass (kg/hectare)		
	N	P	K
Wheat grain	115	22	26
Oat grain	72	13	18
Sugar beet root	86	14	302
Potatoes	109	14	133
Pasture grass	128	14	100

(b) The table below shows the composition of three NPK fertilisers that a farmer uses. Which fertiliser should she use (i) on her pasture (ii) on her wheat and (iii) on her sugar beet? Explain your choices.

Fertiliser	Composition (percentage mass)		
	N	P	K
Fertiliser A	62	13	25
Fertiliser B	22	27	51
Fertiliser C	44	19	37

29.7 pH

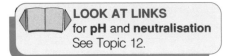

A soil may be acidic or alkaline or neutral. For each type of plant, there is a range of soil pH over which the plant can be grown successfully.

Table 29.2 ● Crop growth and pH

Crop	Potatoes	Swedes	Oats	Wheat	Sugar beet	Kale	Barley
pH	2	4.7–5.6	4.8–6.3	6.0–7.5	7.0–7.5	9–11	12

Farmers have a choice. They can measure the pH of their soil and then choose a crop which will grow well on it. Alternatively, they can alter the pH of the soil to suit the crop which they want to grow. Often soils are too acidic. Rainwater, being weakly acidic, neutralises bases which are present in topsoil. When soils are too acidic, farmers add lime (calcium hydroxide). This neutralises the excess of acid. Occasionally, soils are too alkaline. Then iron(II) sulphate is added. This salt is weakly acidic in solution and neutralises the excess of alkali in the soil.

> **LOOK AT LINKS**
> for **pH** and **neutralisation**
> See Topic 12.

29.8  Herbicides

FIRST THOUGHTS

Farmers and gardeners have to cope with pests. These include weeds which compete with crops for nutrients, animals which eat crops, and disease-causing organisms like harmful fungi which attack plants and livestock. Substances which are used to kill pests are pesticides. There are three main types: herbicides, which kill plants, insecticides, which kill insects (Topic 29.9) and fungicides, which kill fungi (Topic 29.10).

Fertile soil is a good growing ground for weeds as well as for the crops that farmers sow. Weeds are plants that compete with crops for space, light and nutrients. At one time, farmers employed many farmhands, and one of the jobs they did was pulling out weeds by hand. Now, farmers employ fewer people and save labour by using chemical weedkillers – herbicides. By using a **selective herbicide**, they can kill weeds and leave the crop untouched. The most common selective herbicides are 2,4-D and 2,4,5-T. They kill broad-leaved plants, (dandelions, thistles and nettles).

Figure 29.8A ● A crop treated with herbicide and a control

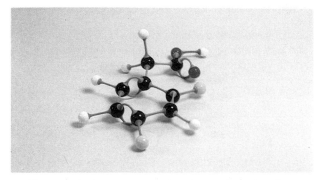

Figure 29.8B ● Models of 2,4-D and 2,4,5-T molecules

They do not harm plants with narrow leaves (grasses and cereals). These selective herbicides are organic compounds which contain chlorine. Other weedkillers, such as paraquat and sodium chlorate(V), are **non-selective**: they kill all plants.

Chemical weedkillers are of great benefit to us. It is difficult to see how scientific farming could continue without them. There have also been some tragic results of using herbicides.

Agent Orange

From 1961–71, there was a war in Vietnam between the USA and the communist Vietcong forces. The US planes found it difficult to carry out bomb attacks on the Vietcong because the Vietcong positions were hidden by dense jungle vegetation. US planes dropped a herbicide called Agent Orange, which killed all the trees and other vegetation where it landed. It destroyed the forests which sheltered Vietcong fighters and their bases and killed their crops. Agent Orange contained 2,4,5-T and an impurity, **dioxin**. At the time, no-one knew that dioxin has teratogenic effects (effects on unborn babies). Terribly deformed babies were born to Vietnamese mothers. Dioxin also has genetic effects. When US servicemen returned home and started families, some of them found to their dismay that they had become the fathers of deformed babies. Agent Orange had affected the fathers' genes. As a result of these tragedies, 2,4,5-T has been banned in many countries. The method of manufacture always produces some dioxin as impurity.

SUMMARY

Selective weedkillers are of enormous benefit to agriculture. Nevertheless, one of them, 2,4,5-T, is now banned because it contains an impurity called dioxin, which has teratogenic effects and genetic effects.

CHECKPOINT

❶ Explain what is meant by (a) herbicide and (b) selective herbicide.

❷ What are the advantages of selective herbicides over non-selective herbicides?

FIRST THOUGHTS

29.9 Insecticides

The previous section dealt with weeds – plant pests. Insects are the major animal pests, although many insects are beneficial. The pesticides which are used to kill insects are called insecticides.

Insects, weeds and plant diseases destroy about one third of the world's crops. Aphids attack the flowers of many fruits and vegetables. Cutworms chew off plants at soil level. Wireworms bore into potatoes and other root vegetables. Grain weevils eat stored grain. Unchecked, insects would consume most of the food we grow, and humans would starve. On the other hand, many insects are helpful. Some pollinate flowers, and some eat harmful insects.

The first chemicals which were used as insecticides were general poisons, such as compounds of arsenic, lead and mercury. These compounds are poisonous to all animals which swallow them. Children and family pets were sometimes accidentally poisoned. These poisons are stable: they remain in the soil for many years.

SUMMARY

Insecticides are used to protect crops against being eaten by insects. Insecticides have played a big part in increasing food supplies worldwide.

The DDT story

In 1939, a scientist called Paul Mueller discovered that a substance called DDT was very poisonous to houseflies. His experiments soon showed

LOOK AT LINKS
for **petrochemicals**
See Topic 25.

that DDT was a very useful substance indeed, killing many insects such as lice and mosquitoes and agricultural pests such as potato-blight and grape-blight. The exciting aspect of DDT was that it left species other than insects unharmed. It is a **broad spectrum insecticide**. That is, it kills many types of insect. DDT is a **petrochemical**. Together with similar compounds, it is often referred to as an 'organochlorine'.

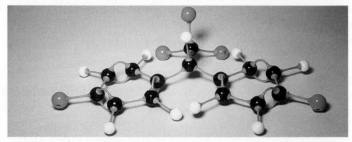

Figure 29.9A ● A model of a DDT molecule (Dichlorodiphenyltrichloroethane)

DDT has saved more lives than any other single substance. It has saved million of lives and prevented billions of illnesses by killing insects which carry disease. It has saved people from hunger and starvation by killing insects which eat crops. Thanks to the use of DDT, 550 million people in tropical countries now no longer live in constant fear of malaria. The World Health Organisation says that DDT has saved five million lives. Cholera is another disease which is carried by insects and which can now be fought with DDT.

═══ SCIENCE AT WORK ═══

DDT came into production in time to be used in the Second World War. In wartime, the breakdown of normal hygienic living conditions means that both troops and civilians may suffer from lice, typhus and other afflictions. In the Italian city of Naples in 1943, the spread of lice caused a massive epidemic of typhus. The whole population of Naples was dusted with DDT, which killed the lice, and the epidemic was halted.

Figure 29.9B ● Crop spraying today. In countries where DDT is banned, safer, often biodegradable chemicals are used.

IT'S A FACT

Ten million tonnes of DDT have been applied to the Earth in the last 25 years.

DDT has also contributed to agriculture. It was used successfully to control tea parasites in Sri Lanka and cotton pests in the USA. In the desert regions of North Africa, crops are often consumed by swarms of locusts. The result is famine, and a whole population will face starvation. DDT has reduced the harm done by locusts.

IT'S A FACT

According to Oxfam, 375 000 people in the Third World are poisoned – 10 000 of them fatally – by pesticides each year. Pesticide controls are lax in some Third World countries. Manufacturers recommend that agricultural workers should wear protective boots, gloves, face masks and hoods while spraying pesticides. Unfortunately, workers often cannot afford this protective clothing.

▽ SUMMARY ▽

Insecticides fight disease by killing disease-carrying insects, e.g. lice and mosquitoes. The insecticide DDT has saved thousands of human lives. Worry over the build up of DDT in animals has led to strict limits on its use.

People predicted that DDT would exterminate all insect pests. In 1948, Paul Mueller was awarded the Nobel Prize for his discovery. No sooner had he received the prize than people began to discover some worrying effects of DDT. In 1962 Rachel Carson wrote a book called *Silent Spring*. She suggested that killing insects would deprive birds of their natural food, and birds would die out, giving a spring without birdsong. Many of her fears have been borne out. Since DDT does not break down easily, an application of DDT will remain in the soil and in water for many years.

In Topic 18, Figure 18.4D shows what happened when Clear Lake in California was sprayed with DDT to combat mosquitoes. People were surprised when aquatic birds, such as grebes, died. The concentration of DDT in the lake water was only 0.02 p.p.m. However, DDT is soluble in fat and therefore difficult to excrete. Along the food chain, the level of DDT built up: micro-organisms 5 p.p.m., fish 250 p.p.m. and at the top of the food chain, grebes 1600 p.p.m., a lethal dose.

DDT has been carried through food chains to the remotest parts of Earth, far from the places where it was used. Figure 29.9C shows how DDT can build up along a food chain on land. A problem is that many insects are becoming immune to DDT. As a result, higher concentrations must be used in spraying.

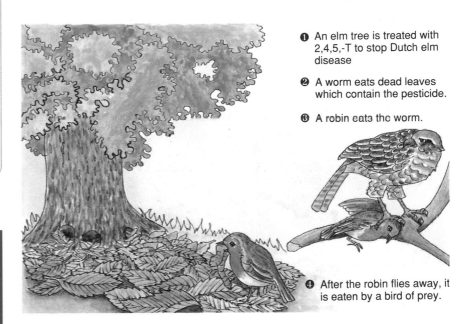

❶ An elm tree is treated with 2,4,5,-T to stop Dutch elm disease

❷ A worm eats dead leaves which contain the pesticide.

❸ A robin eats the worm.

❹ After the robin flies away, it is eaten by a bird of prey.

Figure 29.9C ● The robin's place in a food chain

DDT does not react with air or water. Since it is such a stable compound, the amount of DDT in the world is steadily increasing. We know that it kills insects and that, in sufficiently large amounts, it kills fish and birds. A human being may eat fish and birds contaminated with DDT over a long period of time. We do not know what the results of the buildup of DDT will be. There is no direct evidence that DDT causes illness in humans. It has been found, however, that cancer victims in the USA have twice as much DDT in their fatty tissues as the rest of the population.

In the USA and Europe, DDT is now banned. DDT is still used for essential jobs, for example., fighting malaria. Safer insecticides are used for agricultural purposes. One choice is organic compounds of phosphorus, which are broken down in time to safe compounds.

Other methods of controlling insects

SUMMARY

It would benefit the environment if we could find ways of cutting down on the use of insecticides. Research is being carried out on the use of predator insects, sterilisation and trapping.

Scientists are looking for other ways of controlling insects. Some methods are described as **biological control**. One such method is to breed **predator insects**. These are insects which kill the insects that are eating the crops. Of course, the predator insects must not eat the crops! Another idea is to use radioactivity to sterilise male insects so that they cannot breed. Insects give out substances called **pheromones** when they want to attract a mate. Research workers have worked out a method of extracting pheromones from insects and using the extract to bait a trap. The insects which are lured into the trap are then killed.

CHECKPOINT

❶ In the spring of 1988, North Africa was threatened by the worst plague of locusts for 30 years. The locust control programme financed by the United Nations uses short-lived insecticides such as fenitrothion. These do not last long enough to lie in wait for advancing swarms of locusts. An area sprayed with fenitrothion is safe for locusts after only three days. With long-lasting insecticides, such as dieldrin, strips of desert can be sprayed to prevent the advance of locusts. Experts say that dieldrin would do no harm to a desert environment. However, dieldrin has been shown to cause genetic damage. Which would you use – fenitrothion or dieldrin – if you were in charge of the locust control programme? Explain your views.

❷ The Mediterranean fruit fly is an insect pest. Large numbers of male flies were irradiated by a cobalt-60 source (see Topic 8.3) and then released. They mated but they produced no offspring.
(a) Why were fruit farmers pleased with the use of radioactivity?
(b) What is the advantage of this technique over the use of chemical pesticides?

29.10 Fungicides and molluscicides

IT'S A FACT

The pesticide picture
In 1985 nearly 10 000 tonnes of pesticide were applied to farmland in the UK. More than 800 000 tonnes were used worldwide, at a cost of about £28 billion. The use of pesticide will continue to grow at a rate of 2–4% a year in developing countries and 7–8% a year in developed countries.

Fungi cause the expensive diseases of mildew and potato-blight. To prevent these diseases, crops are sprayed with fungicides. Many copper salts are fungicides, e.g. copper(II) sulphate.

Molluscs, including slugs and snails, eat crops and spread diseases. Molluscicides are being used to fight them. Bilharzia is a disease which affects people in tropical countries. It is spread by water snails.

Advantages of using pesticides
• They kill pests quickly.
• Pesticides increase food production.
• They are easy to store and use.
• There is such a large variety of chemicals available that most pests can be eliminated.

Disadvantages

- Pests can develop resistance to a particular pesticide.
- The spray can be carried by the wind and so affect wildlife.
- Pesticides can seep into the soil and drain into rivers and lakes, where they will harm wildlife.
- Pesticides can enter food chains and build up from one feeding level to the next. They can reach a toxic level at the top of a food chain.

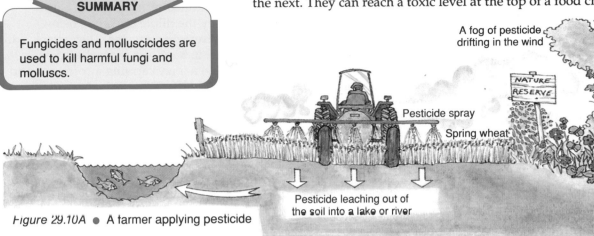

Figure 29.10A ● A farmer applying pesticide

> **SUMMARY**
>
> Fungicides and molluscicides are used to kill harmful fungi and molluscs.

CHECKPOINT

❶ (a) How do chemical insecticides help farmers?
(b) What did farmers use as insecticides before DDT?
(c) In what way is DDT better than the insecticides which were used previously?
(d) Why has the use of DDT on farms been banned?
(e) For what purpose is DDT still used? Why is it in such demand?

❷ Why did Rachel Carson's book get people worried?

❸ Do you think we should give up using chemical insecticides?

❹ DDT was found in the flesh of eagles. They are birds of prey: they do not eat insects. How did the insecticide get into the eagles' flesh?

❺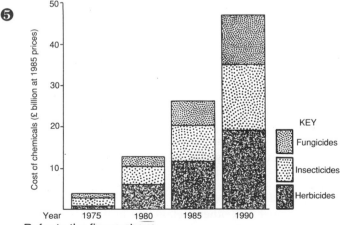

Refer to the figure above.
(a) Give approximate figures for the money spent on (i) fungicides, (ii) insecticides and (iii) herbicides in 1975 and in 1990.
(b) Give the ratio:
Cost of pesticides used in 1990/Cost of pesticides used in 1980
(c) Name crops on which a farmer might use (i) a fungicide (ii) an insecticide and (iii) a herbicide.

29.11 ● Concentrated acids

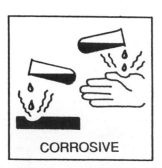

CORROSIVE

Figure 29.11A ● A Hazard sign for a corrosive substance

A concentrated solution is one that contains a large amount of solute per litre of solution. Concentrated acids take part in the same chemical reactions as dilute acids but react more vigorously. In addition, concentrated sulphuric acid and concentrated nitric acid take part in some reactions which are not shared by the dilute acids (see below). Concentrated solutions of acids are extremely corrosive. They attack skin and clothing. Great care must be taken in handling concentrated acids and in transporting them. Occasionally, a lorry carrying a corrosive acid is involved in an accident (see Figure 29.11B). Two methods are employed to deal with a spill. The acid may be diluted with plenty of water and then washed away or it may be diluted and then neutralised with a base.

Figure 29.11B ● Tanker in simulated road accident

Concentrated hydrochloric acid

Concentrated hydrochloric acid has the same reactions as dilute hydrochloric acid. The concentrated acid reacts more vigorously with metals, bases and carbonates.

Concentrated sulphuric acid

Concentrated sulphuric acid takes part in the same reactions as dilute sulphuric acid and has two additional properties. It is an **oxidising agent** and a **dehydrating agent**.

● *Oxidising agent*

1. Copper and other metals low in the reactivity series do not react with dilute sulphuric acid. Concentrated sulphuric acid reacts with these metals. With copper it forms copper(II) sulphate, sulphur dioxide and water. Sulphuric acid has acted as an oxidising agent; it has taken electrons from copper atoms to form copper ions.

Copper + Conc. sulphuric acid ⟶ Copper(II) + Sulphur + Water
                                                        sulphate   dioxide

$Cu(s) + 2H_2SO_4(aq) \longrightarrow CuSO_4(aq) + SO_2(g) + H_2O(l)$

2. Hot, dilute sulphuric acid reacts with iron to form iron(II) sulphate. $FeSO_4$. With concentrated sulphuric acid, the product is iron(III) sulphate, $Fe_2(SO_4)_3$, showing that concentrated sulphuric acid has acted as an oxidising agent, taking electrons from Fe^{2+} ions to form Fe^{3+} ions.

● *Dehydrating agent*

Ask your teacher to let a drop of concentrated sulphuric acid fall on to a piece of cloth. You will see a brown patch develop, which looks exactly as though the cloth has been burnt. **Don't try it** – but people who have accidentally spilt concentrated sulphuric acid on their skins will tell you that the same thing happens: it feels very much like being burnt by fire. The reason is that concentrated sulphuric acid removes water from the cloth, from skin and from other substances. A reagent which removes water from a compound is called a **dehydrating agent** (hudor = water in Greek).

1. Copper(II) sulphate crystallises with water of crystallisation as copper(II) sulphate-5-water, $CuSO_4.5H_2O$. Concentrated sulphuric acid can remove the water of crystallisation.

Conc. sulphuric acid

Copper(II) sulphate-5-water → Copper(II) sulphate + water;
(blue crystals)   (white powder)

Heat is given out

The water of crystallisation combines with the sulphuric acid to leave anhydrous (without water) copper(II) sulphate, which is a white powder. Evidently, the water of crystallisation gives copper(II) sulphate-5-water its colour and its crystalline form.

2. Cane sugar, sucrose, is a carbohydrate (see Topic 29.3) of formula $C_{12}H_{22}O_{11}$. Concentrated sulphuric acid dehydrates sucrose (removes the elements of water from it) to leave a form of carbon known as 'sugar charcoal'.

Conc. sulphuric acid

Sucrose   →   Carbon + Water; Heat is given out
$C_{12}H_{22}O_{11}(s)$ $12C(s)$ + $11H_2O(l)$

3. The reaction between concentrated sulphuric acid and water itself is very exothermic. This makes it difficult to prepare a dilute solution of sulphuric acid from concentrated sulphuric acid. Never attempt to make a dilute solution by adding water to concentrated sulphuric acid. Heat will be generated in a small volume of acid and will make the added water boil and splash out of the container, bringing with it a shower of concentrated sulphuric acid. Instead, add concentrated sulphuric acid slowly, with stirring, to a large volume of water. Then the heat generated will be spread through a large volume of water.

Figure 29.11C ● (a) Don't try it this way (b) This is how to dilute a concentrated acid safely – but wear safety glasses

LOOK AT LINKS
for **soapless detergents**
See Topic 16.10.

● *Sulphonating agent*

Concentrated sulphuric acid is used in the manufacture of soapless detergents. Many of these contain a sulphonic acid group, $-SO_3H$, or a sulphate group, $-SO_4^-$. They are made by the reaction of concentrated sulphuric acid with a number of compounds from petroleum oil. This reaction is called **sulphonation**.

Nitric acid – concentrated and dilute

Concentrated nitric acid takes part in the same reactions as dilute nitric acid and also acts as an oxidising agent. It is an even more powerful oxidising agent than concentrated sulphuric acid.

Dilute nitric acid is an oxidising agent. When heated, it reacts with copper to form copper(II) nitrate. Hydrogen is not formed. The colourless gas nitrogen oxide, NO, and the pungent brown gas nitrogen dioxide, NO_2, are formed. This gas is toxic, and the reaction must therefore be carried out in a fume cupboard.

When you need to work with a concentrated acid or alkali it is essential to wear safety glasses (see Figure 29.11C). Spilling a concentrated acid or alkali on your skin is painful and dangerous, and you should immediately wash off the spill with plenty of cold water. The danger to your eyes is very much greater. Firstly, the tissues of the eye are very delicate and easily damaged. Secondly, if you get something in your eye, you cannot see your way to the cold water tap to wash it out. Concentrated alkalis are dangerous too.

SUMMARY

Concentrated sulphuric acid is an oxidising agent, a dehydrating agent and a sulphonating agent. Nitric acid, both concentrated and dilute, is an oxidising agent. In addition, the concentrated acids take part in the same chemical reactions as the dilute acids.

CHECKPOINT

❶ Name two types of reaction which concentrated sulphuric acid takes part in but dilute sulphuric acid cannot bring about. Give one example of each.

❷ 'Here lies Susan still and placid.
She added water to the acid.'

Suggest which acid Susan could have been using to result in an accident. Explain why her method was dangerous. Describe how she should have diluted the acid.

❸ 'Dilute nitric acid is an oxidising agent.' Give two reactions which support this statement.

❹ Explain why the following reactions are oxidation reactions (see Topic 21).
   (a) Copper is converted into copper(II) sulphate
   (b) Iron is converted into iron(II) sulphate
   (c) Iron is converted into iron(III) sulphate
   (d) Iron(II) sulphate is converted into iron(III) sulphate.

TOPIC 30 PLASTICS AND FIBRES

FIRST THOUGHTS

The toys you see in Figure 30A are made of plastics. The clothes you see in Figure 30B are made of synthetic fibres. Both types of material are polymers. In this topic, you will find out what a polymer is.

Figure 30A ● Plastics

Figure 30B ● Fibres

30.1 Alkenes

LOOK AT LINKS
Alkenes are formed when alkanes are cracked. See Topic 25.1.

Ethene is a hydrocarbon of formula C_2H_4 (see Figure 30.1A). There is a double bond between the carbon atoms.

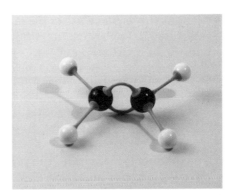

Figure 30.1A ● (a) A model of ethene   (b) The formula of ethene

Ethene and other hydrocarbons which contain double bonds between carbon atoms are described as **unsaturated** hydrocarbons. The double bond will open to allow another molecule to add on. Unsaturated hydrocarbons will add hydrogen to form **saturated** hydrocarbons. For example, ethene adds hydrogen to form ethane, which is an **alkane**. Reactions of this kind are called **addition reactions**; see Figure 30.1B. Saturated hydrocarbons, such as alkanes, contain only single bonds between carbon atoms.

Ethene + Hydrogen → Ethane

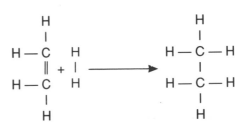

Figure 30.1B ● The addition reaction between ethene and hydrogen

Propene is an unsaturated hydrocarbon which resembles ethene (see Figure 30.1C). Ethene and propene are members of a **homologous series**, that is, a set of similar compounds whose formulas differ by CH_2. They are called **alkenes** (see Table 30.1). The general formula is C_nH_{2n}.

SUMMARY

Alkenes are unsaturated hydrocarbons. They possess a double bond between carbon atoms. They are a homologous series of general formula C_nH_{2n}.

```
H   H   H
|   |   |
C = C — C — H
|       |
H       H
```

Figure 30.1C ● The formula of propene

Table 30.1 ● The first members of the alkene series

Alkene	Formula, C_nH_{2n}
Ethene	C_2H_4
Propene	C_3H_6
Butene	C_4H_8
Pentene	C_5H_{10}
Hexene	C_6H_{12}
Heptene	C_7H_{14}
Octene	C_8H_{16}

Reactions of alkenes

● *Hydrogenation*

Animal fats, such as butter, are solid. Vegetable oils, such as sunflower seed oil, are liquid. More vegetable oil is produced than we need for cooking, and insufficient butter is produced to satisfy the demand for solid fat. It is therefore profitable to convert vegetable oils into solid fats. Manufacturers make use of the fact that fats are saturated, while oils are unsaturated. Hydrogenation (the addition of hydrogen) is used to convert an unsaturated oil into a saturated fat. The vapour of an oil is passed with hydrogen over a nickel catalyst.

Vegetable oil (unsaturated) + Hydrogen $\xrightarrow[\text{nickel catalyst}]{\text{pass over heated}}$ Solid fat (saturated)

The product is margarine. The process can be modified to leave some of the double bonds intact and yield soft margarine.

● *Hydration*

Water will add to alkenes. A molecule of water can add across the double bond. Combination with water is called **hydration**. The product formed by the hydration of ethene is ethanol, C_2H_5OH.

Ethene + Water → Ethanol

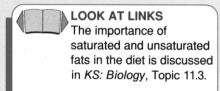

Ethanol is the compound we commonly call **alcohol**. It is an important industrial solvent. The industrial manufacture is

Ethene + Steam $\xrightarrow[\text{(phosphoric acid), under pressure}]{\text{pass over a heated catalyst}}$ Ethanol

LOOK AT LINKS
for **fats** and **oils**
See Topic 31.5.

LOOK AT LINKS
The importance of saturated and unsaturated fats in the diet is discussed in *KS: Biology*, Topic 11.3.

LOOK AT LINKS
for **ethanol**
See Topic 31.1.

Only about 10% of ethene is converted to ethanol. The unreacted gases are recycled over the catalyst.

● *Addition polymerisation*

The double bond in ethene enables many molecules of ethene to join together to form a large molecule.

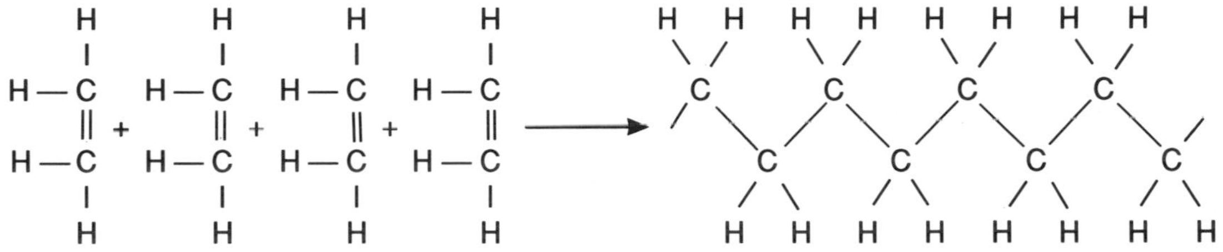

This reaction is called **addition polymerisation**. Many molecules (30 000–40 000) of the **monomer**, ethene, **polymerise** (i.e. join together) to form one molecule of the **polymer**, poly(ethene). The conditions needed for polymerisation are high pressure, a temperature of room temperature or higher and a catalyst.

$$Ethene \xrightarrow[\text{over a heated catalyst}]{\text{pass at high pressure}} Poly(ethene)$$

$$nCH_2{-}CH_2 \xrightarrow[\text{catalyst}]{\text{heat, pressure,}} \text{---}(CH_2{-}CH_2)\text{---}_n$$

Poly(ethene) is better known by its trade name of **polythene**. It is used for making plastic bags, kitchenware (buckets, bowls, etc.), laboratory tubing and toys. It is flexible and difficult to break.

> **SUMMARY**
>
> The double bonds makes alkenes reactive. They take part in addition reactions, e.g. with hydrogen to form alkanes and with water to form alcohols. Hydrogenation is used to turn vegetable oils into saturated fats.

● *Addition of bromine*

A solution of bromine in an organic solvent or in water is brown. If an alkene is bubbled through a solution of bromine, the solution loses its colour. Bromine has added to the alkene to form a colourless compound. The reaction can be shown as

Ethene + Bromine → 1,2-Dibromoethane

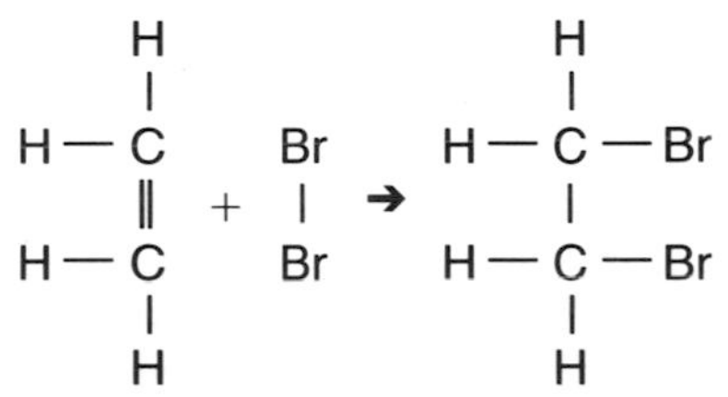

The product has single bonds: it is a saturated compound. With two carbon atoms in the molecule, it is named after ethane. With two bromine atoms in the molecule, it is a dibromo-compound, 1,2-dibromoethane. The numbers 1,2- tell you that one bromine atom is bonded to one carbon atom and the second bromine atom is bonded to the second carbon atom. The decolourisation of a bromine solution is used to distinguish between an alkene and an alkane. Chlorine adds to alkenes in a similar way.

30.2 Plastics

Polymers such as poly(ethene) and other polyalkenes are **plastics**. Plastics are materials which soften on heating and harden on cooling. They are therefore useful materials from which to mould objects. There are two kinds of plastics: **thermosoftening plastics** and **thermosetting plastics**.

Thermosoftening plastics can be softened by heating, cooled and resoftened many times. Thermosetting plastics can be softened by heat only once.

LOOK AT LINKS
The ways in which the uses of materials depend on their properties are mentioned in Topic 3.6.

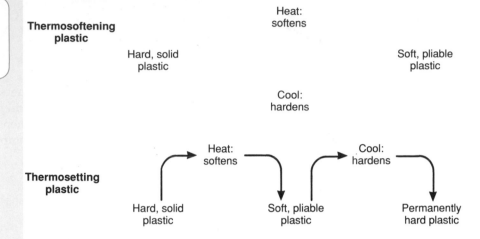

The reason for the difference in behaviour is a difference in structure (see Figure 30.2A).

Thermosoftening plastics consist of long polymer chains. The forces of attraction between chains are weak

(a) Part of the structure of a thermosoftening plastic

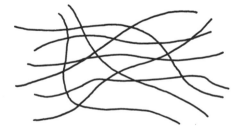

SCIENCE AT WORK

When someone breaks a bone, the broken ends must be held in place by a plaster cast until new bone grows. Sometimes, metal pins are used to join the broken ends together. Now there is an easier way. Poly(methylmethacrylate) can be used to 'glue' the ends of a broken bone together without the need for pins or plaster.

When a thermosetting plastic is softened and moulded, the chains react with one another. Cross-links are formed and a huge three-dimensional structure is built up. This is why thermosetting plastics can be formed only once

(b) Part of the structure of a thermosetting plastic
Figure 30.2A ●

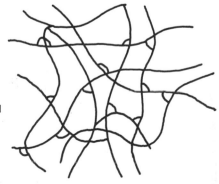

Manufacturers find thermosoftening plastics very convenient to use. They can buy thermosoftening plastic in bulk in the form of granules, then melt the material and press it into the shape of the object they want to make. Thermosoftening plastics are easy to colour. When a pigment is added to the molten plastic and thoroughly mixed, the moulded objects are coloured all through. This is much better than a coat of paint which can become chipped. Thermosetting plastics have important uses too. Materials used for bench tops must be able to withstand high temperatures without softening. 'Thermosets' are ideal for this use.

● *Condensation polymerisation*

The chief thermosetting polymers are not poly(alkenes). Many of them are made by another type of polymerisation called **condensation polymerisation**. For this to occur, the monomer must possess two groups of atoms which can take part in chemical reactions. When the reactive groups in one molecule of monomer react with the groups in other molecules of monomer, large polymer molecules are formed (see Figure 30.2B). In the reaction, small molecules are eliminated, e.g. H_2O, HCl, NH_3. The elimination of water gave the name **condensation** to this type of polymerisation.

SUMMARY

Addition polymerisation is the addition of many molecules of monomer to form one molecule of polymer. Thermosoftening and thermosetting plastics have advantages for different uses.

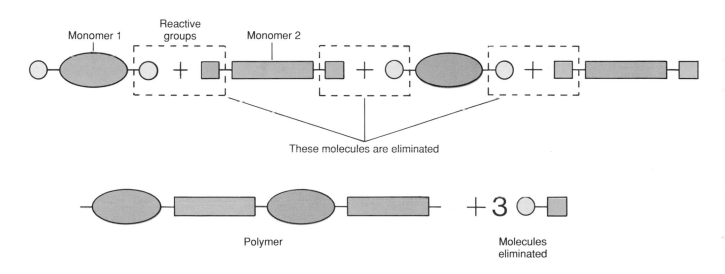

Figure 30.2B ● Condensation polymerisation

LOOK AT LINKS
for **polysaccharides and polypeptides**
See *KS: Biology*, Topic 10.4.

Examples of condensation polymers are:
● epoxy resins, which are used in glues,
● polyester resins, which are used in glass-reinforced plastics,
● polyurethanes, which are used in varnishes.
● See Table 30.3 for nylon, terylene and rayon.

SUMMARY

Condensation polymerisation is the reaction between many molecules of monomer to form one molecule of polymer with the elimination of a small molecule such as a molecule of water.

● *Methods of moulding plastics*

Different methods are used for moulding thermosoftening and thermosetting plastics (see Figures 30.2C, D, E, F).

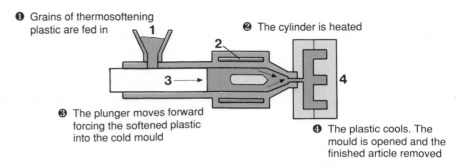

❶ Grains of thermosoftening plastic are fed in

❷ The cylinder is heated

1

2

3 →

4

❸ The plunger moves forward forcing the softened plastic into the cold mould

❹ The plastic cools. The mould is opened and the finished article removed

Figure 30.2C ● For thermosoftening plastics, injection moulding is used for many objects, e.g. milk bottle crates and construction kits

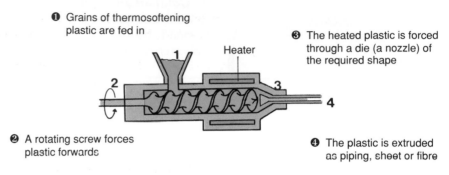

❶ Grains of thermosoftening plastic are fed in

❸ The heated plastic is forced through a die (a nozzle) of the required shape

Heater

1

2

3

4

❷ A rotating screw forces plastic forwards

❹ The plastic is extruded as piping, sheet or fibre

Figure 30.2D ● For thermosoftening plastics, the extrusion method is used to make threads of fabric, pipes and tubes, e.g. insulation for electrical cable

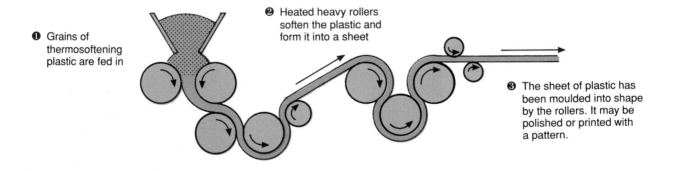

❶ Grains of thermosoftening plastic are fed in

❷ Heated heavy rollers soften the plastic and form it into a sheet

❸ The sheet of plastic has been moulded into shape by the rollers. It may be polished or printed with a pattern.

Figure 30.2E ● For thermosoftening plastics, calendering is used to make large sheets of plastic, e.g. floor coverings and car seat covers

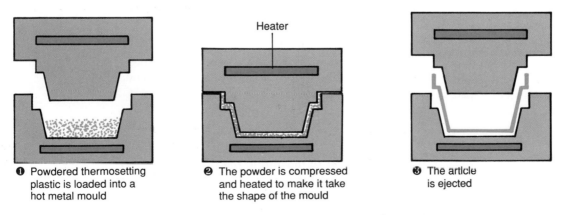

❶ Powdered thermosetting plastic is loaded into a hot metal mould

Heater

❷ The powder is compressed and heated to make it take the shape of the mould

❸ The article is ejected

Figure 30.2F ● The usual method of shaping thermosetting plastics is compression moulding

SUMMARY

Different methods are used for moulding thermosoftening plastics (which can be softened by heat many times) and thermosetting plastics (which can be softened by heat only once before setting permanently).

Some manufacturers mix gases with plastics to make low density plastic foams. These foams are used in packaging, for thermal insulation of buildings, for insulation against sound and in the interior of car seats. Sometimes it is necessary to strengthen plastics by the addition of other materials. Boat hulls, plastic panels in cars, instrument panels and wall-mounted hand-driers are just a few of the articles which are made of GRP (glass-fibre-reinforced plastic).

Polyalkenes

Table 30.2 ● Some poly(alkenes) and their uses

Poly(ethene); trade name Polythene

Monomer	Polymer	

$$\begin{array}{cc} H & H \\ | & | \\ C = C \\ | & | \\ H & H \end{array} \qquad \left(\begin{array}{cc} H & H \\ | & | \\ C - C \\ | & | \\ H & H \end{array}\right)_n$$

Polythene is used to make plastic bags. High density polythene is used to make kitchenware, laboratory tubing and toys.

Poly(chloroethene); trade name PVC

Monomer   Polymer

$$\begin{array}{cc} Cl & H \\ | & | \\ C = C \\ | & | \\ H & H \end{array} \qquad \left(\begin{array}{cc} Cl & H \\ | & | \\ C - C \\ | & | \\ H & H \end{array}\right)_n$$

PVC is used to make wellingtons, raincoats, floor tiles, insulation for electrical wiring, gutters and drainpipes.

Poly(propene); trade name Polypropylene

Monomer   Polymer

$$\begin{array}{cc} H & CH_3 \\ | & | \\ C = C \\ | & | \\ H & H \end{array} \qquad \left(\begin{array}{cc} H & CH_3 \\ | & | \\ C - C \\ | & | \\ H & H \end{array}\right)_n$$

Polypropylene is resistant to attack by chemicals. Since it does not soften in boiling water, it can be used to make hospital equipment which must be sterilised. Polypropylene is drawn into fibres which are used to make ropes and fishing nets.

Poly(tetrafluoroethene); trade names PTFE and Teflon

Monomer   Polymer

$$\begin{array}{cc} F & F \\ | & | \\ C = C \\ | & | \\ F & F \end{array} \qquad \left(\begin{array}{cc} F & F \\ | & | \\ C - C \\ | & | \\ F & F \end{array}\right)_n$$

PTFE is a hard plastic which is not attacked by most chemicals. Few substances can stick to its surface. It is used to coat non-stick pans and skis.

Perspex

Monomer

$$\begin{array}{cc} H & CO_2CH_3 \\ | & | \\ C = C \\ | & | \\ H & CH_3 \end{array}$$

Perspex finds many applications because it is transparent and can be used instead of glass. It is more easily moulded than glass and less easily shattered.

(continued on following page)

RESOURCE —ACTIVITY— PACK

SUMMARY

Some important poly(alkenes) are:
- poly(ethene),
- poly(chloroethene),
- poly(propene),
- poly(tetrafluoroethene),
- perspex,
- polystyrene.

LOOK AT LINKS
Cellulose is a polysaccharide. It forms the walls of plant cells. Over 3000 sugar molecules join to form one cellulose molecule.
See Topic 31.4 and *KS: Biology*, Topic 10.4.

SUMMARY

Some important condensation polymers are nylon, terylene, bakelite, melamine and rayon.

Polystyrene
Monomer

H   H
| |
C = C
| |
H C_6H_5

Polystyrene is a hard, brittle plastic used for making construction kits. Polystyrene foam is made by blowing air into the softened plastic. It is used for making ceiling tiles, insulating containers and packaging materials.

Table 30.3 ● Some condensation polymers and their uses

Polymer	Properties	Uses
Nylon	A polyamide Thermosoftening m.p. 200 °C High tensile strength	Textile fibre used in clothes; also in ropes, fishing lines and nets
Terylene	A polyester Thermosoftening	Important in clothing manufacture
Bakelite	Thermosetting An electrical insulator Insoluble in water and organic solvents Not attacked by chemicals	Electrical appliances, e.g. switches, sockets, plugs. Casings for radios and telephones
Melamine	Similar to bakelite	Kitchen surfaces, bench tops
Rayon	Made from cellulose by reshaping, that is breaking the long chains of sugar molecules in cellulose into shorter chains and then rejoining them	Important in clothing manufacture

● Some drawbacks
Nylon and terylene do not absorb water well. Clothes made of these fibres do not absorb perspiration and allow it to evaporate. Nylon, terylene and other polyesters are usually mixed with natural fibres such as wool and cotton. The mixtures absorb water and are therefore more comfortable to wear.

Many refreshment stands serve coffee and soft drinks in disposable cups. These are often made of polystyrene. All the plastic cups, plates and food containers which people use and throw away have to be disposed of. This is difficult to do because plastics are **non-biodegradable**. They are not decomposed by natural biological processes. Some plastic waste is burned in incinerators and the heat generated is used. Other plastics cannot be disposed of in this way because they burn to form poisonous gases, for example, hydrogen chloride, carbon monoxide and hydrogen cyanide. Much plastic waste is buried in landfill sites. One third of the plastics manufactured is used for packaging: it is made to be used once and thrown away. As the mass of non-biodegradable plastic waste increases, more land is used up to bury the waste.

Chemists are working on the problem. They have invented some **biodegradable** plastics, that is plastics which can be broken down by micro-organisms. One type has starch incorporated into the plastic. When bacteria feed on the starch, the plastic partially breaks down. Other new types of plastic are completely biodegradable.

SUMMARY

Some disadvantages in the use of plastics are:
• Clothes made of synthetic fibres do not absorb perspiration.
• Most plastics are non-biodegradable.
• They ignite easily, and some burn to form toxic products.

There are some dangers in the use of plastics. Plastic foams are used as insulation in many buildings. Furniture is often stuffed with plastic foam. If there is a fire, burning plastics spread the fire rapidly. This is because plastics have lower ignition temperatures than materials like wood, metal, brick and glass. There is another danger too: some burning plastics give off poisonous gases.

SUMMARY

Plastics are petrochemicals obtained from oil. Earth's resources of oil are limited. Should we use oil as fuel or save it for the petrochemicals industry?

Oil: a fuel and a source of petrochemicals

Industry is constantly finding new uses for plastics. The raw materials used in their manufacture come from oil. At first the price of oil was low and plastics were cheap materials. As the price of oil has risen, this is no longer true. The Earth's resources of oil will not last for ever. We should be thinking about whether we ought to be burning oil as fuel when we need it to make plastics and other petrochemicals.

CHECKPOINT

❶ Say what materials the following articles were made out of before plastics came into use. Say what advantage plastic has over the previous material, and state any disadvantage.
(a) gutters and drainpipes (g) electrical plugs and sockets
(b) toy soldiers (h) wellingtons
(c) dolls (i) furniture stuffing
(d) motorbike windscreens (j) electrical cable insulation
(e) lemonade bottles (k) dustbins
(f) buckets

❷ (a) What does the word 'plastic' mean?
(b) Plastics can be divided into two types, which behave differently when heated. Name the two types. Describe how each behaves when heated. Say how the difference in behaviour is related to (i) the use made of the plastics and (ii) the molecular nature of the plastics.

❸ Study the following list of substances.

nylon sucrose ethane styrene olive oil starch
margarine glass silk rubber melamine

(a) List the naturally occurring polymers.
(b) List the synthetic polymers.
(c) Name the substance which is not a polymer but can easily be converted into one.
(d) List the substances which are not polymers and which cannot easily polymerise.

❹ PVC is used in the manufacture of drainpipes and plastic bags.
(a) Calculate the relative molecular mass of the monomer, $CH_2 = CHCl$.
(b) The M_r of the polymer is 40 000. Calculate the number of molecules of monomer which combine to form one molecule of polymer.
(c) What method could be used to mould PVC pipes?
(d) If you needed to mould a straight length of PVC pipe to make it fit round a curve in a drainage system, how would you do this?
(e) Used PVC bags have to be disposed of. Burning PVC bags in an incinerator would cause pollution. Name two pollutants that would be released into the atmosphere.

30.3 **Review of materials**

You have now studied materials of many types. In this section, we compare the properties of different solid materials. We look at how these properties make different materials useful for different purposes. We look at the types of structure which give rise to the different properties.

LOOK AT LINKS
for the **metallic bond** and the **crystalline structure of metals**
See Topic 19.2;
for the **uses of metals and alloys**
See Topic 19.7.

LOOK AT LINKS
for **ceramics**
See Topic 28.3.

IT'S A FACT

The ceramic of formula $YBa_2Cu_3O_7$ is a superconductor. (A superconductor has practically zero resistance to the flow of an electric current.) The ceramic is metallic in appearance and becomes a superconductor at 90 K. It has a potential for use in zero-loss power transmission lines. Schemes for cooling underground superconducting cables in liquid air have been put forward.

LOOK AT LINKS
for **plastics**
See Topic 30.2.

Metals

Metals are crystalline materials. Metals:
- are generally hard, tough and shiny
- change shape without breaking (i.e. are ductile and malleable)
- are generally strong in tension and compression
- are good thermal and electrical conductors
- are in many cases corroded by water and acids.

These properties result from the nature of the metallic bond (see Topic 19.2). Metals are giant structures in which some electrons are free to move. These electrons hold the atoms together and also allow atoms to slide over one another when the metal is under stress and to conduct heat and electricity. The bonds are non-directed in space. Dislocations (see Topic 19.2) allow atoms in one plane of a crystal to move relative to an adjacent plane. This is why metals can be bent so easily.

Ceramics

Ceramics are crystalline compounds of metallic and non-metallic elements. Ceramics:
- are very hard and brittle
- are strong in compression but weak in tension
- are electrical insulators
- have very high melting points
- are chemically unreactive.

The bonds are mainly covalent, but some are ionic. The bonds are directed in space, and the structure of a ceramic is therefore more rigid and less flexible than a metallic structure. As a result of this structure, ceramics are both harder than metals and also more brittle. Ceramics have lower densities and higher melting points than metals.

Glasses

Glasses are similar to ceramics. Glasses:
- are generally transparent
- have lower melting points than ceramics.

Plate glass is transparent because it is non-crystalline so there are no reflecting surfaces, such as grain boundaries, to make the material opaque. It is brittle because of the rigid bonds.

Plastics

Plastics consist of a tangled mass of a very long polymer molecules. Plastics:
- are usually strong in relation to their mass
- are usually soft, flexible and not very elastic
- soften easily when heated and melt or burn
- are thermal and electrical insulators.

The structure is described as *amorphous* (shapeless) because the polymer chains take up a random arrangement. The bonds between chains are weak in thermosoftening plastics and strong in thermosetting plastics. Most polymers are poor conductors due to the lack of free electrons.

IT'S A FACT

Polymers can be made to behave like metals. Polymers which conduct electricity are made by including salts in certain polymers. They have up to 10^{11} times the conductivity of typical polymers.

In some polymers there are regions where the chains are packed together in a regular way (see Figure 30.3A). Many polymers are a mixture of ordered regions and amorphous regions (see Figure 30.3B). In high-density poly(ethene), the chains pack closely together to give a material which is stronger than low-density poly(ethene) and is not as easily deformed by heat. High-density poly(ethene) is used to make water tanks and kitchen equipment, e.g. buckets and food containers. Articles made from high-density poly(ethene) can be heated to sterilise them so they are used for hospital equipment.

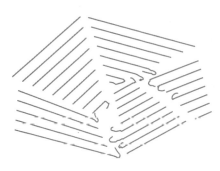

Figure 30.3A ● The arrangement of chains in high-density poly(ethene)

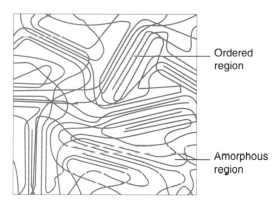

Figure 30.3B ● A polymer with ordered and amorphous regions

Synthetic fibres

Fibres are made by drawing (stretching) plastics. Molten plastic is forced through a fine hole. As the fibre cools and solidifies, it is stretched to align its molecules along the length of the fibre. Fibres therefore have greater tensile strength along the length of the fibre than plastics. Some polymers can be deformed by straightening out the polymer chains. Examples are rubber, polyesters, polyamides and polycarbonates.

Concrete and most types of rock

Concrete and most types of rock are:
● strong in compression (loading)
● weak in tension (stretching).
Reinforced concrete is described below.

Composite materials

Some materials are not homogeneous. They include composite materials, e.g. reinforced concrete, and cellular materials, e.g. coral. Composite materials combine the properties of more than one material. The resulting material is more useful for particular purposes than the individual components.

● Reinforced concrete

Concrete is weak in tension, strong in compression and brittle. It can be reinforced with steel wires or bars. Steel is strong in tension, and the combination is a relatively cheap, tough material. It is essential for the construction of large structures, e.g. high-rise buildings, bridges and oil platforms.

The reason why unreinforced concrete has low tensile strength is that it contains microscopic pores. To reduce porosity and increase tensile strength, one solution is to drive air out of the powder by vibrating it before mixing with water. Another method is to add to the cement–water mixture materials, e.g. sulphur or resin, which will fill the pores. Alternatively, a water-soluble polymer can be added to the cement–water mixture to fill the spaces left between cement particles. Other filler materials, e.g. glass fibre, silicon carbide, aluminium oxide particles or fibres have been used in the polymer–cement mixture.

● Fibre-reinforced plastics

Fibre composites such as epoxy resin and polyester resin contain thermosetting plastics. These are chosen because they are stronger than thermoplastics. However, thermoplastics have the advantage of being easy to mould, and this encouraged chemists to develop fibre-reinforced thermoplastics. These can be used for some applications which were traditionally filled by metals. Examples of fibre-reinforced plastics are:

1. Glass-fibre-reinforced-polyester: The glass fibres are extremely strong, though relatively brittle. The polymer matrix is weaker but relatively flexible. The combination of materials and properties results in a tough, strong material. The composite has many applications, including small boats, skis and motor vehicle bodies.

Figure 30.3C ● Glass-fibre-reinforced polyester (GRP)

2. Carbon-fibre-reinforced epoxy resin: This is used in tennis racquets and in aircraft. It is stronger and lower in density than conventional materials, e.g. aluminium alloys.

● Fibres

The fibres may be glass, carbon, silicon carbide, poly(ethene), kevlar and other substances. The tensile strength of carbon fibres is not high but they are stiff and therefore included in materials used for the frames of aircraft.

● Economics

Composites are expensive. However, a reduction in fuel consumption by lightweight transport vehicles could increase the use of composites.

Figure 30.3D ● Carbon-fibre-reinforced resin in use

Cellular materials

The properties of cellular materials depend on their cellular structure. Examples are:

- cork, which is used as an insulator against heat and sound and as floats
- wood
- aluminium honeycomb structures (see Figure 30.3E)
- paper honeycomb structures, used for packaging and in internal doors.

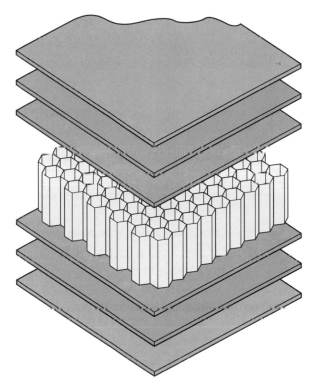

Figure 30.3E ● A honeycomb structure which is part of an aircraft wing. A cellular structure of aluminium is covered by thin strong layers of a metal or a composite. The result is a stiff, lightweight material used in aircraft, skis and door panels

Why do metals bend, polymers bend, stretch or break and ceramics and glasses break?

Metals are strong in tension and in compression because the bonds between atoms are strong. Metals can change shape without breaking because the metallic bond allows layers of atoms to slide over one another when the metal is under stress. Metal crystals contain dislocations which will move under stress and allow the metal to change shape. Metals can withstand smaller loads without changing shape than can glass.

Glass shatters easily. Glass should be a strong material because the bonds between atoms are strong, and it does not contain dislocations. It is able to withstand mechanical loading without breaking. The brittleness of glass is due to surface defects or scratches. Glass breaks easily if a scratch is first made on its surface. Cracks find it harder to grow in metals because of the movement of dislocations.

Glass fibres resist cracking when they are bent because they are free of surface scratches.

SUMMARY

The properties of metals, ceramics, glasses, plastics, fibres and concrete are revised. Composite materials, e.g. reinforced concrete, combine the properties of their components. Cellular materials, e.g. cork and wood, are strong because of their structure. The physical properties of the different materials are related to their structures.

Polymers. If you break a plastic ruler, made of poly(methyl methacrylate), only 10% of the fracturing is due to the breaking of covalent bonds; the rest is due to individual chains being pulled out of the material. In other polymers chain pull-out occurs less because the molecular chains are more densely packed and more entangled. Polymers such as rubber are elastic because the polymer chains can adjust in position and shape relative to one another.

CHECKPOINT

❶ Explain why it is possible to bend a piece of poly(ethene) tubing more easily than a piece of glass tubing.

❷ Explain why a piece of metal is dented by a hammer blow but a piece of pottery shatters.

❸ Explain why a rubber band stretches more than a length of metal wire.

❹ For each of the materials below list the uses to which it can be put. Each material may have more than one use.

Material	Use
1. Metal	A Windows
2. Ceramic	B Ovenware
3. Glass	C Electrical plugs
4. Thermosoftening plastic	D Carrier bags
5. Thermosetting plastic	E Ropes
6. Synthetic fibre	F Bridge
7. Concrete	G Rifle

TOPIC 31 ● ORGANIC CHEMISTRY

31.1  Alcohols

LOOK AT LINKS
For the meaning of
organic,
See Topic 25.1.

FIRST THOUGHTS

Timothy's idea of a good time is to go down to the pub and have a few beers with his friends. Sometimes, however, he wakes up the next day with a throbbing headache and a feeling of nausea. He has a 'hangover'. The substance which has produced these effects is **ethanol**, a liquid which we usually call **alcohol**.

LOOK AT LINKS
The effects of alcohol on behaviour and health are described in *KS: Biology*, Topic 11.3. People who abuse alcohol, that is, drink it in more than moderate quantities, become addicted to alcohol, and their health suffers.

Ethanol

Ethanol is the best-known member of a series of compounds called alcohols. Ethanol is a drug. It is classified as a depressant. This means that it depresses (suppresses) feelings of fear and tension and therefore

Figure 31.1A ● Ethanol and relaxation

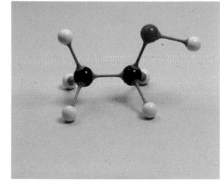

Figure 31.1B ● Ethanol

makes people feel relaxed. The body can absorb ethanol quickly because it is completely soluble in water.

The alcohols

Ethanol is a member of a homologous series of compounds called **alcohols**. Alcohols possess the group

$$
\begin{array}{c}
\mid \\
-\,C\,-\,O\,-\,H \\
\mid
\end{array}
$$

and have the general formula $C_nH_{2n+1}OH$ (see Table 31.1). The formulas are written as CH_3OH, etc. rather than CH_4O to show the —OH group and show that they are alcohols. The members of the series have similar physical properties and chemical reactions. Ethanol is the only alcohol that is not poisonous. Methanol is very toxic: drinking only small amounts of methanol can lead to blindness and death.

Table 31.1 ● The alcohols

Alcohol	Formula, $C_nH_{2n+1}OH$
Methanol	CH_3OH
Ethanol	C_2H_5OH
Propanol	C_3H_7OH
Butanol	C_4H_9OH
Pentanol	$C_5H_{11}OH$

SUMMARY

Alcohols are a homologous series of formula $C_nH_{2n+1}OH$. Ethanol is the only alcohol which is safe to drink in moderate quantities. Regular abuse of alcohol ruins your health. The alcohols have important industrial uses.

Drinking ethanol

Ethanol is the only alcohol that can be safely drunk (in moderation). Methanol, CH_3OH, is very toxic. Drinking only small amounts of methanol can lead to blindness and death. Ethanol dissolves completely in water and is therefore rapidly absorbed through the stomach and intestines. It can take up to 6 hours for the ethanol in a single drink to be absorbed when the stomach is full, but only about 1 hour when the stomach is empty. This is why people feel the effects of a drink faster on an empty stomach than on a full one. After ethanol enters the bloodstream, it diffuses rapidly into the tissues until the concentration of ethanol in the tissues equals that in the blood. This is why measuring the concentration of ethanol in the breath or in urine will indicate the level of ethanol in the blood. As the concentration of ethanol in the blood increases, speech becomes slurred, vision becomes blurred and reaction times increase. This is why it is so dangerous to drive 'under the influence' of alcohol. A driver needs short reaction times. Drinking large amounts of ethanol regularly causes damage to the liver, kidneys, arteries and brain. Many people abuse alcohol; they do not use it properly, that is, in moderation. Such people become addicted to alcohol, and their health suffers.

SUMMARY

Alcohols have the general formula $C_nH_{2n+1}OH$. The group C_nH_{2n+1} is called an alkyl group, e.g. $-CH_3$ the methyl group and $-C_2H_5$, the ethyl group. The group $-OH$ is called the hydroxyl group. This is the group which gives alcohols their reactions: it is the **functional group** of the alcohols.

Uses of alcohols

The use of ethanol is not restricted to drinking. Ethanol is an important solvent. It is used in cosmetics and toiletries, in thinners for lacquers and printing inks. Being volatile (with b.p. 78 °C), the solvent evaporates and leaves the solute behind. Other alcohols also are used as solvents for paints, lacquers, shellacs and industrial detergents. The big advantages of alcohols as solvents is that they are miscible with water and many organic liquids.

CHECKPOINT

❶ (a) Write the structural formulas of (i) methane and methanol (ii) ethane and ethanol (iii) propane and propanol.
  (b) What is the difference in structure between the members of each pair of compounds in (a)?
  (c) What general formula can be written for the members of (i) the alkane series and (ii) the alcohol series?

LOOK AT LINKS
for **carbohydrates**
See Topic 31.4.

Manufacture of ethanol

● **Fermentation**

People have known for centuries how to obtain ethanol from sugars and starches. These substances are carbohydrates. They are compounds of carbon, hydrogen and oxygen, e.g., the sugar glucose, $C_6H_{12}O_6$, and

LOOK AT LINKS
for **fractional distillation**
See Topic 11.6.

LOOK AT LINKS
for **biotechnology**
See *KS: Biology*, Topic 20.

IT'S A FACT

One of the effects of drinking alcohol is to increase reaction times. This is why it is dangerous to drive 'under the influence' of alcohol. A survey of 17–18 year-old drivers involved in road accidents showed that half of them had drunk too much.

SUMMARY

Alcoholic drinks are made from sugars by fermentation. The reaction is catalysed by an enzyme in yeast. Starches can be hydrolysed to sugars and then fermented to give ethanol. Fermentation also takes place in baking when carbon dioxide makes dough rise. Industrial ethanol is made by the hydration of ethene.

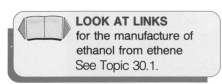

LOOK AT LINKS
for the manufacture of ethanol from ethene
See Topic 30.1.

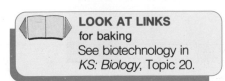

LOOK AT LINKS
for baking
See biotechnology in *KS: Biology*, Topic 20.

starch, $(C_6H_{10}O_5)_n$. Glucose can be converted into ethanol by an enzyme called zymase, which is found in yeast. The conversion of glucose into ethanol is called **fermentation.**

$$\text{Glucose} \xrightarrow{\text{Enzyme in yeast}} \text{Ethanol} + \text{Carbon dioxide}$$
$$C_6H_{12}O_6(aq) \longrightarrow 2C_2H_5OH(aq) + 2CO_2(g)$$

Wine is made by adding yeast to fruit juices, which contain sugars. When the content of ethanol produced by fermentation reaches 14%, it kills the yeast. More concentrated solutions of ethanol (up to 96% ethanol) can be obtained by fractional distillation.

Beer is made from a number of starchy foods, such as potatoes, rice, malt, barley, hops and others. Germinated barley (malt) contains an enzyme which hydrolyses starch to a mixture of sugars. The addition of malt is followed by yeast which ferments the sugars.

$$\text{Starch} + \text{Water} \xrightarrow{\text{Enzyme in malt}} \text{Sugar}$$
$$(C_6H_{10}O_5)_n(aq) + nH_2O(l) \longrightarrow nC_6H_{12}O_6(aq)$$

Ethanol is sold in four main forms.
- Absolute alcohol: 96% ethanol, 4% water.
- Industrial alcohol or methylated spirit: 85% ethanol, 10% water, 5% methanol. The methanol is added to make the liquid unfit to drink.
- Spirits: gin, rum, whisky, brandy, etc. which contain about 35% ethanol.
- Fermented liquors: wines (12–14% ethanol), beers and ciders (3–7% ethanol). These contain flavourings and colouring matter.

Baking, like wine-making and brewing, involves fermentation. In baking, the important product of fermentation is carbon dioxide. In bread-making, flour, water, yeast, salt, sugar and fat are mixed to form a dough. The yeast starts to act on the sugar in the dough. The dough is divided into portions and allowed to stand in a warm container while the carbon dioxide produced makes it 'rise'. Then it is passed on a conveyor belt through a hot oven, from which emerge baked loaves.

Figure 31.1C ● Ethanol in beer, wine, brandy, after-shave lotion

● *Hydration*

Ethanol which is to be used as an industrial solvent is made from ethene by **catalytic hydration.**

$$\text{Ethene} + \text{Steam} \xrightarrow[\text{under pressure}]{\text{Heated catalyst (phosphoric acid)}} \text{Ethanol}$$
$$CH_2=CH_2(g) + H_2O(g) \longrightarrow C_2H_5OH(g)$$

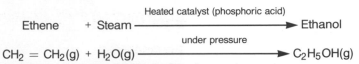

About 10% of the ethene is converted; the unreacted ethene and steam are recycled.

Table 31.2 shows a comparison of the two methods.

Table 31.2 ● A comparison of two methods of manufacturing ethanol

Method	Raw materials	Rate	Quality of product	Type of process
Fermentation	Sugars (a renewable resource)	Slow	Impure ethanol	Batch process
Catalytic hydration	Ethene (from oil, a finite resource)	Faster, but needs heat and pressure	Pure ethanol	Continuous process

Reactions of alcohols

● *Oxidation*

Ethanol is oxidised by air if the right micro-organisms are present. The reason why wine goes sour if it is open to the air is that the ethanol in it is oxidised to ethanoic acid. Vinegar is 3% ethanoic acid.

$$\text{Ethanol} + \text{Oxygen} \xrightarrow{\text{Certain micro-organisms}} \text{Ethanoic acid} + \text{Water}$$
$$C_2H_5OH(aq) + O_2(g) \longrightarrow CH_3CO_2H(aq) + H_2O(l)$$

This reaction is used as an industrial method of making ethanoic acid.

A faster method of oxidising ethanol is to use the powerful oxidising agent, acidified potassium dichromate (VI), $K_2Cr_2O_7$. Dichromate(VI) ions, $Cr_2O_7^{2-}$, which are orange, are reduced by ethanol and other reducing agents to chromium(III) ions, which are blue. As the reaction proceeds, the colour changes from orange through green to blue. This colour change was the basis of the first 'breathalyser' test. A motorist suspected of having too much ethanol in his or her blood had to breathe out through a tube containing some orange potassium dichromate(VI) crystals. If they turned green, or even blue, he or she was 'over the limit' (Figure 31.1D).

Figure 31.1D ● A modern breath analysis

● *Dehydration*

When ethanol is dehydrated, it gives ethene (see Figure 31.1E). Catalysts for the reaction include aluminium oxide, unglazed porcelain, porous pot and pumice stone.

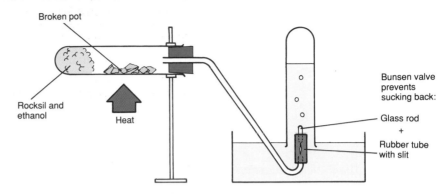

Figure 31.1E ● Making ethene by the dehydration of ethanol

● *Esterification*

Alcohols react with organic acids to form sweet-smelling liquids called **esters** (see Topic 31.3). For example,

LOOK AT LINKS
for **the hydration of
ethene**
See Topic 30.1.

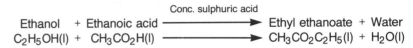

Concentrated sulphuric acid is added to catalyse the **esterification** reaction and to combine with the water that is formed. The structural formula for ethyl ethanoate is:

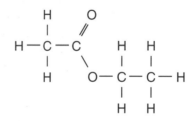

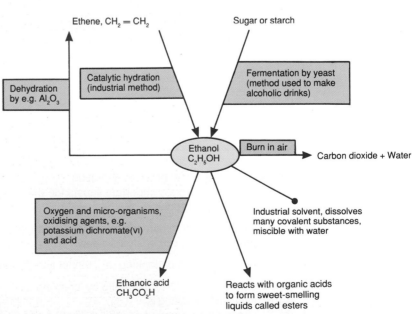

SUMMARY

Ethanol is an important industrial solvent. For this use, ethanol is made from ethene by catalytic hydration. Ethanol is oxidised to ethanoic acid. Alcohols react with organic acids to form esters, e.g. ethyl ethanoate, $CH_3CO_2C_2H_5$.

Figure 31.1F ● Some reactions of ethanol

LOOK AT LINKS
To obtain more agricultural land, Brazil is clearing rain forest. The cost to the environment is discussed in *KS: Biology*, Topic 5.4.

The Ethanol Question
(program)

Use this simulation to study the factors affecting the chemical reactions involved in producing ethanol. Try your hand at producing ethanol yourself, at a competitive selling price.

Gasohol

The cost of petroleum oil has risen since the oil crisis of 1973. Countries which have to import oil want to find alternatives. The major use of petroleum oil fractions is in vehicle engines. Petrol engines are designed to operate over a temperature range at which petrol will vaporise. Any fuel added to petrol must vaporise at the engine temperature and it must dissolve in petrol. Ethanol has the same boiling point as heptane, and it dissolves in petrol. Ethanol burns well in vehicle engines, producing 70% as much heat per litre as petrol does. A petrol engine will take 10% ethanol in the petrol without any adjustments to the carburettor (which controls the ratio of air to fuel in the cylinders). A mixture of petrol and ethanol is described as **gasohol**. The combustion of ethanol produces carbon dioxide and water: there is little atmospheric pollution.

To invest in the production of ethanol by fermentation, a country needs land available for growing suitable crops. It needs plenty of sunshine to ripen the crops quickly and supply sugar or starch for fermentation. Brazil has already started using ethanol as a vehicle fuel. Brazil has very little oil, but plenty of land and sunshine. Most of the petrol sold there contains 10% ethanol. This reduces Brazil's expenditure on oil imports. By the year 2000, Brazil hopes to provide for 75% of its motor fuel needs by using 2% of its land for growing crops for fermentation (Figure 31.1G).

SUMMARY

Ethanol can be added to the petrol in vehicle engines. Brazil is a country with no oil but with plenty of arable land and a sunny climate. Brazil is growing crops which can be fermented to give ethanol. Ethanol is a clean fuel: it burns to form carbon dioxide and water.

Figure 31.1G ● Petrol pumps in Brazil selling 'Alcool' gasohol

CHECKPOINT

❶ (a) Explain what is meant by fermentation.
(b) Name a commercially important substance which is made by this method. Say what it is used for.

❷ (a) Why does wine turn sour when it is left to stand?
(b) Suggest two methods of slowing down the rate at which wine turns sour.

❸ What are the dangers of drinking (a) ethanol and (b) methanol?

❹ Petrol is produced by distillation and by 'cracking'.
(a) Explain what cracking is.
(b) Say where the energy used in (i) distillation and (ii) cracking comes from.
(c) Say where the energy used in fermentation comes from.

⑤ Europe has a surplus of grain. Someone proposes building plants to obtain ethanol from the surplus cereals. Say what advantages this would bring to (a) the environment (b) the farmer (c) the motorist (d) industry (e) the tax payer.
Can you see anything wrong with the idea? Explain your answers.

⑥ Ethanol is made by fermentation and by catalytic hydration.
  (a) Why is it an advantage to make ethanol for solvent use by a continuous process, rather than a batch process?
  (b) Why are alcoholic drinks not made by catalytic hydration?
  (c) Which method is more economical with energy?
  (d) Which method is more economical with the Earth's resources?

31.2 Carboxylic acids

Ethanoic acid has the structural formula:

$$\begin{array}{ccc} H & & O \\ | & & \parallel \\ H-C-C & & \\ | & & \diagdown \\ H & & O-H \end{array}$$

The formula is usually written as CH_3CO_2H. Ethanoic acid is a member of a homologous series called **carboxylic acids** (or alkanoic acids). Other members of the series are
   methanoic acid, HCO_2H
   propanoic acid, $C_2H_5CO_2H$
   general formula, $C_nH_{2n+1}CO_2H$.
The carboxylic acids possess the group $-CO_2H$, which is called the carboxyl group. This is the functional group of the carboxylic acids. The carboxyl group ionises in solution to give hydrogen ions:

Carboxylic acid + Water $\rightleftharpoons$ Hydrogen Ions + Carboxylate Ions
$$RCO_2H(aq) \rightleftharpoons H^+(aq) + RCO_2^-(aq)$$

R is an alkyl group, e.g. CH_3, C_2H_5. Since they are only partially ionised, carboxylic acids are weak acids (see Topic 27.2).
Some of the reactions of ethanoic acid are shown in Figure 31.2A.

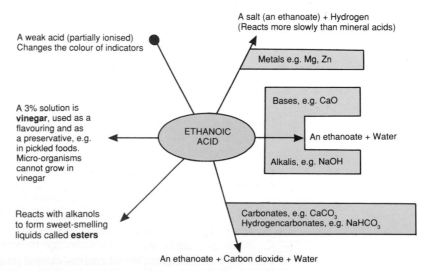

SUMMARY

Ethanoic acid is a carboxylic acid. It is a weak acid. It reacts in the same way as mineral acids but more slowly. Carboxylic acids react with alcohols to form esters.

Figure 31.2A ● Reactions of ethanoic acid

❶ (a) Name three substances which will react with ethanoic acid. Name the products of the reactions.
   (b) Explain why a solution of ethanoic acid is less reactive than a solution of hydrochloric acid of the same concentration.
   (c) Describe an experiment which you could do to demonstrate that ethanoic acid is weaker than hydrochloric acid.

31.3 Esters

Many esters are liquids with fruity smells. They occur naturally in fruits. They are used as food additives to improve the flavour and smell of processed foods. Other esters are used as solvents, for example in glues. People can get 'high' by inhaling esters. Some people enjoy the sensation so much that they become 'glue sniffers'. It is really the solvent, which may be a hydrocarbon or an ester, that they want to inhale. This dangerous habit is called **solvent abuse**. It produces the same symptoms as ethanol abuse. In addition, sniffers who are 'high' on solvents become disoriented. They may believe that they can jump out of windows or off bridges or walk through traffic. Many deaths occur from solvent abuse both through disoriented behaviour and from sniffers passing out and suffocating on their own vomit.

Figure 31.3A ● Esters in thinner, nail varnish remover, UHU glue

Animal fats and vegetable oils are esters. They are liquids or solids, depending on the size and shape of their molecules. They are esters of glycerol, an alcohol which has three hydroxyl groups:

$$CH_2OH$$
$$|$$
$$CHOH$$
$$|$$
$$CH_2OH$$

Each —OH group can esterify with a molecule of a carboxylic acid. The acids hexadecanoic acid, $C_{15}H_{31}CO_2H$, octadecanoic acid, $C_{17}H_{35}CO_2H$ and others combine with glycerol to form fats. These acids are known as 'fatty acids'. (Hexadecane means 16, and octadecane means 18. Count the number of carbon atoms, and you will see how they get their systematic names.)

LOOK AT LINKS
for soaps
See Topic 16.9.

SUMMARY

Esterification is the reaction between carboxylic acids and alcohols to form esters. These compounds are used as food additives and as solvents. Solvent abuse is a dangerous habit. Fats and oils are esters. They can be used for soap-making.

In soap-making, fats and oils are boiled with a concentrated solution of sodium hydroxide. The reaction can be represented by:

$$G\begin{matrix}-A\\-A\\-A\end{matrix} + 3NaOH(aq) \rightarrow G\begin{matrix}-OH\\-OH\\-OH\end{matrix} + 3NaA$$

where $G(OH)_3$ represents glycerol and HA represents the fatty acid. The sodium salt of the fatty acid, e.g. sodium hexadecanoate or sodium octadecanoate, is a soap. This important reaction is called **saponification**.

CHECKPOINT

❶ Explain the terms 'esterification' and 'solvent abuse'.

❷ (a) How are soaps manufactured?
   (b) How do soaps emulsify oil and water?

31.4 Carbohydrates

FIRST THOUGHTS

What do sugar and starch, the cell walls of plants and the exoskeletons of insects have in common? This topic will tell you!

Sugars

The sugar in the sugar bowl is a compound called **sucrose**. There are many other sugars. One of the simplest is **glucose**, $C_6H_{12}O_6$. It is a **carbohydrate**, that is, a compound of carbon, hydrogen and oxygen only, in which the ratio (number of hydrogen atoms/number of oxygen atoms) = 2. A molecule of glucose contains a ring of six atoms, five carbon atoms and one oxygen atom; see Figure 31.4A(a). A shorthand way of writing the formula is shown in Figure 31.4A(b).

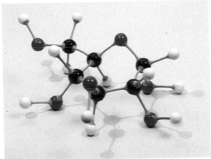

Figure 31.4A ● (a) Model of a glucose molecule   (b) The formula in shorthand form

Fructose, $C_6H_{12}O_6$, is another sugar. It is an isomer (see Topic 24.2) of glucose and, as with glucose, its molecules contain a ring of six atoms. Sugars whose molecules contain one ring structure are called

monosaccharides (mono = one). Monosaccharides such as glucose and fructose are **hexoses**, in which the ring contains six atoms. Others are **pentoses**, in which the ring contains five atoms, e.g. ribose, $C_5H_{10}O_5$.

Sucrose, $C_{12}H_{22}O_{11}$, is a **disaccharide**: its molecules contain two ring structures. It is formed from glucose and fructose:

Glucose   +   Fructose   →   Sucrose   + Water

$C_6H_{12}O_6(aq)$ + $C_6H_{12}O_6(aq)$ → $C_{12}H_{22}O_{11}(aq)$ + $H_2O(l)$

Maltose, $C_{12}H_{22}O_{11}$, is a disaccharide formed from glucose:

Glucose   →   Maltose   + Water

$2C_6H_{12}O_6(aq)$ → $C_{12}H_{22}O_{11}(aq)$ + $H_2O(l)$

The shorthand form of the formula of maltose is shown in Figure 31.4B.

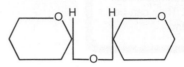

Figure 31.4B ● The formula for maltose in shorthand form

LOOK AT LINKS
for the **reaction of sucrose with concentrated sulphuric acid**
See Topic 29.11.

Carbohydrates have a vital function in all living organisms: they provide energy. When carbohydrates are oxidised, carbon dioxide and water are formed, and energy is released.

Sucrose   + Oxygen   →   Carbon dioxide + Water

$C_6H_{12}O_6(aq)$ + $6O_2(aq)$ → $6CO_2(g)$ + $6H_2O(l)$; Energy is released

LOOK AT LINKS
for **respiration**
See Topic 14.6 and *KS: Biology*, Topic 12.

If you burn sucrose in the laboratory, you will observe a very rapid release of energy. Inside the cells of a plant or animal, however, the oxidation takes place in a controlled manner so that energy is made available to the organism as needed. The process is called cellular respiration. Living organisms also respire fats, oils and proteins.

Starch, glycogen and cellulose

Starch is a food substance which is stored in plant cells. Glycogen is a food substance which is stored in animal cells. Cellulose is the substance of which plant cell walls are composed. All these substances are carbohydrates. They are **polysaccharides**, that is, their molecules contain a large number (several hundred) of sugar rings. These are glucose rings (see Figure 31.4C). Polysaccharides differ in the length and structure of their chains (see Figure 31.4D).

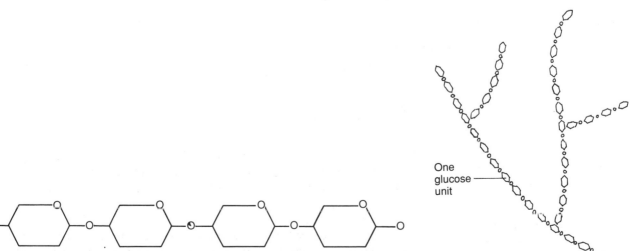

One glucose unit

Figure 31.4C ● Part of a starch molecule

Figure 31.4D ● Structure of a starch molecule

Starch and glycogen are only slightly soluble in water. They can remain in the cells of an organism without dissolving and therefore make good food stores. When energy is needed, cells convert starch and glycogen into glucose, which is soluble, and then oxidise the glucose.

Cellulose is insoluble. The walls of plant cells consist of a tough framework of cellulose fibres (see Figure 31.4E). Another polysaccharide, chitin, forms the exoskeleton that encloses the body of some invertebrates (see Figure 31.4F).

> **SUMMARY**
>
> Carbohydrates contain carbon, hydrogen and oxygen only. Monosaccharides are sugars, including hexoses, with six-atom rings in the molecule, e.g. glucose and fructose, and pentoses, with five-atom rings, e.g. ribose. Disaccharides have two sugar rings in the molecule, e.g. sucrose and maltose. Polysaccharides have molecules with a large number of sugar rings. They include starch and glycogen (food stores), cellulose (in plant cell walls) and chitin (in insect exoskeletons).

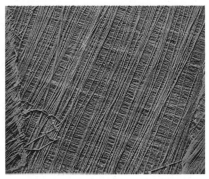

Figure 31.4E ● Plant cell wall (x10 000) showing the framework of cellulose fibres

Figure 31.4F ● An invertebrate exoskeleton of chitin

CHECKPOINT

❶ Runners in the London marathon ate pasta the evening before. Why did they think this would give them energy?

❷ Give examples of carbohydrates which are used as structural materials in plants and animals.

❸ Glucose, maltose and starch are carbohydrates. Answer the following questions about them.
   (a) Which one tastes sweet?
   (b) Which two are very soluble in water?
   (c) List the elements that make up all three.
   (d) The formula of glucose is $C_6H_{12}O_6$. What is the ratio number of atoms of hydrogen: number of atoms of oxygen?
   (e) The formula of maltose is $C_{12}H_{22}O_{11}$. What is the ratio number of atoms of hydrogen: number of atoms of oxygen?
   (f) Using your answers to (d) and (e), try to explain where the name 'carbohydrate' comes from.
   (g) The 'shorthand' formula for glucose is

   Draw the shorthand formulas for maltose and starch.

FIRST THOUGHTS

31.5 Fats and oils

Have you seen brands of margarine and cooking oil described as 'high in polyunsaturates' and wondered what it means? This section will tell you.

Fats and oils are together called **lipids**. They contain carbon, hydrogen and oxygen only. At room temperature, most fats are solid and most oils are liquid. Fats and oils are insoluble in water. Their functions are as follows:

Source of energy: Lipids are important stores of energy in living organisms. The oxidation of fats and oils gives carbon dioxide and water with the release of energy. Lipids provide about twice as much energy per gram as do carbohydrates.

SUMMARY

Fats and oils are together called lipids. As foods, they provide energy and dissolved vitamins. They also provide thermal insulation and protection for delicate organs.

Thermal insulation: Mammals have a layer of fat under the skin. This acts as a thermal insulator.
Protection: Delicate organs, e.g. kidneys, are protected by a layer of fat.
Food: Some vitamins, e.g. vitamins A, D and E, are insoluble in water and soluble in lipids. Foods containing lipids provide these vitamins.
Cell membranes: Lipids are incorporated in cell membranes.

LOOK AT LINKS
for the **importance of fats and oils in the diet,**
See *KS: Biology,* Topic 11.3.

Saturated and unsaturated fats and oils

Do you watch your diet? Many people are concerned about having too much fat in their diet, particularly **saturated fats**. Let us look at what the term 'saturated' means here.

Figure 31.5A ●

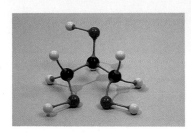
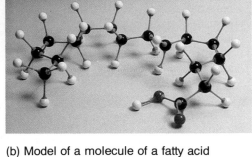

(a) Model of a molecule of glycerol

(b) Model of a molecule of a fatty acid (hexadecanoic acid)

(c) The formula of glycerol

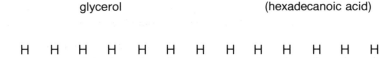

(d) The formula of hexadecanoic acid

LOOK AT LINKS
for **esters**
See Topic 31.3;
for **unsaturated compounds**
See Topic 30.1.

Fats and oils contain esters of glycerol and fatty acids (carboxylic acids with long alkyl groups) (see Topic 31.3). The fatty acid may be saturated or unsaturated. If there is one carbon–carbon double bond in the molecule, the compound is described as **unsaturated**. If there is more than one carbon–carbon double bond, the compound is described as **polyunsaturated**. Below is part of the carbon chain in (a) a saturated fatty acid, (b) an unsaturated fatty acid and (c) a polyunsaturated fatty acid.

(a)
$$-C-C-C-C-C-C-C-C-C-$$

(b)
$$-C-C-C-C-C-C=C-C-C-$$

(c)
$$-C-C=C-C-C-C=C-C-C-$$

SUMMARY

Lipids (fats and oils) are mixtures of esters of glycerol with carboxylic acids (called fatty acids). The carboxylic acid may be saturated or unsaturated or polyunsaturated. Unsaturated fats and oils can be converted into saturated fats and oils by catalytic hydrogenation.

Fats and oils are mixtures of esters. Fats which contain esters of glycerol and saturated fatty acids are called saturated fats. Fats which contain esters of glycerol and unsaturated fatty acids are called unsaturated fats or polyunsaturated fats, depending on the number of double bonds in a molecule of the acid. Animal fats contain a large proportion of saturated esters and are solid. Plant oils contain a large proportion of unsaturated esters and have lower melting points. Many scientists believe that eating a lot of saturated fats increases the risk of heart disease.

Unsaturated oils can be converted into saturated fats by catalytic hydrogenation. The vapour of the oil is passed with hydrogen over a nickel catalyst (see Topic 30.1).

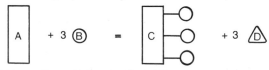
CHECKPOINT

❶ (a) Say which are the fats and which are the oils.
soft margarine, butter, lard, hard margarine, cooking oil
(b) What is the main physical difference between fats and oils?

❷ The manufacturers of *Flower* margarine claim that their product encourages healthy eating because it is 'low in saturated fats'.
(a) What do they mean by 'low in saturated fats'?
(b) Name a product which contains much saturated fat.
(c) What particular benefit to health is claimed for products such as *Flower*?

❸ The following equation represents the formation of a fat.

$$A + 3\,⒝ = C + 3\,△D$$

(a) Name the substances A and D.
(b) Name one substance which could be B.
(c) Name one fat and one oil which could be C.

31.6 Proteins

Proteins are compounds of carbon, hydrogen, oxygen, nitrogen and sometimes sulphur. They are vitally important in living things for the following reasons:

1. Proteins are the compounds from which new tissues are made. When organisms need to grow or to repair damaged tissues, they need proteins.

2. Enzymes are proteins which catalyse reactions in living things.

3. Hormones are proteins which control the activities of plants and animals.

4. Proteins can be used as a source of energy in respiration.

369

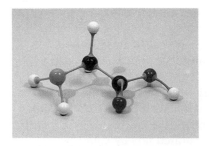

Figure 31.6A ● Model of a glycine molecule

Proteins have large molecules. A protein molecule consists of a large number of amino acid groups. There are about 20 **amino acids**. The simplest is glycine, of formula

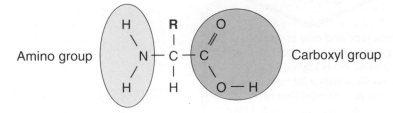

The general formula of an amino acid is

Amino group

Carboxyl group

R is a different group in each amino acid; it can be —H, —CH_3, —CH_2OH, —SH and many other groups. The carboxyl group (—CO_2H) is acidic, and the amino group (—NH_2) is basic (like ammonia, NH_3). The amino group of one amino acid can react with the carboxyl group of another amino acid. When this happens, the two amino acids combine to form a **peptide** and water.

$$H_2NCH_2CO_2H + H_2NCH_2CO_2H \rightarrow H_2NCH_2CONHCH_2CO_2H + H_2O$$

The bond which has formed between the two amino acids, —CONH—, is called the **peptide link**. Its structure is

$$-\overset{O}{\overset{\|}{C}}-\overset{H}{\overset{|}{N}}-$$

The peptide which has been formed has an amino group at one end and a carboxyl group at the other end so it can form more peptide links. We can show this by an equation, using a different shape, e.g. ○ and □, for each different amino acid.

$$□ + ○ \rightarrow □—○ + H_2O$$
$$□—○ + ▲ + ○ \rightarrow ▲—□—○—○ + 2H_2O$$
$$▲—□—○—○ + ▲ + □ \rightarrow ▲—▲—□—○—○—□ + 2H_2O$$

In this way, many amino acids can combine to form a long chain. Each link in the chain is a peptide bond.
Peptides have molecules with up to 15 amino acid groups.
Polypeptides have molecules with 15–100 amino acid groups.
Proteins have still larger molecules.

SUMMARY

Many tissues in living organisms are made of proteins. The molecules of proteins are long chains of amino acid groups. Peptides and polypeptides have smaller molecules.

IT'S A FACT

Fifteen million children a year die of starvation and disease. You will have seen pictures of children with swollen abdomens. They are suffering from a disease called kwashiorkor. They are starved of protein. In parts of the world where rice is the staple diet, kwashiorkor is common.

Diet

An animal can break down the proteins in its food into amino acids and then use the amino acids to build the proteins which it needs. Animals are able to make some amino acids in their bodies. The other amino acids which they need must be supplied in the diet. These are called **essential amino acids**. A protein which contains all the essential amino acids is called a **first class protein**. Meat, fish, cheese and soya beans supply first class protein. A protein which lacks some essential amino acids is called a **second class protein**. Such proteins are in flour, rice and oatmeal.

Structure

A protein molecule is twisted into a three-dimensional shape (see Figures 31.6B and C). It is kept in this shape by bonding between groups in different parts of the chain.

Figure 31.6B ● Insulin. In 1953, Dr Frederick Sanger of the UK reported the complete sequence of amino acids in the protein insulin. This was the first time that a protein structure had been worked out. Insulin controls blood sugar levels. Its molecule is one of the smallest protein molecules.

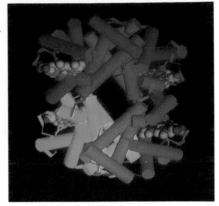

Figure 31.6C ● Haemoglobin. In 1959, Dr Max Perutz worked out the shape of the haemoglobin molecule. Haemoglobin is the red pigment in blood which transports oxygen round the body. The haemoglobin molecule is larger than the insulin molecule.

Dr Max Perutz began work on the structure of haemoglobin in 1936. It took over 20 years to find out that the molecule contains 574 amino acid groups, arranged in four chains. With a total of 10 000 atoms, the molecule has a mass 65 000 times that of a hydrogen atom.

Enzymes

The reactions which take place inside the cells of living organisms are catalysed by **enzymes**. Enzymes are proteins. An enzyme is **specific** for a certain reaction or type of reaction, that is, it will catalyse only that reaction or type of reaction. For example, the enzyme maltase catalyses the reaction of maltose and water to form glucose. The substance which the enzyme helps to react, e.g. maltose, is called the **substrate**. The enzyme amylase catalyses the reaction between starch and water to form sugars. Figure 31.6D shows how the enzyme is thought to interact with its substrate.

LOOK AT LINKS
A substance which increases the speed of a chemical reaction without being used up in the reaction is called a **catalyst**.
See Topic 23.7.

The enzyme has a group of atoms called the **active site**. Part of the substrate molecule fits into the active site. This fit has been described as 'like a lock and key'.

The starch molecule is attacked by a water molecule. Weakened by its bonds to the enzyme, the starch molecule is hydrolysed to form two molecules of sugar.

SUMMARY

Protein molecules are twisted into three-dimensional structures. Enzymes are proteins. They catalyse reactions which take place in living organisms.

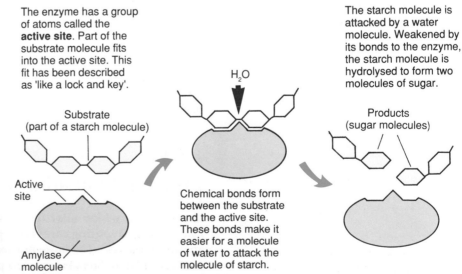

Substrate (part of a starch molecule)

Active site

Amylase molecule

H_2O

Chemical bonds form between the substrate and the active site. These bonds make it easier for a molecule of water to attack the molecule of starch.

Products (sugar molecules)

Figure 31.6D ● How the enzyme amylase catalyses the breakdown of starch. Notice that the enzyme is unchanged at the end of the reaction and able to catalyse the breakdown of more starch

CHECKPOINT

❶ What is the difference between (a) a peptide and a polypeptide (b) a polypeptide and a protein?

❷ (a) To what group of chemical compounds do enzymes belong?
(b) What vital job do enzymes do?
(c) Give the names of three substrates on which enzymes work and the names of the products that are formed.

❸ What compounds are formed when molecules of the following compounds combine? (a) monosaccharides (b) amino acids (c) glycerol and fatty acids?

❹ (a) What type of compound is this?

$$H_2N - \overset{\overset{\displaystyle H}{|}}{\underset{\underset{\displaystyle CH_2 - SH}{|}}{C}} - CO_2H$$

(b) Draw the structure of the compound (other than water) that is formed when two molecules of the above substance react.
(c) What type of bond is formed in the reaction?
(d) What type of compound is the substance in (b)?
(e) What further reaction or reactions can this substance undergo?

31.7 Food additives

FIRST THOUGHTS

We eat some foods exactly as they are harvested, e.g. apples and cucumbers. Most foods, however, go to the food industry to be processed, that is changed in some way. Three quarters of the food eaten in industrialised countries is processed.

Why use additives?

In food processing, many substances are added. They are called **food additives**. A food additive is defined as a substance which is not normally eaten or drunk as a food either by itself or as a typical ingredient of food. Salt and sugar are added to foods but are not called additives. No substance may be used as an additive unless the food manufacturer can give a good reason for its use. The reasons for using additives are:
• to flavour food
• to colour food
• to alter the texture of food
• to preserve food.
Additives are chemicals, but so are proteins, vitamins and all the other substances which occur naturally in foodstuffs. In 1950, 50 additives were in general use; now 3500 additives are used in our food.

Types of food additives

● Additives which alter the taste of food
Flavourings Flavourings are the largest group of food additives, with about 3000 in use. The large number is not so surprising when it takes a mixture of up to 50 substances to produce a natural flavour such as apple or peach.
Sweeteners The commonest sweetener is sucrose (sugar); this is a food, not an additive. Some people want to cut down on sucrose either

because it causes tooth decay or because they are overweight. Diabetics cannot cope with sucrose. Substitutes are saccharin, sorbitol and mannitol.

Flavour enhancers Flavour enhancers are not flavourings; they are substances which make existing flavours seem stronger. The best known is monosodium glutamate, MSG. It stimulates the taste buds.

Figure 31.7A ● They all contain MSG

● Additives which alter the colour of food

When food is processed, it may lose some of its colour; then the manufacturer will want to restore the original colour of the food. Different countries vary widely in the number of colourings allowed. Colourings are not added to baby foods.

● Additives which alter the texture of food

Emulsifiers and stabilisers When oil and water are added, they form two layers. Some substances, called **emulsifiers**, can make oil and water mix. The mixture is called an emulsion. Any substance which helps to prevent the emulsion from separating out again is called a **stabiliser**. Margarine, ice cream and salad dressings all use emulsifiers and stabilisers.

Figure 31.7B ● All these contain emulsifiers and stabilisers

Thickeners You will see that 'modified starch' appears on many labels. It is used to thicken foods. It can be a main ingredient of instant soups and puddings.

Figure 31.7C ● These contain modified starch

Anti-caking agents and humectants Anti-caking agents are substances which can absorb water without becoming wet. They are added to powdery or crystalline foods, such as cake mixes and table salt, to prevent lumps from forming. Humectants keep products moist. They are added to products such as bread and cakes.

Figure 31.7D ● Which products contain an anti-caking agent? Which contain a humectant?

Gelling agents To make jams, desserts etc. set, a gelling agent is added. Pectin is the commonest.

● *Anti-oxidants*

LOOK AT LINKS
for **sulphur dioxide and sulphites as reducing agents**
See Topic 21.7.

Foods which contain fats and oils can turn rancid on exposure to the air. The fats and oils are oxidised to unpleasant-smelling acids. Anti-oxidants are added to prevent oxidation. Two common ones are BHA and BHT (butylated hydroxyanisole and butylated hydroxytoluene). Sulphur dioxide and sulphites are widely used as anti-oxidants. They have two effects: as well as preventing oxidation of fats and oils, they also deprive micro-organisms of the oxygen they need and delay their growth. Sulphur dioxide destroys the vitamin thiamine, and is not used in foods which are important sources of thiamine, e.g. meat. It is, however, used in products such as salami which are processed at temperatures above room temperature and could otherwise become infected by bacteria.

Figure 31.7E ● Preserved by sulphur dioxide and sulphites

● *Preservatives*

Food is stored in the warehouse, in the shop and in the home before it is eaten. Chemical preservatives are added to stop the growth of micro-organisms. Preservatives increase the food's shelf-life, i.e. the length of time the food will keep before it deteriorates. Longer shelf-lives mean less wastage on the shelves. This allows the shopkeeper to charge lower prices and to stock a wider range of foods. Some people are doubtful about whether the customer receives all the benefit of the lower prices. Some people believe that the savings from the widespread use of additives go to the food manufacturers rather than the customers.

Controls on additives

It is illegal to put anything harmful into food. Before an additive may be used, it must be approved by the Government. By law, all additives must be safe, that is, safe for almost everyone. Some people are made ill by some additives, but then some people are made ill by some natural foods.

E Numbers

On the labels of some foods you will see E numbers. E numbers are a device of the European Community (EC). The EC has drawn up a list of 314 safe additives. This is the numbering system:
Colourings: E number begins with 1, e.g. E150, caramel
Preservatives: E number begins with 2, e.g. E221 sodium sulphite
Anti-oxidants: E numbers 310–321, e.g. E320, butylated hydroxyanisole, BHA
Texture controllers: E numbers 322–494, e.g. E461, methylcellulose.
Additives which have been passed in the country of origin but not yet in the EC have a number only, e.g. 107 yellow 2G; 524 sodium hydroxide; 925 chlorine.

Are additives good for you?

Some foods make some people ill. When a person reacts to a food by becoming ill, the reaction is called an **intolerance reaction** or an **allergic reaction**. Allergic reactions can take the form of asthma (breathing difficulty), eczema (a skin complaint), digestive troubles, rhinitis (like hay fever), headaches, migraines and hyperactivity. Putting hyperactive children on a diet free from additives often produces a dramatic improvement. Tartrazine (E102), a yellow dye, is the one that is most

> **IT'S A FACT**
>
> We eat between 3 and 7 kg of additives a year. Some people think that over a long period some additives could be a threat to health. Medical doctors believe that additives make little, if any, contribution to serious illness.

under suspicion. It is used in sweets, fizzy drinks and packet desserts. If you know that you are allergic to tartrazine, you can read the labels on the foods you fancy and reject any which list E102. Many people prefer to buy additive-free foods. Many of the large supermarket chains are reducing the number of additives in their products, and some firms are offering additive-free items.

CHECKPOINT

❶ (a) Select a number of food items from the list. Visit your local supermarket, and find the number of additives in each item you have chosen.
List of food items: Soft margarine, Orangeade, Vanilla ice cream, Sweet pickles, Tomato soup, Salmon paste, Fish fingers, Irish stew, Canned garden peas, Cheese snacks, Pork pies, Strawberry jam, Caramel dessert

(b) Compare your survey with surveys done by other members of your class. Make a list of the items and the number of additives which each contains.
How many have no additives?
What percentage of the foods contain additives?
What is the food with the maximum number of additives?

❷ Which types of additives are present in this Apricot Pie?

APRICOT PIE
Ingredients: Flour, Sugar, Apricot, Animal and Vegetable fat, Dextrose, Modified starch, Sorbitol syrup, Glucose syrup, Salt, Citric acid, Preservative E202, Flavouring, Emulsifiers E465, E471, E475, Whey powder, Colours E102, E110

❸ A ham manufacturer wants to increase the weight of his hams by injecting water into them. He needs an additive to keep the water and other ingredients well mixed. What type of additive does he use? Who benefits from the use of this additive, the manufacturer or the consumer?

❹ The labels on two soft drink bottles are shown below.

Coca-cola
Soft drink with vegetable extracts

Ingredients: Carbonated water, Sugar, Colour (caramel), Phosphoric acid, Flavourings, Caffeine

**BEST SERVED
– ICE COLD**

A

diet Coke
Low calorie soft drink with vegetable extracts

Ingredients: Carbonated water, Colour (caramel), Artificial sweetener (Aspartame), Phosphoric acid, Flavourings, Citric acid, Preservative (E211) Contains phenylamine

**JUST FOR THE TASTE OF IT
– ICE COLD!**

B

(a) Which sweetener is present (i) in **A** (ii) in **B**?
(b) Which preservative is present in both **A** and **B**?
(c) Why is a preservative needed?
(d) Why do the labels recommend serving the drinks ice-cold?
(e) What is the difference between **A** and **B** that enables one to describe 'diet Coke' as 'low calorie'?
(f) What pH would you expect the two drinks to have?

31.8 Chemistry helps medicine

Chemotherapy is the treatment of illness by chemical means. Chemicals can cure diseases and relieve pain. Advances in chemotherapy have made life safer, longer and more free from pain than it was a century ago. Some of the chemicals which are used in medicine are mentioned in this section.

IT'S A FACT

Many of the victims of AIDS are drug addicts who used infected needles to inject themselves with heroin. Never be tempted to give drugs a try. The dangers of addiction and AIDS are very real. It's not worth the risk to satisfy your curiosity.

LOOK AT LINKS
for **adrenalin**
See *KS: Biology*, Topic 16.5.

Laughing gas was discovered by Sir Humphrey Davy (see Theme C, Topic 13.3). At first it was used only as a curiosity at parties. One of these parties was attended by a dentist, Horace Morton. He decided to try laughing gas as an anaesthetic on his patients. Morton obtained better results by using ether. The use of chloroform spread rapidly after Queen Victoria took it during the birth of one of her children in 1853.

Painkillers

Aspirin is the most popular pain-killer. It was first sold in 1899. Each person in the UK swallows an average of 200 aspirins a year. Many other medicines, e.g. APC, contain aspirin. When you swallow aspirin, there is slight bleeding of the stomach wall. To reduce irritation of the stomach wall, you should always swallow plenty of water with aspirin. **Codeine** is a stronger pain-killer, used in headache tablets and in cough medicines.

Morphine is a substance with an amazing ability to relieve intense pain. The problem with morphine is that it is **addictive**. It is only used in cases of dire necessity, for example when soldiers are wounded in battle. Morphine can only be obtained by doctors. **Heroin** is a pain-reliever which is even more potent and more addictive than morphine. Heroin is not used in medicine because it is so addictive. Drug-dealers like to get their customers to try heroin because they will soon become 'hooked' and come back to buy again and again.

Tranquillisers and sedatives

One person in twenty takes **tranquillisers** every day. Tranquillisers are substances which relieve tension and anxiety. With so many people taking tranquillisers over long periods of time, some doctors are worried about long-term effects. Research workers are now investigating whether there is any danger in taking tranquillisers for a long time.

Many people take **sedatives** (sleeping tablets). Drugs called **barbiturates** are used for this purpose. They are habit-forming. People who rely on barbiturates sometimes kill themselves accidentally by taking an overdose. These drugs are also used in the treatment of high blood pressure and mental illness.

Stimulants

Adrenalin is a substance which the body produces when it needs to prepare for action: for 'fight or flight'. It is a **stimulant**: it makes the heart beat faster and makes a person ready for strenuous action.

Amphetamines (pep pills) have similar effects on the body. Some people take amphetamines to keep themselves awake; other people want to pep themselves up and make themselves more entertaining. People can become addicted to amphetamines. A person taking amphetamines becomes excitable and talkative, with trembling hands and enlarged pupils.

Anaesthetics

Anaesthetics made modern surgery possible. Before the days of anaesthetics, a surgeon asked his patient to drink some brandy or smoke some opium to deaden the pain. Then the surgeon did the operation as quickly as he could. Sometimes patients died from pain and shock. After anaesthetics came into use, surgeons were able to take time to do the best operation they could, rather than the fastest. They were able to explore new techniques.

The first anaesthetics to be used were chloroform (1846), ether (1847)

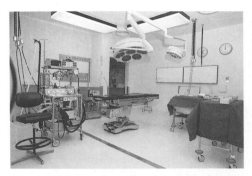

Figure 31.8A ● A modern operating theatre

Who's behind ⁖ the science

The power of sulphonamide drugs was discovered by Gerhardt Domagh. The first human being on whom he tested the drug was his own daughter. She was dying of 'child-bed fever', a bacterial infection which can attack women who have just had a baby. Happily, the drug worked. Domagh received the Nobel prize in 1939.

Who's behind ⁖ the science

The old treatment for syphilis was to apply a paste of heavy metal salts (salts of arsenic, lead, mercury and antimony) to the skin. The metal salts killed the micro-organism that caused the disease, but many patients experienced metal poisoning. Paul Ehrlich reasoned that, if he could attach one of these metals to a dye which was able to stain the micro-organism, the metal–dye compound might target the micro-organism 'like a magic bullet' without attacking body tissues. Ehrlich and his team of research workers set to work. At the 606th attempt they were successful. They called their new compound salvarsan. Ehrlich received the Nobel prize in 1908.

and dinitrogen oxide (laughing gas). There were some drawbacks to using these gases. In large doses, chloroform is harmful. Ether is very flammable, and it has been the cause of fires in hospitals in the past. Dinitrogen oxide (laughing gas) does not produce a very deep anaesthesia; however it is suitable for use in dentistry. Research workers have found a better anaesthetic, **fluothane**, which was first used in 1956. It has been so successful and free from side-effects that most operations are now done under fluothane.

Antiseptics

Another advance in surgery came with antiseptics. A century ago, many patients survived operations but died later in the wards. A surgeon called James Lister realised that the patients' wounds were becoming infected. He sprayed the operating theatre and the wards with a mist of phenol and water. His experiment produced immediate results: the death rate fell. Phenol had killed micro-organisms that would otherwise have infected surgical wounds. Phenol is an **antiseptic**. It is unpleasant to use. Solid phenol will burn the skin, and its vapour is toxic. Research chemists made other compounds which work as well as phenol and are safer to use. TCP® and Dettol® are antiseptics which contain trichlorophenol.

Antibiotics

Infectious diseases, such as tuberculosis and pneumonia, used to kill thousands of people every year. The grim picture changed in 1935 with the discovery of the **sulphonamides**. They are antibiotics: substances which fight disease carried by bacteria. Infectious diseases are no longer a serious threat. In surgery too, powerful antibiotics have made having an operation safer than it was at the beginning of this century.

Alexander Fleming was a research bacteriologist in a London hospital. When the First World War began, he joined the Medical Corps. He was distressed by what he saw in his field hospital. Soldiers who did not seem to be mortally wounded when they arrived in the hospital died later from infections. Bacteria in mud and dirty clothing had infected their wounds, and gangrene set in. Watching men die a slow, painful death, Fleming wished that his work as a bacteriologist would enable him to discover a substance that would kill bacteria: a **bactericide**. After the war, Fleming went back to his research. One day, he found a mould called *Penicillium* on a dish of bacteria which he was culturing. To his amazement, he saw that the mould had killed bacteria. From the mould, he prepared an extract which he called **penicillin**. Would this be the powerful bactericide he had been hoping for? Sadly, although penicillin worked in the laboratory, it did not work in patients. Substances in the blood made the bactericide inactive.

In 1940, work on penicillin recommenced. The Second World War had started, and a powerful bactericide was needed urgently. Two chemists, Howard Florey and Ernst Chain, succeeded in making a stable extract of penicillin. They tested it, first on mice and then on human patients. The tests were successful, and the USA built a plant for the mass-production of penicillin. In 1942, penicillin was used in hospitals on the battle field. The results were spectacular. No longer did soldiers die from minor wounds. In 1944, Fleming was knighted. Later, Fleming, Florey and Chain shared the Nobel Prize for medicine.

Penicillin has been widely used to treat a variety of infections. One disadvantage is that penicillin cannot be taken by mouth. It is broken down by acids, in this case the hydrochloric acid in the stomach. A more recent antibiotic, **tetracycline**, does not share this drawback. Tetracycline is a 'broad spectrum' antibiotic which can be used against many kinds of bacteria.

CHECKPOINT

❶ Your Uncle Bert is always swallowing aspirins. Explain to him (a) why he should not take too many aspirins and (b) why he should take water with aspirins.

❷ (a) A singer you know feels so tired that she is thinking of taking pep pills before giving a performance. Explain to her why this is not a good idea in the long run.
(b) What is the body's natural stimulant? How does it work?

❸ Briefly explain how surgery has changed as a result of (a) the discovery of anaesthetics and (b) the discovery of antiseptics.

❹ Morphine is a very powerful pain reliever. Why do doctors prescribe it for so few patients? For what types of patient is morphine prescribed?

❺ Why do doctors never prescribe heroin as a pain-killer?

❻ Someone you know takes barbiturates to help her to get to sleep. Suggest to her what else she could do, instead of taking pills, to get to sleep.

❼ (a) Before the time of James Lister, surgeons did not change their operating gowns between patients. What was wrong with this practice?
(b) What did James Lister do to improve surgery?

❽ (a) What did Sir Alexander Fleming do to merit a Nobel prize for medicine?
(b) He had a piece of luck in his research, but his success was not due to luck. What else went into his discovery?

31.9 The chemical industry

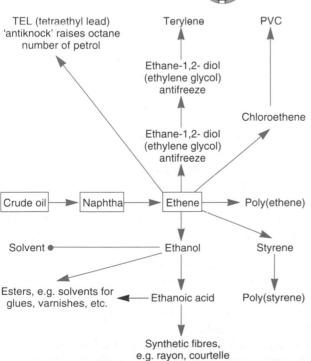

Figure 31.9A ● Some petrochemicals

The chemical industry can be divided roughly into ten sections.
1. **Heavy chemical industry:** oils, fuels, etc. (see Topic 25).
2. **Agriculture:** fertilisers, pesticides, etc. (see Topic 29).
3. **Plastics:** poly(ethene), poly(styrene), PVC, etc. (see Topic 30.2).
4. **Dyes**
5. **Fibres:** nylon, rayon, courtelle, etc. (see Topic 30.2).
6. **Paints, varnishes,** etc.
7. **Pharmaceuticals:** medicines, drugs, cosmetics, etc. (see Topic 31.8).
8. **Metals:** iron, aluminium, alloys, etc. (see Topic 19).
9. **Explosives:** dynamite, TNT, etc. (see Topic 29.3).
10. **Chemicals from salt:** sodium hydroxide, chlorine, hydrogen, hydrochloric acid, etc. (see Topic 9.6 and Topic 13.1).

Figure 31.9A shows some of the petrochemicals which are obtained by the route

Petroleum oil → Naphtha →
Ethene → Petrochemical

Make a Million
(program)

Could you make a financial success in industry? Use the program to find out. The industrial processes you can choose from include:
*   manufacturing acid,
*   electrolysis,
*   fractionating,
*   generating electricity.

Is the chemical industry the place for you? Have you a legal brain or a financial brain? Are you good at relating to other people? Have you a flare for advertising and marketing? Are you an engineer? Are you a scientist who wants to spend his or her time in the laboratory inventing new products? Do you want to be part of the workforce that runs the plant and is responsible for maintaining a high-quality product? These questions span a huge variety of people. It takes a different type of ability to work at finance from the ability to do scientific research. If you add up the numbers of people who can answer 'Yes' to just one of these questions, the sum total is the number of people employed by the chemical industry. Figure 31.9B shows how they work together.

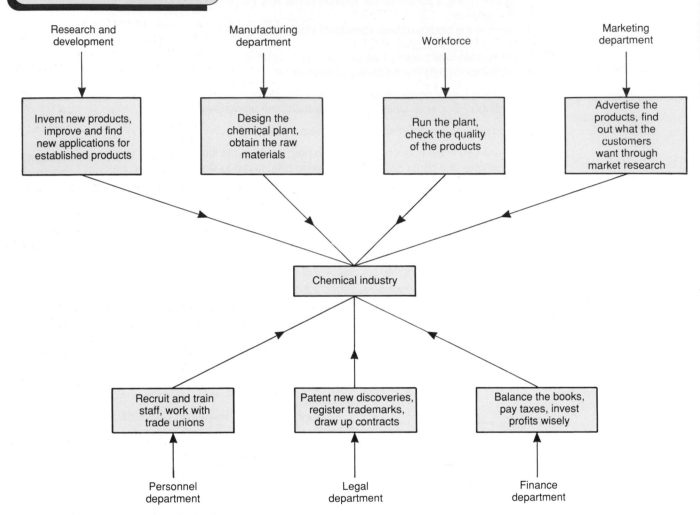

Figure 31.9B ● How different parts of the chemical industry work together

? THEME QUESTIONS

1 For each material, give the letter which describes its properties.

Material	Properties
1 Aluminium	A High tensile strength, brittle
2 Iron	B Tough, low density, electrical insulator
3 Concrete	C Good electrical conductor, fairly low density
4 Glass	D High tensile strength, corrodes easily
5 Plastic	E Hard, brittle, very high melting point
6 Ceramic	G High compressive strength, easily made on site

2 The main raw materials used by ICI at their fertilizer plant at Billingham are natural gas, water, air and sulphur, which is obtained in bulk from the USA. To make ammonia, nitrogen and hydrogen are combined together at high temperature and pressure in the presence of the catalysts platinum and rhodium.

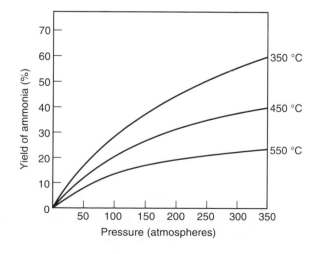

(a) Suggest **four** reasons for the expansion of the chemical industry on Teesside.

(b) The diagram below shows the yields of ammonia at various temperatures and pressures, using the Haber process. The reaction is reversible.

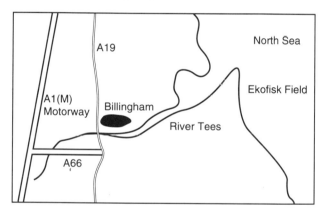

(i) What is the yield of ammonia at a temperature of 450 °C and 200 atmospheres pressure?
(ii) Find the best conditions on the graph for producing the largest yield of ammonia.
(iii) The plant actually operates at 350 °C and 150 atmospheres pressure. Compare these conditions with your answer to (ii) and suggest a reason for any difference.

(c) (i) Give **two** reasons why some ammonium compounds are suitable as fertilisers.
(ii) Discuss the advantages and disadvantages of applying fertilisers as liquids or solids.

(d) Ammonium nitrate (NH_4NO_3) is a basic ingredient in compound fertilisers. Calculate the percentage of nitrogen in the compound (H = 1, N = 14, O = 16).

(e) (i) State **two** ways in which catalysts such as platinum and rhodium are different from biological catalysts such as pepsin.
(ii) Pepsin acts on proteins to produce peptones and proteoses. They are later broken down to amino acids in the body. What happens to these amino acids?

(f) Discuss the statement "Plants can survive without animals but animals could not survive without plants". (LEAG)

3 Concentrated sulphuric acid has some properties which are not shared by dilute sulphuric acid. It is a dehydrating agent and an oxidising agent.
(a) Give two examples of the action of concentrated sulphuric acid as a dehydrating agent.
(b) Give one example of concentrated sulphuric acid acting as an oxidising agent.
(c) Give two examples of the reactions of dilute sulphuric acid.
(d) Name three substances which are attacked by the sulphuric acid in acid rain.

4 Calcium carbonate is quarried as limestone. It is heated in lime kilns to form calcium oxide (quicklime) and a second product.
(a) What is the second product?
(b) State two uses of this product.
(c) Explain why quicklime is spread by some farmers on their fields.
(d) Explain the advantage of using powdered limestone in place of quicklime.
(e) Most of the limestone which is quarried is used in the manufacture of concrete. Briefly explain the process involved.
(f) What are the properties of concrete that make it a useful building material?
(g) When all the limestone has been extracted from a quarry, what should be done to the quarry?

5   The table shows the average prices paid to UK farmers for some products. All prices are shown in 1980 money values. (From *CAS Report 9*, 1985)

Crop	1900	1920	Year 1940	1960	1980
Wheat (£/tonne)	155	161	179	136	99
Barley (£/tonne)	172	241	284	141	93
Potatoes (£/tonne)	123	105	78	66	51
Milk (p/litre)	—	23	23	17	13

(a)   On graph paper, plot the price of each product (on the vertical axis) against the year (on the horizontal axis).

(b)   In which year was barley (i) most expensive (ii) cheapest?

(c)   In which year were potatoes (i) most expensive (ii) cheapest?

(d)   By what percentage did the price of wheat decrease between (i) 1900 and 1980 (ii) 1940 and 1980?

(e)   By what percentage did the price of potatoes drop between 1940 and 1980?

(f)   What was the percentage fall in the price of milk between 1940 and 1980?

(g)   Explain why farm produce is cheaper now than it was in 1940.

(h)   Name other people, besides the general public, who have benefited from the changes in agriculture since 1940.

6   (a)   What type of plants can use atmospheric nitrogen as a nutrient?

(b)   Why can't most plants use atmospheric nitrogen in this way?

(c)   Name the nitrogen-containing compounds that are built by plants.

(d)   Name one natural process that produces nitrogen compounds that enter the soil and are used by plants.

(e)   Explain why there is not enough nitrogen from natural sources to support the growth of crop after crop on the same land.

7   (a)   Explain why ammonium salts are used as fertilisers even though plants cannot absorb them (see Topic 14.5).

(b)   Describe how you could make ammonium sulphate crystals (see Topic 13.5).

8   (a)   Explain what is meant by the **nitrogen cycle**.

(b)   Are we likely to run out of nitrogen?

(c)   Say how the human race alters the natural nitrogen cycle (i) by taking nitrogen out of the cycle and (ii) by adding nitrogen.

(d)   Explain how fertilisers have created a problem concerning nitrogen compounds.

(e)   Suggest two ways in which this problem could be attacked.

9   Farmer Short had a field in which his crops did not grow well. One year, he added fertiliser to the soil and his crops grew better. However, a pond near the field became stagnant and full of algae, and the fish in the pond died. When the pond was tested, it was found to contain fertiliser.

Farmer Long did not use fertiliser. She rotated her crops between fields so that every few years each field grew peas, beans or clover.

(a)   Explain how fertiliser got into the pond.

(b)   Why did the algae in the pond increase?

(c)   Why did the fish die?

(d)   How do crops of peas, beans and clover make up for the lack of fertiliser?

(e)   Give one advantage and one disadvantage of Farmer Long's system compared with Farmer Short's.

10   Explain each of the following statements:

(a)   Prolonged use of artificial fertilisers is bad for the soil.

(b)   Extensive use of fertilisers on arable land can lead to the pollution of waterways.

(c)   Growing clover improves the fertility of the soil.

11   The diagram shows the nitrogen cycle. The labelled arrows represent different processes.

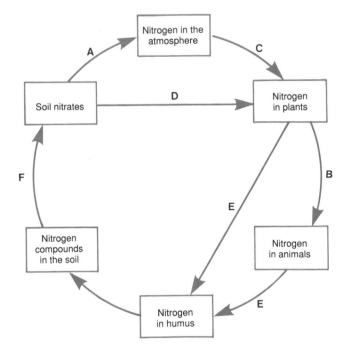

(a)   Name the processes A–F, e.g. A = denitrification.

(b)   (i) Why is process C useful to plants?
        (ii) Process C takes place in clover but not in grass. Why is this so?

(c)   What organisms are responsible for process E?

(d)   In what form is nitrogen passed in process B?

12   Many pesticides are potentially harmful because they are not biodegradable and can build up two lethal limits in animals. The food chain shows the build up of a particular pesticide.

Aphid → Lacewing larva → Blue tit → Buzzard
  5   500   5000   2500

(Pesticide levels shown in relative amounts)

(a)   How did the pesticide enter the food chain in the first place?

(b) Which animal in the food chain is likely to receive a lethal amount first?

(c) The most widespread insecticide use in Britain used to be DDT. Why do you think that its use has been banned?

(d) One of the sub-lethal effects of DDT was that it affected the shell-producing gland in many birds of prey such as the peregrine falcon. It resulted in eggs being laid with thinner than normal shells. How did this lead to the death of many young?

(e) An alternative method of keeping down pest numbers is biological control. Explain how it works.

13 The flow chart below shows the main stages in the production of sugar from sugar cane plants.

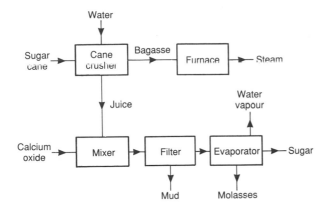

(a) (i) Suggest why water is added to the cane-crusher.
(ii) What do you think 'bagasse' is?
(iii) Suggest **two** reasons why the bagasse is used as a fuel for the furnace.
(iv) Explain the reason for the filter.
(v) Explain what happens inside the evaporator.

(b) Sugar cane plants are grown in Brazil to make sugar and to make fuel for cars. To make fuel the sugar is fermented to make ethanol. Fermentation involves a micro-organism like yeast respiring in an oxygen-free atmosphere.
(i) Apart from ethanol name **one** other substance produced in this reaction.
(ii) Suggest **one** effect this reaction would have on its environment.
(iii) Give **one** way in which this reaction is different from aerobic respiration.

(c) The ethanol produced is used in cars and other road vehicles instead of petrol. It contains fewer impurities than petrol and burns more easily. This means that it has several advantages over petrol. Suggest **two** of these advantages and explain each one.

(d) (i) In Britain research is being carried out on cars powered by electricity. Suggest **one** reason why, apart from the cost, Brazil has **not** spent money on this research. Explain your answer.
(ii) Draw up a table to show **three** advantages and **three** disadvantages of electric-powered vehicles compared with petrol-powered vehicles.

(SEG)

14 Match up the type of glass with the use.

Type of glass
A Heat-resistant glass, which can be heated or cooled quickly without cracking
B Light-sensitive glass, 'Reactolite', which darkens in bright light
C Slow-dissolving glass which slowly dissolves in water
D Float glass (sheets of glass)
E Glass fibres (thin strands)

Use
1 Windowpanes
2 Oven dishes and laboratory glassware
3 Glass-reinforced plastic, GRP
4 Spectacle lenses
5 Pellets of glass containing a substance which kills the snails which spread the disease Bilharzia and which live in canals

15 (a) What is meant by 'recycling' glass?
(b) The value of used glass from a Bottle Bank just pays for the cost of its collection. Is it worth the trouble of collecting glass?
(c) Why do Bottle Banks collect brown, green and colourless glass separately?
(d) Why are people asked to remove metal bottle tops before placing bottles in the bottle banks?
(e) In which section of the Periodic Table would you be likely to find metal oxides to colour glass blue, green, brown and other colours?

16 This flow diagram shows the Haber process for making ammonia gas.

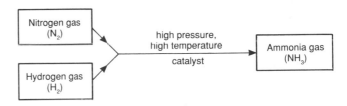

(a) A catalyst is used because it makes the process cheaper.
(i) What effect does a catalyst have on a chemical reaction?
(ii) Explain how using a catalyst can make the process cheaper.
(b) Write a balanced chemical equation for the reaction which takes place during the Haber process.
(c) Most of the ammonia gas from the Haber process is made into fertilisers such as ammonium sulphate. Ammonia gas can be compressed to form liquid ammonia. In some countries this liquid is injected into the soil as a fertiliser.
(i) Give **one** reason why fertilisers are added to soil.
(ii) Suggest which acid is used to make ammonium sulphate.
(iii) Suggest and explain **two** reasons for injecting liquid ammonia *into* the soil rather than spraying it *onto* the surface.
(iv) The world use for fertilisers has been increasing for several years. Suggest an explanation for this increase.

(SEG)

17 The USA produces 8 million tonnes of sulphuric acid a year as a by-product in the smelting of sulphide ores.

(a) Calcium oxide is used to neutralise the acid by the reaction:

Calcium oxide + Sulphuric acid → Calcium sulphate + Water

Write the chemical equation for the reaction. Calculate the mass of calcium oxide needed to neutralise 8 million tonnes of sulphuric acid.

(b) Calculate the mass of calcium carbonate (limestone) that must be heated in a lime kiln to make the calcium oxide used in (a).

18 Borosilicate glass has very good resistance to heat and to sudden temperature changes. It does not react with many chemicals and is strong. Identical pieces of glass were placed in different boiling liquids for eight hours. The graphs in the figure show how the mass of each piece of glass changed.

(a) (i) Draw a labelled diagram to show the laboratory apparatus you could use to boil the acid.

(ii) State **two** safety precautions you would need to take when using this apparatus.

(b) The change in mass is given in milligrams. What is the symbol for milligrams?

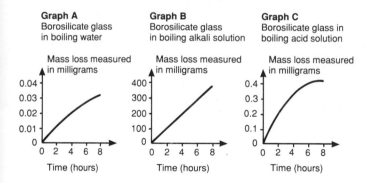

Graph A
Borosilicate glass in boiling water

Graph B
Borosilicate glass in boiling alkali solution

Graph C
Borosilicate glass in boiling acid solution

(c) (i) What pattern connects the mass change of the glass pieces in each liquid?

(ii) Describe **two** ways in which the mass loss shown in graph B is different from that shown in graph A and graph C.

(d) (i) The test using acid was carried out at 105 degrees Celsius (°C). Using your understanding of particle behaviour, explain why the rate of mass loss is greater at this temperature than at room temperature.

(ii) Suggest **one** other method of increasing the rate of mass loss.

(SEG)

19 (a) The diagram at the top of the next column shows some of the uses of limestone.

(i) Give one use of limestone in the building industry.

(ii) Give one use of limestone in the farming industry.

(iii) Explain how limestone can be turned into concrete.

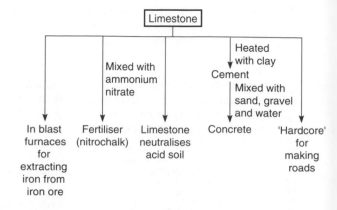

(b) This diagram shows a blast furnace.

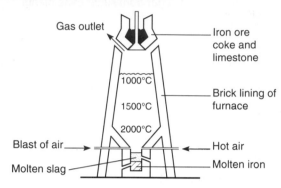

(i) Iron ore and limestone are used in making iron. Name **two** other substances which are needed.

(ii) Suggest why molten slag floats on top of the molten iron.

(iii) The blast of hot air is at a temperature of a few hundred degrees Celsius. Why is the temperature in the blast furnace 2000 °C?

(iv) The gas escaping from the top of the blast furnace contains nitrogen. Explain why.

(v) Brick is a poor conductor of heat. Why does this make it a suitable material for the lining of the blast furnace?

(vi) Why could limestone not be used for the lining of the blast furnace?

(c) A hundred years ago there were three blast furnaces in the village of Glaisdale. The site is now named "Green Acres" but older people in the village still call it "Slag Yard". Some of the statements below could be used to explain why this site is now quite fertile.

A. Microbes which produce nitrates from organic material in the soil require oxygen.

B. Some types of grass (e.g. Rye Grass and Bent) grow best on acid soil.

C. Slag is alkaline.

D. Worms pull leaves and other organic material into the soil.

E. Nitrates are needed for plant growth.

F. The site is surrounded on three sides by the River Esk.

G. Worms prefer an alkaline soil.

H. Worms let air into the soil.

Which **two** of these statements could **not** be used to explain why the site is now quite fertile?

(LEAG)

Numerical Answers

CHECKPOINT 1.2
❷ 2 mm

CHECKPOINT 1.4
❶ (a) Philippine or Caribbean   (b) Andes
(c) Basaltic   (d) Sliding
❷ (a) A = Continental plate, B = Oceanic
trench, C = Subduction zone   D = Sedimentary
(c) (i) W to E   (ii) E to W   (iii) W to E
(d) Mantle, sufficiently fluid to transport
plates   (e) Sea-floor spreading
❸ (a) 4 m   (b) 1.5 m–1.9 m   (c) a > 2b
❻ (a) (i) 44 million years   (ii) 7 million years
(iii) 1 million years

CHECKPOINT 2.5
❶ (a) D   (b) C   (c) F

CHECKPOINT 3.3
❶ 2.7 g/cm³
❷ 7500 g (7.5 kg)
❸ 2720 g (2.72 kg)
❹ A 0.86 g/cm³: floats   B 4.3 g/cm³: sinks

CHECKPOINT 3.6
❽ (a) (i) 55 g   (ii) 100 g   (b) 45 g crystallise
(c) 200 g   (d) 300 g   (e) (i) 3 g   (ii) 25 g

CHECKPOINT 4.4
❸ (a) E   (b) C   (c) D   (d) A, B

THEME B QUESTIONS
6 (b) 48 g/100 g   (c) 50 °C   (d) 55 g KNO₃
crystallise

CHECKPOINT 7.2
❷ 9p, 9e, 10n
❸ (a) 17, 35   (b) 25, 59   (c) 50, 119

CHECKPOINT 7.4
❸ N 7, 7, 7;   Na 11, 11, 12;   K 19, 20, 19;
U 92, 143, 92

CHECKPOINT 8.1
❷ (a) 29.5 c.p.m.   (b) 355.7 c.p.m.
(c) β-radiation
❸ (a) 92p, 146n   (b) 91p, 143n
(c) 89p, 138n
❹ (a) ²³⁴₉₀Th; ²³⁸₉₂U → ⁴₂He + ²³⁴₉₀Th
(b) ²³⁴₉₂U; ²³⁴₉₁Pa → ⁰₋₁e + ²³⁴₉₂U

CHECKPOINT 8.2
❶ (a) 4.0 mg   (b) 2.0 mg   (c) 0.031 mg
❷ (a) 5 years   (b) 25 years
❸ 19.5 hours
❹ (c) 1.55 hours

CHECKPOINT 8.3
❶ (b) 4000 years
❷ β, strontium
❸ Sodium-24
❹ Use cobalt-60

CHECKPOINT 8.9
❶ (a) 1860 units   (b) 521 units   (c) medical
(d) 2381 units   (i) 0.125%   (ii) 0.420%

THEME C QUESTIONS
13 11p, 12n
14 82p, 124n, No
17 (a) 375 c.p.m.   (b) 6 c.p.m.
18 Sodium-24, γ-emitter with suitable half life
20 (c) 12 minutes

CHECKPOINT 13.3
❸ (a) 4   (b) 9   (c) 3   (d) 11   (e) 18
❹ (a) C   (b) A   (c) B

THEME D QUESTIONS
5 (b) 45%
13 (a) blue   (b) orange   (c) orange
(d) 4–6   (e) 7.5   (f) purple   (g) orange

CHECKPOINT 16.7
❶ 200 g KCl crystallises
❷ 20 g NaCl crystallises
❸ 12 g KCl crystallises
❹ 10 g K₂SO₄ + 10 g KBr crystallise

CHECKPOINT 17.8
❻ (a) (i) 80–100 km/h   (ii) 30 km/h
(iii) 100 km/h
(b) (i) 80–100 km/h   (ii) 50–60 mph

THEME E QUESTIONS
13 (a) 3.75 × 10⁵ tonnes   (b) 0.075
19 2400 tonne
20 1600 tonne

CHECKPOINT 19.2
❷ 2PbO(s) + PbS(s) → 3Pb(s) + SO₂(g)

CHECKPOINT 21.3
❶ (a) Zn(s) → Zn²⁺(aq) + 2e
      Sn²⁺(aq) + 2e → Sn(s)
(b) Zn(s) + Sn²⁺(aq) → Zn²⁺(aq) + Sn(s)
(c) Loss of electrons = oxidation   (d) No,
zinc is reduced.
❷ (a) Al(s) → Al³⁺(aq) + 3e
      Cu²⁺(aq) + 2e → Cu(s)
(b) 2Al(s) + 3Cu²⁺(aq) → 2Al³⁺(aq) + Cu(s)
(c) Al loses electrons; Al is oxidised; Al is the
anode

THEME F QUESTIONS
6 (i) Alkaline manganese dry cell gives 400
flashes/£; zinc–carbon cell gives
approximately 100 flashes/£.
(ii) Zinc–carbon cell gives 35 hours/£;
alkaline manganese cell gives 31 hours/£

CHECKPOINT 24.2
❶ 28, 44, 64, 80, 40, 58.5, 56, 58, 106, 159.5,
249.5, 162

CHECKPOINT 24.3
❶ (a) 20%   (b) 70%   (c) 80% C, 20% H
(d) 40% S, 60% O   (e) 2.5% H, 97.5% F
(f) 20% Mg, 27% S, 53% O
❷ (a) 36%   (b) 67.5%

CHECKPOINT 24.4
❶ (a) 27 g   (b) 96 g   (c) 50 g   (d) 14 g
(e) 8 g   (f) 64 g
❷ (a) 2.5 mol   (b) 0.33 mol   (c) 1.0 mol
(d) 0.25 mol
❸ (a) 98 g   (b) 23 g   (c) 100 g   (d) 10 g

CHECKPOINT 24.5
❶ 8.0 g
❷ (b) 3.2 g
❸ 10.3 g
❹ 107 g

CHECKPOINT 24.6
❶ (a) 8.0 g   (b) 95%
❷ (a) 34 g   (b) 10%
❸ 98%
❹ (b) 132 g   (c) 91%
❺ (a) 25 g   (b) 88%

CHECKPOINT 24.7B
❶ A CH₄O, B C₃H₆, C C₄H₈O₂, D C₂H₆O₂,
E C₂H₆, F C₂H₄Cl₂, G C₆H₆, H C₆H₃N₃O₆
❷ (i) (a) CH₃   (b) C₂H₆   (ii) (a) CH₂   (ii) C₂H₄
❸ (a) CH₂Cl   (b) C₂H₄Cl₂

CHECKPOINT 24.8
❶ 6 dm³
❷ 500 cm³ O₂, 250 cm³ CO₂
❸ 21 million dm³
❹ 3 dm³
❺ 0.43 dm³
❻ 25 g

CHECKPOINT 24.9
❶ (a) 15.9 g   (b) 14.8 g   (c) 52.0 g   (d) 20.0 g
(e) 29.8 g
❷ (a) 6.00 dm³H₂, 6.00 dm³ Cl₂
(b) and (c) 6.00 dm³ H₂, 3.00 dm³ O₂
❸ 1.035 g
❹ 120 cm³
❺ +2
❻ 64.8 g
❼ (a) 965 C   (b) 5 ×10⁻³ mol   (c) 2 mol
(d) Ni²⁺(aq) + 2e   Ni(s)
❽ (a) 12 cm³   (b) 24 cm³   (c) 0.0270 g
❾ (b)
❿ +3

CHECKPOINT 24.10A
❶ (a) 1.00 mol/l   (b) 0.10 mol/l
(c) 2.0 mol/l   (d) 0.40 mol/l
❷ (a) 0.02 mol   (b) 1.00 mol   (c) 0.06 mol
(d) 0.025 mol
❸ (a) 0.144 mol/l
❹ 2 mol/l
❺ (a) 0.2 M   (b) 0.02 M   (c) 0.2 M   (d) 8 M
❻ (a) 0.25 mol   (b) 0.010 mol
(c) 0.05 mol   (d) 0.0025 mol

CHECKPOINT 24.10B
❸ (a) 61   (b) 11
❹ Weight out 768 g, dissolve, make up to 2.00 l
❺ 140 cm³
❻ (a) 1090 cm³ made up to 10.0 l
❼ Weight out 0.42 g, dissolve, make up to
500 cm³
62.5 cm³ of 4.00 M acid in a burette, made up
❽ 1.00 l
❾ 235 cm³ of glacial ethanoic acid in a
measuring cylinder, made up to 2.00 l

CHECKPOINT 24.11
❶ 0.15 M
❷ (a) 0.25 M   (b) 10.0 cm³
❸ (a) 2   (b) 0.24 M
❹ (a) Paingo   (b) speed of action, taste

CHECKPOINT 26.4
❶ diagram similar to Figure 26.4B; ΔH positive
❷ 50450 kJ
❸ 20.2 g
❹ −1790 kJ/mol
❺ (a) −57 kJ/mol   (b) the same

CHECKPOINT 26.5
❶ −124 kJ/mol
❷ −484 kJ/mol
❸ −56 kJ/mol
❹ −95 kJ/mol

THEME G QUESTIONS
2 (b) 0.14 mol/l
3 (i) 1.25 × 10⁻³ mol   (ii) a, b, d, e
4 (a) Al³⁺ + 3e → Al(l)   (b) 162 kg
5 95.5%
8 (b) 11 cm³   (c) 90 cm³   (d) 35 cm³

CHECKPOINT 29.6
❼ (a) 65 tonnes   (b) 18 tonnes
❽ 2 months

CHECKPOINT 30.2
❹ (a) 61.5   (b) 650

THEME H QUESTIONS
2 (b) (i) 30%   (ii) 350 °C, 350 atm
(iii) The cost of a plant which can withstand
high pressure.
5 (d) (i) 59%   (ii) 45%   (e) 35%   (f) 43%
17 (a) 4.6 million tonnes   (b) 8.2 million
tonnes

Index